Emergence of Pharmaceutical Industry Growth with Industrial IoT Approach

Emergence of Pharmaceutical Industry Growth with Industrial IoT Approach

Edited by

Valentina Emilia Balas
"Aurel Vlaicu" University of Arad, Arad, Romania

Vijender Kumar Solanki
CMR Institute of Technology (Autonomous), Hyderabad, India

Raghvendra Kumar
LNCT Group of College, Jabalpur, India

ELSEVIER

ACADEMIC PRESS
An imprint of Elsevier

Academic Press is an imprint of Elsevier
125 London Wall, London EC2Y 5AS, United Kingdom
525 B Street, Suite 1650, San Diego, CA 92101, United States
50 Hampshire Street, 5th Floor, Cambridge, MA 02139, United States
The Boulevard, Langford Lane, Kidlington, Oxford OX5 1GB, United Kingdom

Notices
Knowledge and best practice in this field are constantly changing. As new research and experience broaden our understanding, changes in research methods, professional practices, or medical treatment may become necessary.

Practitioners and researchers must always rely on their own experience and knowledge in evaluating and using any information, methods, compounds, or experiments described herein. In using such information or methods they should be mindful of their own safety and the safety of others, including parties for whom they have a professional responsibility.

To the fullest extent of the law, neither the Publisher nor the authors, contributors, or editors, assume any liability for any injury and/or damage to persons or property as a matter of products liability, negligence or otherwise, or from any use or operation of any methods, products, instructions, or ideas contained in the material herein.

British Library Cataloguing-in-Publication Data
A catalogue record for this book is available from the British Library

Library of Congress Cataloging-in-Publication Data
A catalog record for this book is available from the Library of Congress

ISBN: 978-0-12-819593-2

For Information on all Academic Press publications
visit our website at https://www.elsevier.com/books-and-journals

Publisher: Andre Gerhard Wolff
Acquisition Editor: Erin Hill-Parks
Editorial Project Manager: Pat Gonzalez
Production Project Manager: Punithavathy Govindaradjane
Cover Designer: Mark Rogers

Typeset by MPS Limited, Chennai, India

Contents

List of contributors xiii
About the editors xv
Preface xix

1. A neoteric swarm intelligence stationed IOT–IWD algorithm for revolutionizing pharmaceutical industry leading to digital health 1

Neha Sharma, Usha Batra and Sherin Zafar

1.1 Introduction 1
1.2 Related work 4
1.3 Proposed neoteric swarm intelligence stationed IOT–IWD algorithm for ZRP optimization for digital health 7
1.4 Simulation results 7
1.5 Conclusion and discussions 15
References 17
Further reading 18

2. A survey on Internet-of-Thing applications using electroencephalogram 21

Debjani Chakraborty, Ahona Ghosh and Sriparna Saha

2.1 Introduction 21
2.2 Electroencephalogram acquisition techniques 22
2.2.1 Invasive 23
2.2.2 Noninvasive 23
2.3 Channel selection techniques 24
2.3.1 Filtering 24
2.3.2 Wrapper 24
2.3.3 Embedded 24
2.3.4 Hybrid 25
2.3.5 Human-based technique 25
2.4 Different brain signals 25
2.5 Preprocessing of electroencephalogram signals 26
2.6 Feature extraction and classification from electroencephalogram signal 27

2.7 Classification of electroencephalogram signals based on features 31
2.7.1 *k*-Nearest neighbor 31
2.7.2 Linear discriminant analysis 32
2.7.3 Decision tree 32
2.7.4 Adaptive Boosting 33
2.7.5 Multilayer perceptron 33
2.7.6 Naive Bayes 34
2.8 Applications of electroencephalogram-based Internet-of-Thing applications 35
2.8.1 Seizure detection 36
2.8.2 Brain injury detection 37
2.8.3 Object controlling 37
2.8.4 Mental state recognition 38
2.8.5 Rehabilitation and human assistance 38
2.8.6 Neuro-marketing studies 39
2.9 Conclusion 40
References 40

3. A case study: impact of Internet of Things devices and pharma on the improvements of a child in autism 49
Muhammad Javaid Afzal, Shahzadi Tayyaba, Muhammad Waseem Ashraf, Farah Javaid and Valentina Emilia Balas

3.1 Introduction 49
3.1.1 Internet of Things devices 49
3.1.2 Pharma in autism 51
3.1.3 Autism 56
3.2 Parent history 60
3.2.1 Patient history 60
3.2.2 Herbal treatment 61
3.3 General behavior up to 2.5 years 61
3.4 Diagnosis 61
3.5 Autism spectrum disorder therapies, pharma, and Internet of Things 61
3.6 Assessment taken by Lahore Children Center at the age of 5.8 years 62
3.6.1 Introduction 62
3.6.2 Relevant medical and developmental history 63
3.6.3 Language and family background 63
3.6.4 Current functioning 63
3.6.5 Summary and recommendations 65
3.7 Improvements up to 8 years of age 66
3.8 Changes in behavior after 8 years of age and use of Internet of Things 66
3.9 Improvements up to 10.5 years of age 68

3.10 Applied behavior analysis therapy assessment report 68
3.11 Autism schools in Pakistan 75
3.12 Cost analysis of some schools with autism 76
3.13 Recommendations for autistic child discipline with Internet of Things and pharma 76
3.13.1 Recommendation of nutritional interventions in autism spectrum disorder 78
3.14 Conclusion 79
References 79

4. Internet of Things–based pharmaceutics data analysis 85
Pranshu Dhingra, N. Gayathri, S. Rakesh Kumar, Vijayakumar Singanamalla, C. Ramesh and B. Balamurugan

4.1 Introduction 85
4.1.1 The era of Internet of Things 85
4.1.2 Intermittent connectivity 88
4.1.3 Connectivity technologies 88
4.1.4 Quirky machine-to-machine communication 89
4.1.5 Revolutionization of Internet of Things in pharma industry 90
4.1.6 Revolution of Industry 4.0 92
4.1.7 Big data in pharmaceutical industry 92
4.1.8 Linking Internet of Things with big data 94
4.1.9 Analysis of pharma data 96
4.2 Related works 97
4.2.1 Interoperability of Internet of Things devices 97
4.2.2 Pharma logistics: helping hand from Internet of Things 100
4.2.3 Data and its analysis—a way to optimize pharmaceutical processes 111
4.2.4 Big data handling of pharma data 113
4.2.5 Visual analytics in pharma industry 118
4.3 Proposed work 120
4.3.1 Proposed framework for Internet of Things–based pharmaceutical data analysis 120
4.4 Implementation 124
4.5 Results and discussion 127
4.6 Conclusion 128
References 128
Further reading 130

5. Reliable pharma cold chain monitoring and analytics through Internet of Things Edge 133
S. Balachandar and R. Chinnaiyan

5.1 Introduction 133
5.1.1 Statement of the problem 134
5.1.2 Objectives 134

5.2 Cold chain logistics 135
5.3 Literature review 135
5.3.1 Edge technologies 137
5.3.2 Common sensors 138
5.4 Internet of Things edge design—conceptual framework 138
5.5 Implementations 146
5.6 High-level approach 146
5.6.1 Methodology—experiments and results 148
5.7 Role of containers in Internet of Things edge 150
5.8 Pharma—cold chain analytics 153
5.8.1 Product demand forecasting 154
5.8.2 Track and trace 157
5.8.3 Conditional monitoring and predictive maintenance of containers 158
5.9 Deployment considerations and issues 158
5.9.1 Deployment considerations 158
5.9.2 Known issues 159
5.10 Conclusion 159
References 159
Further reading 161

6. The growing role of Internet of Things in healthcare wearables 163

R. Indrakumari, T. Poongodi, P. Suresh and B. Balamurugan

6.1 Introduction 163
6.2 Impact of Internet of Things–based wearables in healthcare 163
6.3 Taxonomy of wearables 166
6.4 Wearable sensors for physiological parameters measurement 166
6.4.1 Physical parameters 167
6.4.2 Biochemical parameters 168
6.5 Types of wearable sensors 169
6.5.1 Invasive sensors 169
6.5.2 Noninvasive wearable sensors 169
6.6 Working principles of wearable sensors 169
6.7 Challenges in the fabrication of wearable sensors 172
6.8 Small wearable antennas for healthcare system 173
6.9 Functions of wearable sensors 175
6.10 Wearable devices in pharmaceutical industry 178
6.10.1 Wireless body area network 179
6.10.2 Respiratory rate sensors 180
6.10.3 Body temperature sensor 182
6.10.4 Blood pressure monitoring sensor 182
6.10.5 Pulse oximetry sensors 183
6.11 Wearable devices revolutionize the entire paradigm in drug dispensing 184
6.11.1 Remove hurdles and offer rewards 185

6.11.2 Form and functions 185
6.11.3 Making the data relevant 186
6.11.4 Market trend 186
6.11.5 Glucose monitoring 187
6.12 Safety and security issues related to wearable health care devices 187
6.13 Wearable devices for women safety 188
6.14 Open challenges and future directions 190
6.15 Conclusion 191
References 191

7. Internet of Things in pharma industry: possibilities and challenges 195

Mohan Singh, Smriti Sachan, Akansha Singh and Krishna Kant Singh

7.1 Introduction 195
7.1.1 Internet of Things 196
7.1.2 Applications of Internet of Things 196
7.2 Internet of Things road map in pharma 198
7.3 Internet of Things in pharma industry 200
7.4 Applying Internet of Things in pharma industry 202
7.4.1 Manufacturing 202
7.4.2 Monitoring of production flow 203
7.4.3 Controlling of environmental factors in drugs manufacturing 204
7.4.4 Quality control 204
7.4.5 Packaging optimization 205
7.4.6 Warehouse operations 206
7.4.7 Facility management 206
7.4.8 Supply chain 206
7.4.9 Inventory management 209
7.5 Role of Internet of Things in challenges of pharma industry 209
7.5.1 Plant safety and security 211
7.5.2 To overcome the short supply of drugs 212
7.5.3 Security of supply chain 212
7.5.4 Theft of drugs during transportation 213
7.6 Conclusion and future scope 213
References 214
Further reading 216

8. Internet of Things technologies for elderly health-care applications 217

Jinesh Padikkapparambil, Cornelius Ncube, Krishna Kant Singh and Akansha Singh

8.1 Introduction 217

8.2 Elderly population distribution 219
8.3 Societal adaptions 220
8.4 Connected homes 224
8.5 What is Internet of Things? 225
8.6 Ambient assistive living systems 226
8.7 Requirements of activity recognition 227
8.8 Internet of Things–based technologies 229
8.8.1 Activity recognition 230
8.8.2 Wearable systems 231
8.8.3 Ready-to-use products 233
8.9 Existing systems 234
8.10 Conclusion 241
References 241
Further reading 243

9. An insight of Internet of Things applications in pharmaceutical domain 245
Sushruta Mishra, Anuttam Dash and Brojo Kishore Mishra

9.1 An overview of Internet of Things 245
9.2 Characteristics features of Internet of Things 247
9.3 Advantages of Internet of Things 248
9.4 Architectural framework of Internet of Things 249
9.5 Application areas of Internet of Things 252
9.6 Potential of Internet of Things in the pharmaceutical industry 255
9.7 Literature review of Internet of Things in pharmacy 257
9.8 Benefits of using Internet of Things in the pharmaceutical industry 258
9.9 Patient-centric Internet of Things 260
9.9.1 Patient-centric versus patient-centered information 261
9.10 Body area network overview 261
9.10.1 Challenges faced by body area network 262
9.11 Internet of Health Things 264
9.11.1 Advantages of Internet of Health Things 264
9.12 Analysis of medical nursing system using Internet of Things in the pharmaceutical domain 265
9.12.1 Discussed work 266
9.12.2 Identity management system 266
9.12.3 Environmental-sensing system 267
9.12.4 Biomedical system 267
9.12.5 Medication system 268
9.12.6 Personal orientation system 271
9.13 Conclusion 272
References 272

10. Smart pills: a complete revolutionary technology than endoscopy 275

Subhashree Sahoo, Amiya Bhusan Bagjadab and Sushree Bibhuprada B. Priyadarshini

10.1 Introduction 275
10.2 Introduction to endoscopy 276
10.3 Why endoscopy? 277
10.3.1 Investigating signs and symptoms 277
10.3.2 Diagnosing 277
10.3.3 Treating 277
10.4 Types of endoscopy 278
10.4.1 Upper Gastro-Intestinal (GI) endoscopy 278
10.4.2 Colonoscopy 279
10.4.3 Endoscopic retrograde cholangiopancreatography 281
10.4.4 Bronchoscopy 282
10.4.5 Percutaneous Endoscopic Gastrostomy (PEG) 284
10.4.6 Flexible sigmoidoscopy 285
10.4.7 Cystoscopy 287
10.4.8 Transbronchial endoscopy 289
10.4.9 Hysteroscopy 291
10.4.10 Endoscopic ultrasound 295
10.5 Smart pills 297
10.6 Purpose of WCE 298
10.6.1 Accuracy of WCE 300
10.6.2 Technology of WCE 300
10.6.3 Preparation of WCE 302
10.6.4 Working of WCE 302
10.7 Conclusions 302
References 303
Further reading 303

11. BioSenHealth 2.0—a low-cost, energy-efficient Internet of Things–based blood glucose monitoring system 305

Vikram Puri, Raghvendra Kumar, Dac Nhuong Le, Sandeep Singh Jagdev and Nidhi Sachdeva

11.1 Introduction 305
11.2 Related studies 308
11.3 Methodology 309
11.3.1 Architecture 309
11.3.2 Circuit diagram 315
11.4 Result and discussion 319
11.5 Conclusion 321
References 322

Index 325

List of contributors

Muhammad Javaid Afzal Government Islamia College Civil Lines, Lahore, Pakistan

Muhammad Waseem Ashraf Government College University, Lahore, Pakistan

Amiya Bhusan Bagjadab Sambalpur University Institute of Information Technology, Burla, India

S. Balachandar Shell India Market Private Limited, Bangalore, India

B. Balamurugan School of Computing Science and Engineering, Galgotias University, Greater Noida, India

Valentina Emilia Balas "Aurel Vlaicu" University of Arad, Arad, Romania

Usha Batra SOE, GD Goenka University, Sohna, India; CSE, SEST, Jamia Hamdard, New Delhi, India

Debjani Chakraborty Department of Computer Science and Engineering, Maulana Abul Kalam Azad University of Technology, Kolkata, India

R. Chinnaiyan Department of Information Science and Engineering, CMR Institute of Technology, Bangalore, India

Anuttam Dash KIIT Deemed to be University, Bhubaneswar, India

Pranshu Dhingra Galgotias University, Greater Noida, India

N. Gayathri Anna University, Chennai, India

Ahona Ghosh Department of Computer Science and Engineering, Maulana Abul Kalam Azad University of Technology, Kolkata, India

R. Indrakumari School of Computing Science and Engineering, Galgotias University, Greater Noida, India

Sandeep Singh Jagdev Ellen Technology (Pvt). LTD, Punjab, India

Farah Javaid Government APWA College for Women, Lahore, Pakistan

Raghvendra Kumar CSE Department, LNCT Group of College, Jabalpur, India

Brojo Kishore Mishra GIET University, Gunupur, India

Sushruta Mishra KIIT Deemed to be University, Bhubaneswar, India

Cornelius Ncube British University in Dubai, Dubai, United Arab Emirates

Dac Nhuong Le Faculty of Information Technology, Haiphong University, Haiphong, Vietnam

Jinesh Padikkapparambil Higher College of Technology, Dubai, United Arab Emirates

T. Poongodi School of Computing Science and Engineering, Galgotias University, Greater Noida, India

Sushree Bibhuprada B. Priyadarshini Institute of Technical Education and Research, Bhubaneswar, Siksha 'O' Anusandhan Deemed to be University, Bhubaneswar, India

Vikram Puri Duy Tan University, Da Nang, Vietnam

S. Rakesh Kumar Anna University, Chennai, India

C. Ramesh Bannari Amman Institute of Technology, Sathyamangalam, India

Smriti Sachan Department of ECE, GL Bajaj Institute of Technology and Management, Greater Noida, India

Nidhi Sachdeva Fairleigh Dickinson University, Vancouver, BC, Canada

Sriparna Saha Department of Computer Science and Engineering, Maulana Abul Kalam Azad University of Technology, Kolkata, India

Subhashree Sahoo Sambalpur University Institute of Information Technology, Burla, India

Neha Sharma SOE, GD Goenka University, Sohna, India; CSE, SEST, Jamia Hamdard, New Delhi, India

Vijayakumar Singanamalla Galgotias University, Greater Noida, India

Akansha Singh School of Computing Science and Engineering, Galgotias University, Greater Noida, India

Krishna Kant Singh Department of ECE, GL Bajaj Institute of Technology and Management, Greater Noida, India

Mohan Singh Department of ECE, GL Bajaj Institute of Technology and Management, Greater Noida, India

P. Suresh School of Mechanical Engineering, Galgotias University, Greater Noida, India

Shahzadi Tayyaba The University of Lahore, Lahore, Pakistan

Sherin Zafar SOE, GD Goenka University, Sohna, India; CSE, SEST, Jamia Hamdard, New Delhi, India

About the editors

Valentina Emilia Balas, Ph.D., is currently a full-time professor in the Department of Automatics and Applied Software at the Faculty of Engineering, "Aurel Vlaicu" University of Arad, Romania. She also holds a Ph.D. in the Applied Electronics and Telecommunications from Polytechnic University of Timisoara. Dr. Balas is the author of more than 300 research papers in refereed journals and international conferences. Her research interests are in intelligent systems, fuzzy control, soft computing, smart sensors, information fusion, modeling, and simulation. She is the Editor-in-Chief to the International Journal of Advanced Intelligence Paradigms (IJAIP) and to the International Journal of Computational Systems Engineering (IJCSysE), a member in the editorial board of several national and international journals, and an evaluator expert for national and international projects. She served as the General Chair of the International Workshop Soft Computing and Applications in seven editions 2005–18 held in Romania and Hungary. Dr. Balas participated in many international conferences as an organizer, session chair, and member in the International Program Committee. Now she is working in a national project with EU funding support: BioCell-NanoART = Novel Bio-inspired Cellular Nano-Architectures—For Digital Integrated Circuits, 3M Euro from the National Authority for Scientific Research and Innovation. She is a member of EUSFLAT, ACM and a Senior Member IEEE, member in TC—Fuzzy Systems (IEEE CIS), member in TC—Emergent Technologies (IEEE CIS), and member in TC—Soft Computing (IEEE SMCS). Dr. Balas was the vice president (Awards) of IFSA (International Fuzzy Systems Association) Council (2013–15) and is a joint secretary of the Governing Council of Forum for Interdisciplinary Mathematics (FIM)—A Multidisciplinary Academic Body, India.

Vijender Kumar Solanki, Ph.D., is an associate professor in Computer Science & Engineering, CMR Institute of Technology (Autonomous), Hyderabad, TS, India. He has more than 10 years of academic experience in network security, IoT, Big Data, Smart City, and IT. Prior to his current role, he was associated with the Apeejay Institute of Technology, Greater Noida, UP, KSRCE (Autonomous) Institution, Tamil Nadu, India, and Institute of Technology & Science, Ghaziabad, UP, India. He has attended an orientation program at UGC—Academic Staff College, University of Kerala, Thiruvananthapuram, Kerala, and refresher course at the Indian Institute of

Information Technology, Allahabad, UP, India. He has authored or coauthored more than 40 research articles that are published in journals, books, and conference proceedings. He has edited or coedited 10 books in the area of Information Technology. He teaches graduate and postgraduate-level courses in IT at ITS. He received Ph.D. in Computer Science and Engineering from Anna University, Chennai, India, in 2017 and ME, MCA from Maharishi Dayanand University, Rohtak, Haryana, India in 2007 and 2004, respectively, and a bachelor's degree in Science from JLN Government College, Faridabad Haryana, India, in 2001. He is the Book Series Editor of Internet of Everything (IoE): Security and Privacy Paradigm, CRC Press, Taylor & Francis Group, United States, Artificial Intelligence (AI): Elementary to Advanced Practices Series, CRC Press, Taylor & Francis Group, United States & IT, Management & Operations Research Practices, CRC Press, Taylor & Francis Group, United States also Book Series Editor of Bio-Medical Engineering: Techniques and Applications with Apple Academic Press, United States. He is an editor to the International Journal of Machine Learning and Networked Collaborative Engineering (IJMLNCE) ISSN 2581-3242; coeditor to the Ingenieria Solidaria Journal ISSN (2357-6014); and associate editor to the International Journal of Information Retrieval Research (IJIRR), IGI-GLOBAL, United States, ISSN: 2155-6377 | E-ISSN: 2155-6385. He is the guest editor with IGI-Global, United States, InderScience and many more publishers.

Raghvendra Kumar is working as an associate professor in the Computer Science and Engineering Department at L.N.C.T Group of College Jabalpur, M.P. India, and serving as a Director of IT and Data Science Department, Vietnam Center of Research in Economics, Management, Environment (VCREME)—Branch VCREME One Member Company Limited, Vietnam. He received B. Tech. in Computer Science and Engineering from SRM University Chennai (Tamil Nadu), India, M. Tech. in Computer Science and Engineering from KIIT University, Bhubaneswar, (Odisha), India, and Ph.D. in Computer Science and Engineering from Jodhpur National University, Jodhpur (Rajasthan), India. He serves as a Series Editor to the Internet of Everything (IOE): Security and Privacy Paradigm publishes by CRC Press, Taylor & Francis Group, United States, and Bio-Medical Engineering: Techniques and Applications, Publishes by Apple Academic Press, CRC Press, Taylor & Francis Group, United States. He also serves as an acquisition editor for Computer Science by Apple Academic Press, CRC Press, Taylor & Francis Group, United States. He has published a number of research papers in the international journals (SCI/SCIE/ESCI/Scopus) and conferences including IEEE and Springer as well as serves as organizing chair (RICE-2019), volume editor (RICE-2018), keynote speaker, session chair, cochair, publicity chair, publication chair (NGCT-2017), advisory board, technical program committee member in many international and national conferences, and serves as a guest editor in many special issues from reputed journals (Indexed By: Scopus, ESCI). He also published 11 chapters in edited book published by IGI Global, Springer and Elsevier. He also received best paper award in IEEE Conference 2013 and Young Achiever Award-2016 by IEAE Association for his research work in the field of distributed database. His research areas are computer networks, data mining, cloud computing and secure multiparty computations, theory of computer science, and design of algorithms. He authored and edited 17 computer science books in the field of Internet of Things, Data Mining, Biomedical Engineering, Big Data, Robotics, Graph Theory, and Turing Machine by IGI Global Publication, United States, IOS Press Netherland, Springer, Elsevier, CRC Press, United States, S. Chand Publication, and Laxmi Publication. He is the managing editor in the International Journal of Machine Learning and Networked Collaborative Engineering (IJMLNCE) ISSN 2581-3242.

Preface

The purpose of this edited book is to inform and educate its audience about the power of Internet of Things and pharma industry. The enormous growth of the Internet of Things (IoT) has urged a vast majority of the companies in the manufacturing sector to make use of this technology to unlock limitless potential. Pharmaceutical manufacturing companies are no exception to this, as IoT has the potential to revolutionize all aspects of the pharmaceutical manufacturing process from drug discovery to manufacturing. IoT in pharma manufacturing coupled with Big Data and advanced analytics can scrutinize massive amounts of data that can be harnessed to improve manufacturing efficiency. The wide variety of topics it presents offers readers multiple perspectives on a variety of disciplines including number of chapters in the edited book. The book is organized into 11 chapters.

Chapter 1, A neoteric swarm intelligence stationed IOT–IWD algorithm for revolutionizing pharmaceutical industry leading to digital health, discussed about the exploration work portrayed around enhancing the execution of zone steering convention by lessening the measure of receptive traffic which is fundamentally in charge of debased system execution, if there should be an occurrence of extensive systems. The methodology is structured to such an extent that the zone sweep of the system stays unaffected while accomplishing better QOS (quality of service) execution alongside effective memory utilization. This is actualized by utilizing the swarm optimization–stationed IWD calculation that plans to accomplish worldwide enhancement which is very hard to accomplish due to nonlinearity of capacities and multimodality of calculations. Different customary streamlining strategies, such as inclination-based procedures and tree-based calculations, need to manage such issues so that this exploration-based work uses the metaheuristic calculation; it takes points of interest of firefly calculation to upgrade QOS of MANET. For evaluation and validation of the performance of the proposed algorithm, a set of benchmark functions is being adopted, such as throughput and packet loss ratio, packet delivery ratio, and delay. Simulation results depict better performance of proposed neoteric Intelligent Water Drop Algorithm (IWD) algorithm when compared to ZRP and its modified Route Conglomeration/Aggregation (RA/RC) methodology.

Chapter 2, A survey on Internet-of-Things applications using electroencephalogram, gives an idea about how these signals can be acquired and what are the signal parameters needed to be measured for successful implementation of electroencephalography-based human–computer interactive devices. Then, the chapter focuses on different feature extraction methods for the signals. Lastly, the chapter gives a thorough understanding of different application domains related to brain–computer interfacing.

Chapter 3, A case study: impact of Internet of Things devices and pharma on the improvements of a child in autism, discusses about how the Internet of Things makes smart devices into ultimate building blocks for the improvement of healthcare and other fields. This field is also remodeling the ignored area of children with autism with favorable social, technical, economical, and prospects, and a case study of triangle of IOT, pharma, and autism works very well for Mujtaba.

Chapter 4, Internet of Things–based pharmaceutics data analysis, aims to provide the various techniques related to the IoT for pharmaceutical-related data and further assist in the analysis of the data generated from pharmaceutical field. The various pharmaceutical concepts based on IoT are being investigated and the clinical data is being investigated for the body movements and the analysis is done based on that. The analysis of the proposed system shows that it is 4% better in reducing the error rate of the results.

Chapter 5, Reliable pharma cold chain monitoring and analytics through Internet of Things edge, discusses about the medical institution or patients wait for such drugs from the supplier and it's important to deliver the drugs with recommended temperature or cold conditions. The cold chain logistics follows specific sensitivity of the medicines and adjusts the temperature and humidity settings as per norm. The logistic companies should ensure that the potency of the product remains intact and safeguards the products until it reaches the consumer or dealer. The role of edge computing for IOT (Internet of Things) will play a pivotal role on monitoring and controlling sensor, devices installed on the refrigerators, storage containers, or boxes inside the trucks or vans. It should also help to predict the equipment or device failure at the edge and recommends the resiliency plan.

Chapter 6, The growing role of Internet of Things in healthcare wearables, begins with the introduction of Wearable Internet of Things, its attributes, and footprint of wearable technology in pharmaceuticals by considering the ethical issues and safety measures.

Chapter 7, Internet of Things in pharma industry: possibilities and challenges, discusses that the most recent innovation, made accessible with the coming of IoT, can be utilized to help this change in outlook in the elements of the pharmaceutical segment. The associated innovation can be sent for covering diverse verticals, for example, producing, checking, appropriation, and control in-travel. With the assistance of the continuous accessibility of information, pharmaceutical organizations can guarantee appropriate quality, while limiting or totally maintaining a strategic distance from any odds of pilferage, wastage, or creation. In this chapter the areas of application where

IoT can play a significant role are discussed. But the technology brings with it some challenges as well. Thus the challenges in bringing IoT in the pharma industry are also discussed.

Chapter 8, Internet of Things technologies for elderly health-care applications, aims to enable elderly people to independently live longer in their own homes, to enhance living qualities, and to reduce costs for society and public health systems. Assisted living systems can help support elderly persons with their daily activities in order to help them maintain health and safety while living independently. In this chapter, IoT technologies for elderly healthcare will be detailed.

Chapter 9, An insight of Internet of Things applications in pharmaceutical domain, discusses about the working in the pharmaceutical sector are also supposed to ensure the secure and safe transfer of drugs, better-planned shipment, and delivery, clinical consequences. In order to facilitate speedy operations, it is required to harvest data in a way that will be both effective and well organized, supplemented by obligatory analytics. We have briefly described the IoT trends and methodology that are being used in the pharmaceutical sector in this chapter. Various aspects revolving around the role of IoT in the pharmaceutical industry have been discussed here. A sample case study has also been highlighted in the subsequent sections of the chapter. In this case study a smart system for medical nursing based on WSN, NFC, and RFID technology has been discussed. This system not only promotes nursing home conditions but also upgrades the drug supply accuracy.

Chapter 10, Smart pills: a complete revolutionary technology than endoscopy, focuses for WCEs is on effective localization, steering, and management of capsules. Device development depends on the investment study and technologies for higher system performance, instead of utterly. The term "smart pills" refers to the miniature electronic devices that are formed and designed within the mildew of pharmaceutical capsules, however perform extremely advanced functions, such as sensing, imaging, and drug delivery. They will include biosensors or image, hydrogen ion concentration, or chemical sensors. Once they're swallowed, they travel along the gastrointestinal tract to capture information that is otherwise tough to get then these pills are simply eliminated from the system. Their classification as ingestible sensors makes them distinct from implantable or wearable sensors.

Chapter 11, BioSenHealth 2.0—a low-cost, energy-efficient Internet of Things–based blood glucose monitoring system, presents a prototype of IoT-based glucose testing meter which is able to connect with cloud services. Our proposed system is embedded with ESP8266 which contains Wi-Fi connectivity and low power microcontroller which helps to save the power consumption. This device can be used as a wearable band and handheld use. The experimentation results show that proposed prototype is well efficient for patient risk extraction as well as energy efficient.

There have been several influences from our family and friends who have sacrificed lot of their time and attention to ensure that we are kept motivated to complete this crucial project. The editors are thankful to all the members of Elsevier, United States, Private Limited especially Pat Gonzalez and Narmatha Mohan for the given opportunities to edit this book.

Valentina Emilia Balas, Vijender Kumar Solanki and Raghvendra Kumar

Chapter 1

A neoteric swarm intelligence stationed IOT–IWD algorithm for revolutionizing pharmaceutical industry leading to digital health

Neha Sharma[1,2], Usha Batra[1,2] and Sherin Zafar[1,2]
[1]*SOE, GD Goenka University, Sohna, India,* [2]*CSE, SEST, Jamia Hamdard, New Delhi, India*

1.1 Introduction

Pharmaceutical manufacturers previously enjoyed high profit margins, but due to the decrease in the number of patients and overseas competition, the profit ratios have declined, so the manufacturers have started experimenting with Internet of Things (IOT) technology to improve communication, identification, and interaction, that is, quality of service (QOS), which will increase efficiency and avoid costly mistakes as depicted in Fig. 1.1. The adoption of computers to imitate and analyze nature to enhance the practice of computers has become an asserting and amusing area of research for achieving digital health. Out of many newly developed algorithms "intelligent water drop" (IWD) is one of such amusing algorithms that have been developed lately. It is inspired from nature that follows the behaviors of natural water drops, which alter their surroundings in order to locate the shortest route to their destination.

The course of actions happens in-between the water drops of a river and the soil of the river's bed. These two aspects are the basis of this algorithm. The IWD algorithm has been categorized as a "swarm-based optimization algorithm." At first, it was brought up by Dr. Shah Hosseini in 2007 [1]. Swarm-inspired algorithms gave rise to a neoteric IWD algorithm. It belongs to the class of population constructive optimization utilized for combinatorial optimization. IWD methodology depends on some of the essential elements of a natural water drops and also the actions as well as reactions that tend to

Emergence of Pharmaceutical Industry Growth with Industrial IoT Approach.
DOI: https://doi.org/10.1016/B978-0-12-819593-2.00001-7

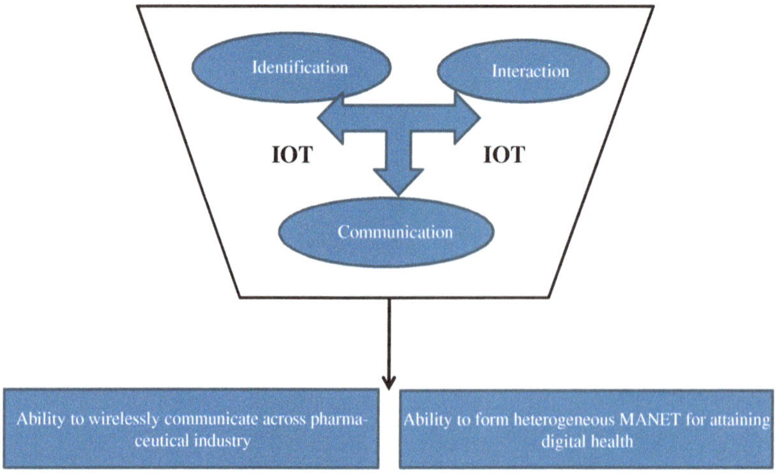

FIGURE 1.1 Attaining digital health through IOT in MANET. *IOT*, Internet of Things; *MANET*, mobile ad-hoc network.

occur between river's bed (soil) and the drops of water that flow within. It has several artificial water drops that cooperate for changing the environment by revealing the optimal path, the one with the lowest soil on its links. The two main factors/variables in IWD velocity of the water drop and the amount of soil are removed from the path that becomes the specific selection criterion. This algorithm until recent times has solved "the traveling salesman problem," (TSP) "n-queen puzzle," "multidimensional knapsack problem" (MKP) [1], "smooth trajectory planning" [2], "robot path planning" [3], "vehicle routing problem" [4], and "economic load dispatch problem" [5].

In nature, water drops are everywhere and mostly gushing in rivers. The rivers can be acknowledged as a giant moving group of water drops. The route of the river is actually designed by this group of moving water drops. Along the path, as the water drop moves, they attempt to alter the surroundings. On the other side the flow of the group of moving water drops gets impacted by the surroundings as well. In other words, we can say that both the surroundings and the group of moving water drops are affected by one another. The surroundings here refer to the soil on the river's bed. As the group of water drops moves faster on the river's bed, it changes the soil on the river's bed more as compared to the slow moving water drops. Also, when the surroundings have hard soils, they resist the movement of the group of water drops [6].

The route that the group of water drops takes is not a smooth ride always, but still they somehow manage to reach their destination, and the overall route taken by them is considered the optimal path. The earth's gravitational force that pulls everything toward the center of the earth also helps the group of water drops reach their destination following the ideal route. The gravitational pull also enhances the speed of water drops [7].

The velocity of a water drop is another aspect of a water drop that is flowing in a river, and it is believed that the each water drop is capable of carrying some soil from a place to another while flowing through its path [1]. As the group of water drops moves from their source point to the destination point, three obvious changes happen during the transition:

1. There is a rise in the velocity of the water drop.
2. The amount of soil carried by the water drop increases.
3. The amount of soil on any of the two points on river's bed decreases.

A high-speed water drop carries a larger amount of soil than the water drop with slow speed. It clearly means that the high-speed water drop removes more soil from the river's bed. The route where the amount of soil on the river's bed is less, the velocity of water drop increases in that region. It means that the velocity of water drop is inversely proportional to the amount of soil on the river bed, and because of this the water drop always chooses a route that has a less amount of soil whenever it has to select a path from several routes that exist in-between the source to the destination. The IWD algorithm works in an organized way in order to locate an optimal solution to a given problem [8].

Zone Routing Protocol (ZRP) advances as a productive half and half directing convention with great possibility inferable from the joining of two drastically unique plans: proactive and receptive, so that a harmony between control overhead and inactivity is accomplished. Different conditions of a system such as span of the zone, versatility, and organization measurement affect ZRP execution. The exploration of this chapter centers for a goal of enhancing the execution of ZRP by controlling receptive traffic, which determines system execution, in the case of extensive systems. The proposed methodology targets a structure to an extent that ZRP stays unaffected and also accomplishes optimized QOS execution along with effective utilization of memory. Achieving better QOS performances is actualized through the swarm optimization stationed IWD methodology that accomplishes worldwide enhancement. The optimization framework is very hard to accomplish as different optimization methodologies have different types of nonlinearity of capacities and multimodality. Different customary streamlining strategies, such as inclination-based procedures and tree-based calculations, need to manage accretion issues, so the proposed exploration-based work utilizes meta-heuristic calculation, by taking points of interest of firefly calculation to upgrade QOS of mobile ad-hoc network (MANET). For the evaluation and validation of the performance of the proposed algorithm, a set of benchmark functions is being adopted such as throughput and packet loss ratio (PLR), packet delivery ratio (PDR), and delay. Simulation results depict a better performance of proposed neoteric IWD algorithm when compared to

ZRP and its modified route conglomeration/aggregation (RA/RC) methodology. Upcoming sections of this chapter will exhibit the following: Section 1.2 contains elaborated literature survey that has been done in order to learn and draw conclusion for the topic. Section 1.3 consists of a swarm intelligence stationed IWD algorithm for ZRP optimization that reveals the general process of attaining optimization by implementing IOT–IWD for digital health. Further, Section 1.4 comprises the distinct simulation results, and the last section concludes the entire topic and showcases the future scope.

1.2 Related work

The related work presents the different specifications of IWD and digital health scenarios. Various authors have focused on experimenting IWD algorithm in order to find the optimal solutions for various mathematical problems such as TSP, n-queen puzzle, and MKPs. One of the dewy swarm-based optimization algorithms, which has been encouraged from the natural drift of water drops in a river, is IWD algorithm. The natural drift of a river discovers the best route betwixt ample of possible routes till the destination from its source. The actions and reactions that take place betwixt the water drops and water drops with the soil at the river's bed contribute in attaining the near optimal routes [1]. Also experimentations of the IWD algorithm onto the four distinct mathematical problems, such as TSP, n-queen problem, MKP, and the AMT, have been performed. The authors have concluded that the test conducted on various problems showed the caliber of IWD to locate optimal or closer optimal solutions. Also, IWD algorithm can be improved by inlaying the mechanism of any other technique, and better results can be obtained for the given problem. The IWD algorithm exhibits that the novel nature influenced optimization algorithms can be excellently designed and conceived by the guidance of nature [2]. Various surveys on the optimization areas are also performed, beginning from the swarm intelligence–based algorithms that exhibit the functioning of different optimization algorithms through the behaviors of distinct groups of insects such as ants, bees, and fishes. Furthermore, the authors have focused on other nature-inspired optimization algorithms. One of such nature-inspired algorithms is IWD, which is based on the dynamic behavior of the river. It is believed that the water in the river follows the optimum path to reach its destination. The water drops and the soil on the river's bed are the two major contributing factors [3].

Various researchers have also focused on MANET and the issues with routing in MANET. The mobile devices in MANET do not require any preinstalled infrastructure, which makes the routing process quite challenging. The major objective of a routing protocol is to locate a path betwixt two communicating nodes along with the optimization of the whole network performance. Further, a neoteric routing protocol influenced from the nature that tackles the dynamic nature of MANET is also focused upon on the basis of IWD that employs the events that occur in the natural river, and it has been witnessed that the water drops in river follow the best route among all the possible paths to their destination [4]. Various applications of IWD with UCAV (unmanned combat aerial vehicle) with some improvisation in IWD are also highlighted by various researchers. The water drops in the river follow some actions that lead to the optimization of route to reach its destination. An updated IWD approach is discussed in order to give a solution to "single UCAV smooth trajectory planning" problems in distinct combating situations. In the process of locating the optimal UCAV trajectory, the water drops can behave as an agent. Various experiments performed by the authors have suggested that the IWD is highly flexible and works outstandingly for dynamic environments [5]. The newly proposed algorithm by the researchers called IWD is inspired by the actions and reactions of water drops that take place in a river in order to reach its destination. It has been noticed that a river mostly selects an optimal or shortest path among all the possible paths. This algorithm is intended to solve various mathematical problems and implemented it on TSP. The implementation of IWD on man-made and real-time problem exhibits quite impressive results, and the author has concluded that IWD can be very promising in future in order to give solutions for other problems as well [6]. Various authors have addressed a universal problem of path planning of air robots, by implementing through IWD algorithm and have configured more effective and feasible results. The IWD is inspired from the dynamic behavior of rivers, and it is observed that in order to reach their destination the rivers follow an optimum route. Through this chapter the authors have proposed an improved IWD to solve the air robot path planning problem in a distinct environment [7]. ELD problem, which is "economic load dispatch" problem, has been solved through the proposed swarm-based nature-inspired algorithm called IWDs. The ELD is a technique of extracting the most capable, profitable, and trustworthy working of a power system by an expedition of all feasible electricity generation resources to provide load on the system. The basic goal of ELD is to reduce the total cost of production to the lowest, while considering all the production resources–related constraints. The overall results depict the effectiveness of the proposed method by giving better quality results [8].

Various works have enlightened the swarm-based nature-inspired optimization algorithm "IWDs," utilizing IWD to solve various problems, and the IWD is coming out to be quite promising. The algorithm has given outstanding contribution in the field of computational intelligence. IWD has been considered one of the best nature-inspired algorithms as the level of imitation of river in IWD is quite high. IWD has been used to solve various mathematical problems such as TSP, AMT, and MKP [9]. The new and extended version of IWD is utilized to solve complex computing problems such as workflows scheduling in cloud computing, which is attracting huge consideration nowadays because of the emergence of workflows as a paradigm to exhibit the complex computing problems. The classical working of IWD has been altered with an enhanced IWD that has been tested in distinct simulated cloud environments [10].

In order to figure out the requirements of web-based applications, web data extracted from the usage pattern is required. This usage pattern is revealed by the web usage mining. In a website the browsing pattern of a web user is exposed by the usage of data. Such information helps various businesses to improve their functionality. Now, biclustering method is used in order to group different browsing patterns to categorize the users. The IWDs algorithm has been used in this study to identify the optimal bicluster. It has been revealed through the experimental studies that the biclusters generated by the implementation of IWD possess great quality results in aspect of objective values and average correlation values [11].

Cloud computing is quite beneficial; as it provides resource scalability, and one has to pay as per the use only. The execution in a cloud computing scenario is carried out by dividing the application into small workflows and scheduling them in order to achieve the expected outcomes. Researchers have applied IWDs algorithm to schedule the various workflows with an intention of cost minimization [12]. Internet of Things (IOT) has become an emerging area in the field of computer science and technologies. The IOT aims at a world that is well interconnected with various physical and virtual devices, objects, etc. via Internet services. IOT can connect distinct technologies together with the standard communication solutions. As a result, it leads to the improvisation of conventional communication solutions. As IoT mainly connects different aspects of computer science through the Internet, thus this chapter offers a review on various protocols related to distinct aspects and their applicability toward an IOT realization [13]. The major issues that are affecting the health-care sector to the most have been targeted by the IOT, ranging from the medical inventions in order to deal with the incurable medical conditions to taking care of the aging population while being distant. The number of things involved in the abovementioned issues are quite huge, which leads to another challenge of connecting them safely together. So, researches offer the examination of the identification management issues of the IOT and suggest a framework to target the same issues. While performing the abovementioned examination, it has been assumed that the most of the health-care devices are cordlessly connected with each other in a mobile environment [14].

Modern communication platforms have been built for distinct IOT applications. The various applications that fall in the IOT domain have sensors, and these sensors are capable of sensing environment and sending data as well. The two of the IOT applications have been focused in this chapter that have been supervised by the cordless sensor nodes. Various IOT applications are stimulated with an intention of checking their working abilities and feasibility to give a wide-ranging resolution [15].

The link of the physical things with the digital world has been made possible with IOT. Different related works penlight the contribution of IOT in today's world of technology and collaboration of IOT with MANET by addressing a new system named MANET–IOT that provides better mobility to the user. Utilizing the combination of wireless sensor network (WSN) and MANET, a routing solution is suggested by different scientists for the same. Various results depict better outcomes in terms of energy consumption of various MANET–IOT collaborations [16].

1.3 Proposed neoteric swarm intelligence stationed IOT–IWD algorithm for ZRP optimization for digital health

The proposed approach discussed and presented in this section utilizes IOT–IWD approach with the following objectives:

The approach will connect IOT sensors to pump up machinery of pharmaceutical industry. The collected information will then be sent through a MANET platform. Historical data will be crunched in MANET, and the platform will predict breakdown and sent notifications to the pharmaceutical industry to repair or replace the equipment before the production fails. Since QOS plays a very important role, the IWD algorithm will optimize the performance like a modern platform and combines sensor information to provide visibility across the industry. Bottlenecks would be quickly spotted, and equipment could be identified as underused to make changes for attaining efficiency.

IWD-enabled approach will provide real-time data to manufacturers so that correct temperature is maintained while the drugs are transported. Data analytics will also be provided, which will enable the companies to check pattern of stock damage during temperature fluctuations. This will provide the visibility of supply chain management.

To achieve a "digital health" scenario the approach will provide medication reminders for patients that will help make patients "project managers" of their own health. The digital health platform will enable one to send information to physicians that can change the frequency of medications and track the results of drugs and their combinations.

1.4 Simulation results

Since the approach focuses on achieving efficiency through the IOT–IWD paradigm, QOS parameters are simulated to validate the methodology. Achieving high QOS parameters for different parameters will support the medical fraternity to achieve "digital heath" objective; this will not only fasten, revolutionize, and regulate the pharmaceutical industry but will also help make patients "project managers" of their own health. IWD approach will be a neoteric methodology in the case of various emergency situations

TABLE 1.1 Simulation setup.

Parameter	Value
Network area	1000×1000 m^2
Velocity	0–20 m/s
No. of nodes	50
Packet size	512 B
Traffic type	CBR
Number of connection	20
Packet rate	4 P/s
Simulation time	100, 200, 300, 400, 500, 600 s
Pause time	0

CBR, Constant bit rate.

where infrastructure will be negligible. The X-graph simulator, which is based upon NS2, authorizes the interpretation of the analysis of accomplishment of the recommended algorithm. In an NS2 simulator the X-graph is utilized to plot the distinct network parameters such as end-to-end delay (E2D), throughput, and normalized routing load (NRL). The X-graphs are believed to provide effective results including interactive plotting and graphing. In this work a random network scenario is created using NS2. An n number of nodes are dispensed over a network area possessing dimensions as 1000×1000 m^2. The considered nodes are varied, and their performance is judged by varying simulation time. The type of traffic considered for this network scenario is constant bit rate. The number of connections considered is 20. The simulation time for this scenario ranges from 100 to 600 seconds, with a node velocity of 0–20 m/s. Each data packet has a size of 512 B, and the packet rate is 4 P/s as listed in Table 1.1.

The PDR can be described as the ratio of data packets that are actually received at the receiver end to those which were originally sent by sender. So, in other terms, it can also be defined as PDR $= R_i/S_i$, where R_i is the number of nodes received by the receiver, and S_i is the number of nodes actually sent by the sender. Table 1.2 showcases the comparison betwixt the ZRP, RA-ZRP, and intelligent water drop routing algorithm (IWDRA)-ZRP for varying simulation time. Fig. 1.2 presents the same comparison in the form of a bar graph, and Fig. 1.3 presents the same in the form of an X-graph.

TABLE 1.2 PDR results for varying simulation time.

	PDR					
Simulation time (s)	100	200	300	400	500	600
ZRP	79	68.3	75.7	72.9	69.32	65
RA-ZRP	82.3	78.4	81.1	75.6	71.2	72
IWDRA-ZRP	83.1	81.6	84.2	79.9	78.3	79

PDR, Packet delivery ratio; *RA*, route aggregation; *ZRP*, Zone Routing Protocol; *IWDRA*, intelligent water drop routing algorithm.

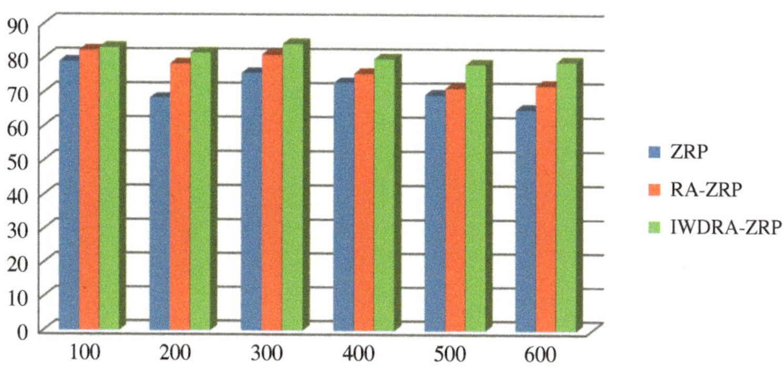

FIGURE 1.2 Bar graph for PDR. *PDR*, Packet delivery ratio.

2. E2D:

The E2D comprises all the possible delays that can be caused by temporary storage of the data during path discovery, latency, and retransmission of the data by intermediate nodes, processing delay, and propagation delay. It can be calculated as $D_i = (T_r - T_s)$, where T_r represents the receive time and T_s represents the sent time of the packet. Table 1.3 displays the comparison results of ZRP, RA-ZRP, and IWDRA-ZRP for varying simulation time. Fig. 1.4 presents the same comparison in the form of a bar graph, and Fig. 1.5 presents the same in the form of an X-graph.

3. Throughput:

Throughput generally gives the average at which over a communication network a data packet is delivered successfully from one node to another. It can be measured in bits per second with the given function:

$\text{Throughput} = \frac{D_p \times P_s}{\text{total duration of simulation}}$ where D_p symbolizes the number of delivered packets, and P_s stands for the size of a packet. Table 1.4 presents the comparison analysis of ZRP, RA-ZRP, and IWDRA-ZRP for varying

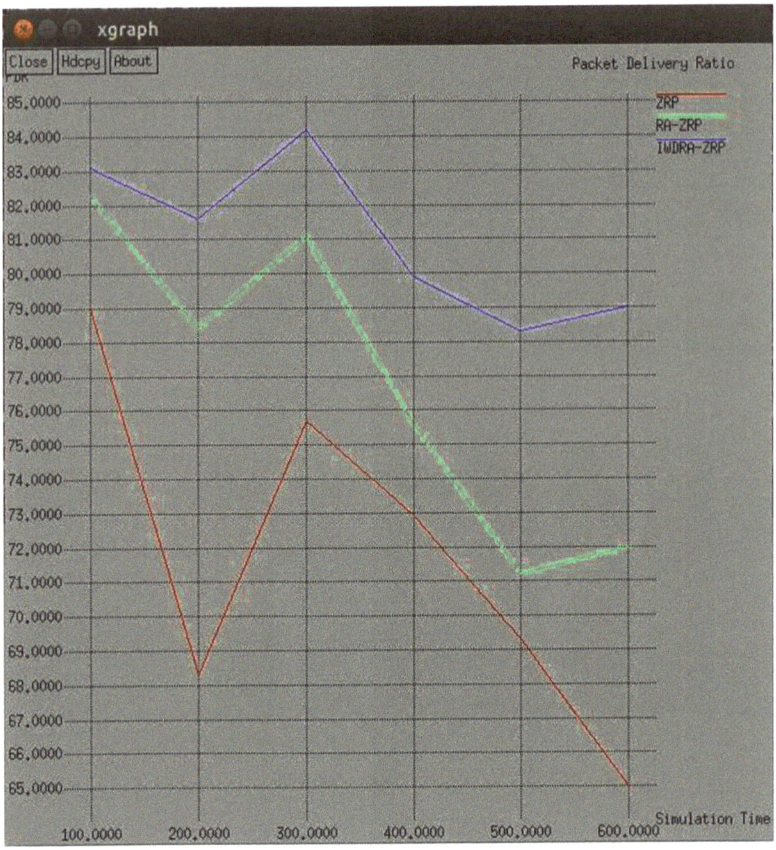

FIGURE 1.3 X-graph for PDR. *PDR*, Packet delivery ratio.

TABLE 1.3 E2D results for varying simulation time.

	E2D					
Simulation time (s)	100	200	300	400	500	600
ZRP	1.7	1.5	1.8	1.2	1.4	1.2
RA-ZRP	1.2	1.1	0.7	0.73	0.82	0.9
IWDRA-ZRP	0.9	0.7	0.5	0.3	0.71	0.51

E2D, End-to-end delay; *RA*, route aggregation; *ZRP*, Zone Routing Protocol.

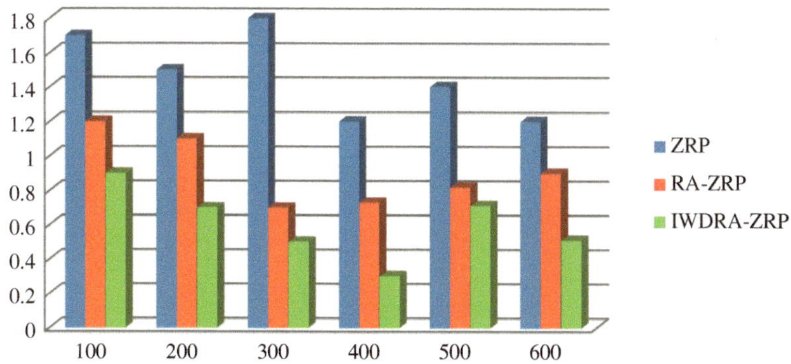

FIGURE 1.4 Bar graph for end-to-end delay.

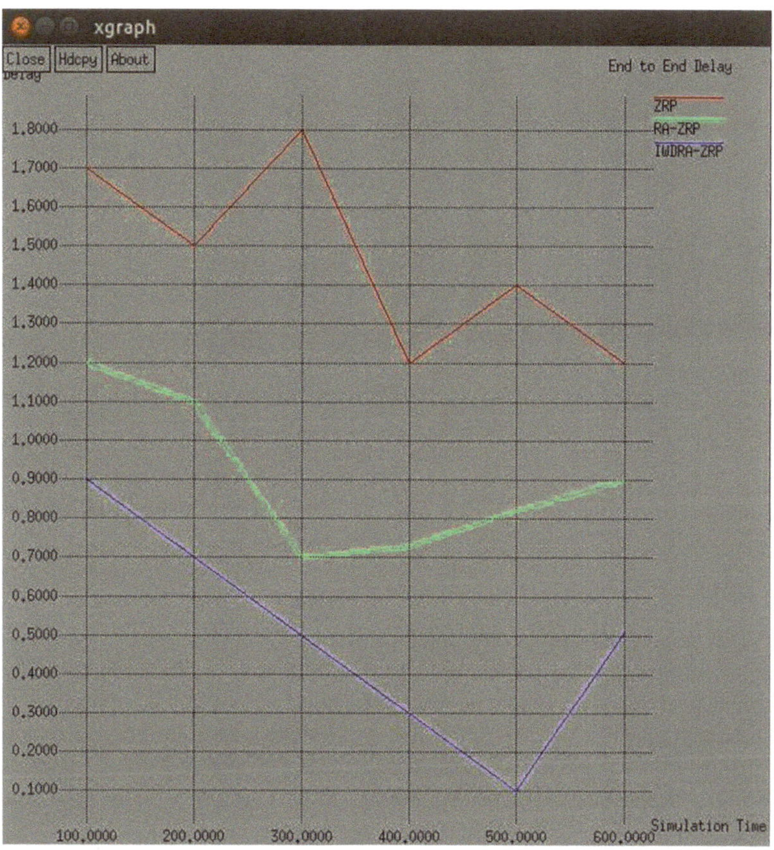

FIGURE 1.5 X-graph for end-to-end delay.

TABLE 1.4 Throughput results for varying simulation time.

	Throughput					
Simulation time (s)	100	200	300	400	500	600
ZRP	85.3	81.7	82.9	79.4	78.9	78.4
RA-ZRP	102.65	91.4	95.7	90.8	87.3	84.67
IWDRA-ZRP	107.41	102.8	105.7	99.3	95.4	87.7

RA, Route aggregation; *ZRP*, Zone Routing Protocol.

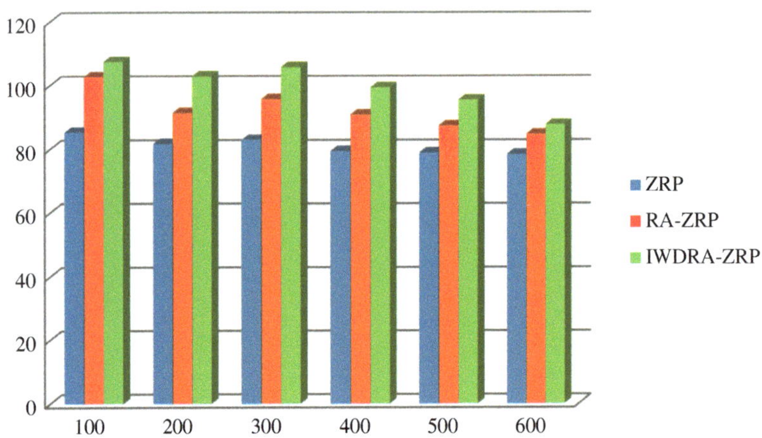

FIGURE 1.6 Bar graph for throughput.

simulation time. Fig. 1.6 presents the same comparison in the form of a bar graph, and Fig. 1.7 presents the same in the form of an X-graph.

4. NRL:

NRL is actually depicts the number of routing packets that are transmitted per data packet that has been sent to the destination. Also each of the forwarded packets is counted as one transmission. Table 1.5 presents the comparison analysis of ZRP, RA-ZRP, and IWDRA-ZRP for varying simulation time. Fig. 1.8 presents the same comparison in the form of a bar graph, and Fig. 1.9 presents the same in the form of an X-graph.

5. PLR:

PLR can be measured as the percentage of packets lost with respect to packets sent. It can be calculated using the following function:

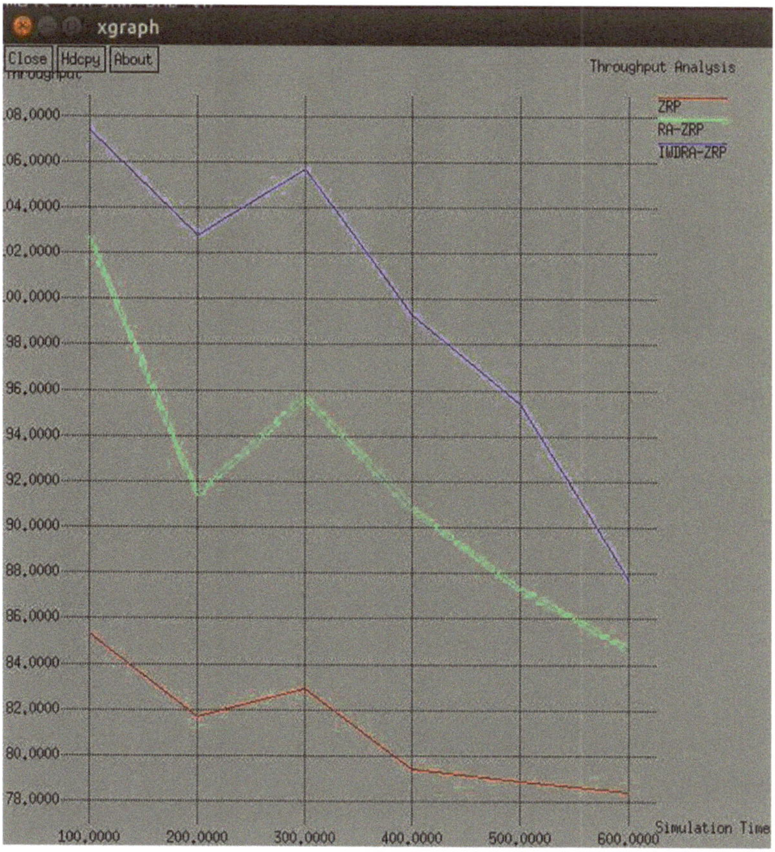

FIGURE 1.7 X-graph for throughput.

TABLE 1.5 NRL results for varying simulation time.

	NRL					
Simulation time (s)	100	200	300	400	500	600
ZRP	5.641	6.684	6.983	7.095	7.148	7.382
RA-ZRP	4.023	4.897	5.038	6.073	6.372	6.842
IWDRA-ZRP	4.009	4.031	4.154	4.927	4.625	4.923

NRL, Normalized routing load; *RA*, route aggregation; *ZRP*, Zone Routing Protocol.

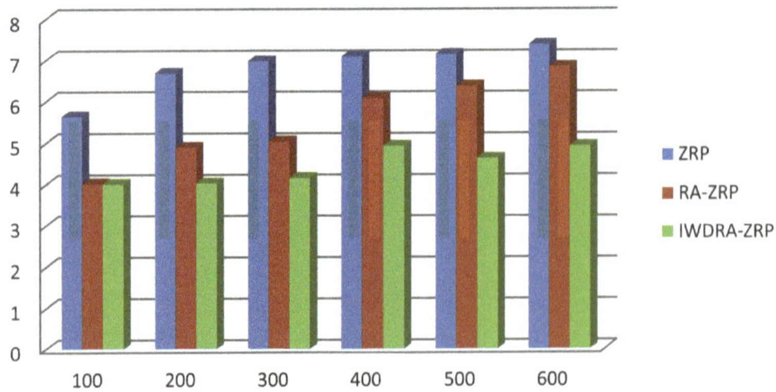

FIGURE 1.8 Bar graph for normalized routing load.

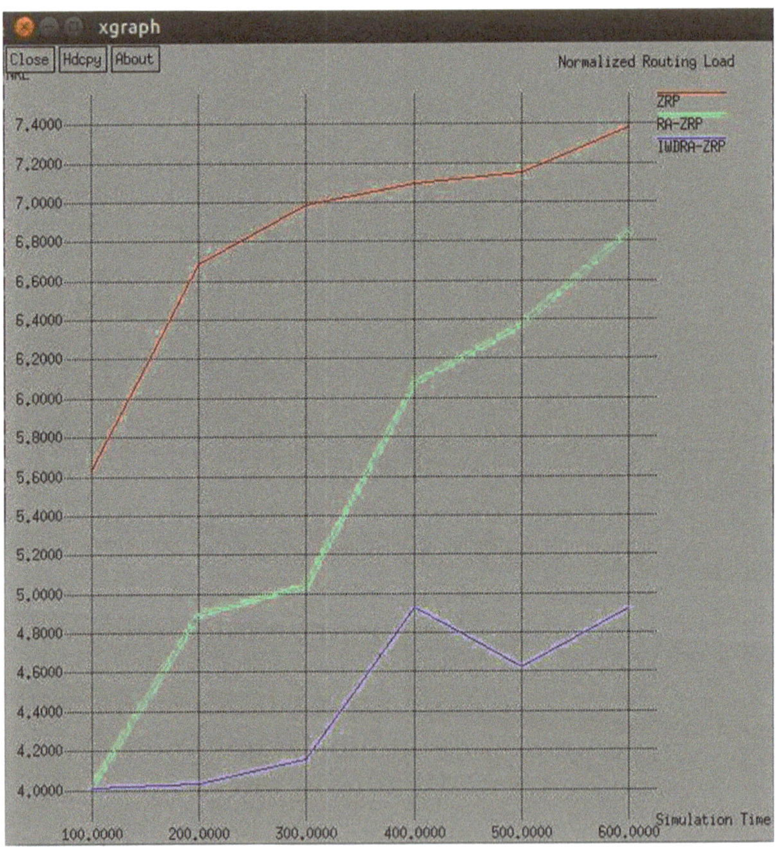

FIGURE 1.9 X-graph for normalized routing load.

TABLE 1.6 PLR results for varying simulation time.

	PLR					
Simulation time (s)	100	200	300	400	500	600
ZRP	21	31.7	24.3	27.1	30.68	35
RA-ZRP	17.7	21.6	18.9	24.4	28.8	28
IWDRA-ZRP	16.9	18.4	15.8	20.1	21.7	21

PLR, Packet loss ratio; *RA*, route aggregation; *ZRP*, Zone Routing Protocol.

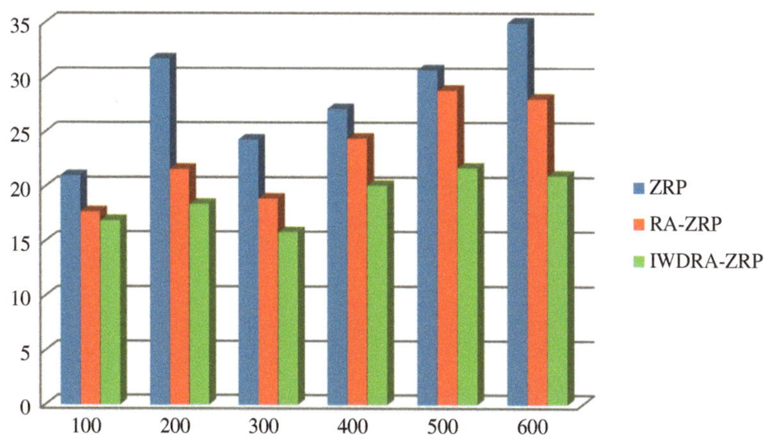

FIGURE 1.10 Bar graph for packet loss ratio.

Packet loss ratio $= (P_s - P_l) \times \frac{100}{P_s}$ where P_s is the number of packets sent, and P_l is the number of packets lost during transmission. Table 1.6 presents the comparison analysis of ZRP, RA-ZRP, and IWDRA-ZRP for varying simulation time. Fig. 1.10 presents the same comparison in the form of a bar graph, and Fig. 1.11 presents the same in the form of an X-graph.

1.5 Conclusion and discussions

The abovementioned simulation results and analysis lead to the conclusion that there is a paramount and radical growth required for pharmaceutical and health-care industry. Owing to the needs of these industries, the result analysis of the proposed study emphasizes the fact that IOT–IWD approach using the five QOS parameters emphatically plays a huge role in revolutionizing the pharmaceutical and health-care sector as all the QOS parameters

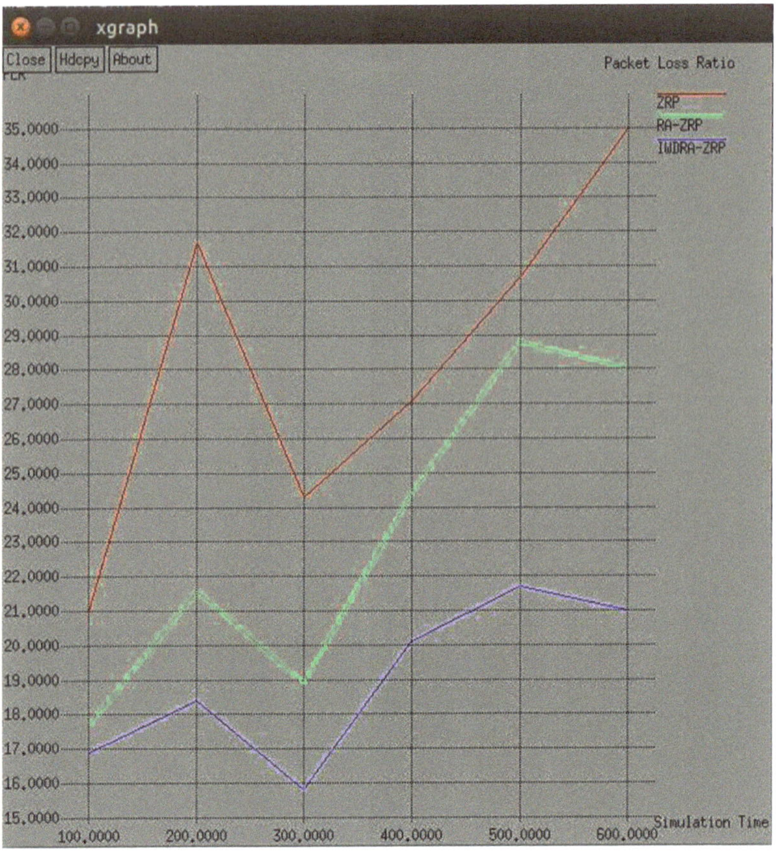

FIGURE 1.11 X-graph for packet loss ratio.

influence on the development of this revolution as depicted in Tables 1.1–1.6 and Figs. 1.2–1.11.

In the digital world today, it is very important that MANET converges in the smart environment for different applications of IOT domain. Sensors that are used in the IOT applications will sense through the environment and will send the data though the node gateway that will send that data to the node of MANET for pharmaceutical analysis and health-care operations. The most challenging part of this chapter is to converge MANET with IOT as different nodes have different levels of power and the various traditional-based heterogeneous protocols apart from the IWD suffers from cochannel interference. Also the simulation results of the proposed study depicted five QOS parameters agree upon the related literature for the course of this neoteric methodology. The validated results of the proposed approach can be adopted by other research analysts to further explore the details related to the subject matter of pharmaceutical and health-care industry leading to "digital health platform." Every patient and individual industrially requires revolutionized health-care and pharmaceutical sector for gaining maximum benefits. Latest technologies are required to be applied to health-care and pharmaceutical operations and management for alleviating and eradicating the problems and malpractices prevalent in the industry. The abovementioned simulation results conclude that IOT–IWD approach will be significant for health-care and pharmaceutical sectors development and growth. Hence, these industries and sectors should focus on the neoteric approaches for proper adoption and diffusion. The researchers recommend top management as well as the stakeholders of health-care and pharmaceutical industries to provide maximum contribution and support for realizing the potentials of the proposed approach.

The proposed approach has presented and overall solution of presenting a network protocol–based IWD solution, routing through MANET and implementation of IOT application simulated through an NS simulator showing QOS and performance stability. The pharmaceutical and health-care industries should focus on providing sufficient training to their staff and stakeholders to be abreast for the latest information communication technology (ICT)-based paradigm and approaches. Government as well as the private sector should evolve for the global revolution in the ICT for enabling as well as promulgating policies and encourage on the best practices for digital health revolution. Pharmaceutical and health-care industries should harness benefits of the proposed approach for remote patient monitoring, detection and prevention of frauds, prediction and control of epidemics and telemedicine technologies.

References

[1] H.S. Hosseini, The intelligent water drops algorithm: a nature-inspired swarm-based optimization algorithm, Int. J. Bio-Inspired Comput. 1 (1/2) (2009) 71. Available from: https://doi.org/10.1504/ijbic.2009.022775.

[2] H.S. Hosseini, Optimization with the nature-inspired intelligent water drops algorithm, in: W.P. dos Santos (Ed.), Evolutionary Computation, I-Tech, Vienna, Austria, 2009 p. 572. ISBN: 978-953-307-008-7.

[3] B. Aldeeb, et al., A survey on intelligent water drop algorithm, Int. J. Comput. Technol. 13 (10) (2014) 5075–5084.

[4] L. Sayad, L.B. Medjkoune, D. Aissani, IWDRP: an intelligent water drops inspired routing protocol for mobile ad hoc networks, Wireless Pers. Commun. (2017). Available from: https://doi.org/10.1007/s11277-016-3692z.

[5] H. Duan, S. Liu, J. Wu, Novel intelligent water drops optimization approach to single UCAV smooth trajectory planning, Aerosp. Sci. Technol. 13 (2009) 442–449. Available from: https://doi.org/10.1016/j.ast.2009.07.002.

[6] H.S. Hosseini, Problem solving by intelligent water drops, in: Proceedings of the IEEE Congress on Evolutionary Computation, IEEE Xplore Press, Singapore, 2009, Sept. 25–28, doi:10.1109/CEC.2007.4424885.

[7] H. Duan, S. Liu, X. Lei, Air robot path planning based on intelligent water drops optimization, in: Proceedings of the International Joint Conference on Neural Networks, IJCNN, 2008, doi:10.1109/IJCNN.2008.4633980.
[8] S.R. Rayapudi, An intelligent water drop algorithm for solving economic load dispatch problem, Int. J. Energy Power Eng. 5 (10) (2011).
[9] H.S. Hosseini, Optimization With the Nature-Inspired Intelligent Water Drops Algorithm, 2009, pp. 572, ISBN 978-953-307-008-7.
[10] S. Elsherbiny, et al., An extended intelligent water drops algorithm for workflow scheduling in cloud computing environment, Egypt. Inf. J. 19 (1) (2018) 33–55. ISSN 1110-8665.
[11] Kavitha, Biclustering using intelligent water drops algorithm with leader clustering approach for web usage mining, Int. J. Pure Appl. Math. 10 (6) (2016).
[12] C.B. Sivaparthipan, J. Rathinaraja, S.S. Kumar, Intelligent water drop algorithm for workflow scheduling in cloud computing, Int. J. Adv. Inf. Eng. Technol. 9 (9) (2015).
[13] D.G. Reina, et al., The role of ad hoc networks in the internet of things: a case scenario for smart environments, in: Internet of Things and Inter-Cooperative Computational Technologies for Collective Intelligence, SCI, vol. 460, 2014. 89–113, doi:10.1007/978-3-642-34952-2_4.
[14] C. Chibelushi, A. Eardley, A. Arabo, Identity management in the Internet of things: the role of MANETs for healthcare applications, Comput. Sci. Inf. Technol. 1 (2) (2013) 73–81. Available from: https://doi.org/10.13189/csit.2013.010201. Available from: http://www.hrpub.org.
[15] S. Mukherjee, G.P. Biswas, Networking for IoT and applications using existing communication technology, Egypt. Inf. J. 19 (2018) 107–127.
[16] R. Bruzgiene, L. Narbutaite, T. Adomkus, MANET network in Internet of Things system, 2013, doi:10.5772/66408.

Further reading

A. Dorri, R.A. Seyed, E. Kheyrkhah, Security challenges in mobile ad hoc networks: a survey, Int. J. Comput. Sci. Eng. Surv. 6 (1) (2015) 15–29.
C. Mbarushimana, A. Shahrabi, Comparative study of reactive and proactive routing protocols performance in mobile ad hoc networks, in: 21st International Conference on Advanced Information Networking and Applications Workshops AINAW '07, Niagara Falls, 2007, pp. 679–684.
D. Mehta, I. Kashyap, S. Zafar, Routing optimization in cloud networks, Int. J. Adv. Res. Comput. Sci. 8 (2) (2017) 16–18.
C.E. Perkins, E.M. Belding-Royer, S.R. Das, Ad hoc on demand distance vector (AODV) routing, Internet Draft, Internet Engineering Task Force (IETF), Cornell University, 2002.
N. Sharma, S. Zafar, U. Batra, Trust based hybrid routing approach for securing MANET, in: INDIACom-2, 2018; ISSN 0973-7529; ISBN 978-93-80544-28-1.
N. Sharma, S. Zafar, U. Batra, Catechize global optimization through leading edge firefly based Zone Routing Protocol, Recent Patents on Computer Science, 12, Bentham Science Publishers, 2019, pp. 1–11.
H. Tian, Hermes: a scalable sensor network architecture for robustness & time energy awareness, A Dissertation Proposal Dept. of Computer Science, University of Virginia, 2013.

S. Zafar, Cyber secure corroboration through CIB approach, Int. J. Inf. Technol. 9 (2) (2017) 167–175.

S. Zafar, M.K. Soni, Secure routing in MANET through crypt-biometric technique, in: Proceedings of the Third International Conference on Frontiers of Intelligent Computing: Theory and Applications (FICTA), 2014, pp. 713–720.

S. Zafar, M.K. Soni, M.M.S. Beg, An optimized genetic stowed approach to potent QOS in MANET, Procedia Comput. Sci. 62 (2015) 410–418.

Chapter 2

A survey on Internet-of-Thing applications using electroencephalogram

Debjani Chakraborty, Ahona Ghosh and Sriparna Saha
Department of Computer Science and Engineering, Maulana Abul Kalam Azad University of Technology, Kolkata, India

2.1 Introduction

Being a very complex part in human body, the human brain can be mainly segregated into four lobes, namely, frontal, parietal, temporal, and occipital. Fig. 2.1 shows the different parts of human brain. The elementary part of the brain consists of different types of neurons. These neurons generate electrical signal while transferring information (e.g., to indicate pain in hand of a subject) from one part of human body to the brain. To record these small value electrical signals, an electronic device, electroencephalogram (EEG), is used. Hans Berger in the year of 1929 invented EEG device for human subjects [1]. By the help of that device the electrical activity of human brain's neuron can be monitored and recorded in a noninvasive way. In this

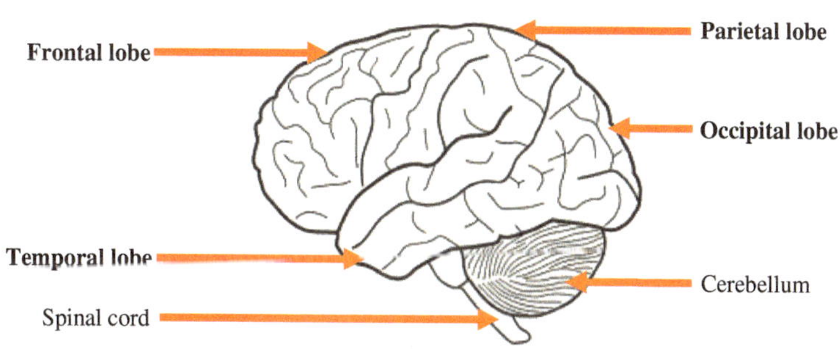

FIGURE 2.1 Depiction of different lobes of human brain.

Emergence of Pharmaceutical Industry Growth with Industrial IoT Approach.
DOI: https://doi.org/10.1016/B978-0-12-819593-2.00002-9

technique, electrodes are placed on the human scalps and those electrodes are connected to an amplifier and then to a recording machine, which measure these impulses in microvolt after amplification. The recording machine changes these electrical signals into a series of wavy lines on a moving sheet of graph paper.

These EEG signals can be used for several biomedical engineering applications [2], by measuring the difference between EEG signals of patients with some brain abnormalities with healthy subjects. The wide applications of biomedical engineering deal with treating epilepsy [3], encephalitis [4], drug effect [5], sleep disorder [6], dementia [7], brain death [8], memory loss [9], etc.

In the later parts of this chapter, different types of acquisition techniques are discussed. Next, we briefly enlightens on feature extraction, feature classification algorithms used for EEG-based biomedical applications. Then, how acquired EEG signals can be used in different Internet-of-Thing (IoT) applications are also elaborated.

2.2 Electroencephalogram acquisition techniques

There are two types of techniques that are generally used in EEG acquisition: invasive and noninvasive [10]. Fig. 2.2 describes the working process of EEG acquisition. At first, the EEG signal is received by placing the electrodes on the brain. After the signal acquisition the signal gets processed and according to the commands placed on it, applied in IoT domain applications.

While acquiring the EEG signals from human brain, the following challenges are faced:

- Getting the right codes for desired acquisition of signals
- Avoiding the damage to the human brain
- Preventing attacks of virus while acquiring the signals

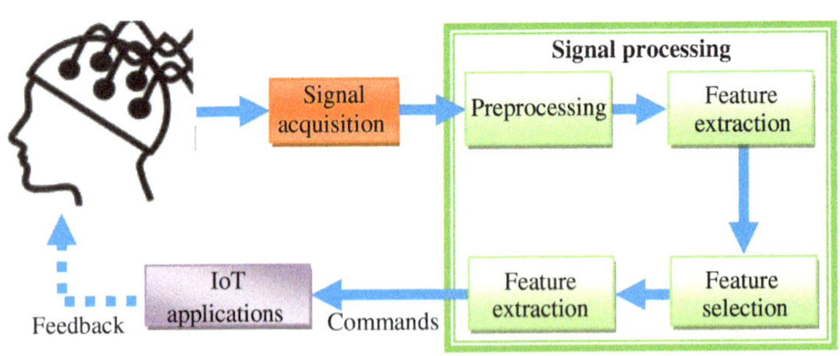

FIGURE 2.2 Working process of EEG acquisition technique. *EEG*, Electroencephalogram.

- Ethically recording the data following proper human rights
- Requirement of extensive training of the subjects before taking the data

The pros and cons of the already stated acquisition methods are described in respect of the challenges are given in the following subsections.

2.2.1 Invasive

This method is also known as deep brain recording [11]. To implement EEG acquisition a skilled person needs to place electrodes inside the scalp using surgery. As the electrodes for invasive technique are placed on the exposed brain and hence, the quality of signal is much better than noninvasive technique, that will be described later on. This type of invasive technique is often used for severe epileptic patients to identify location of seizures [12]. Though this process produces high-quality EEG signal, sometimes scar tissues can built up, which, in turn, make the signal weaker and, in the worst case, signal may get lost [13].

2.2.2 Noninvasive

For noninvasive technique the electrodes are placed outside of the brain scalp; hence, no surgery is required. This type of method lacks in quality of EEG signals. As the recording of the signal is taken far away from the source, and so signals may get distorted. In addition, the strength of obtained signals is of lesser amplitude that those for invasive ones [14].

The acquisition systems can be segregated based on electrodes type—gel, water, or dry [15]. Short circuit noise level is low in case of water-based system. The highest P300 spelling accuracies are obtained in hydrogel-based system. The least inconvenience is available by using dry electrode-based system. Fig. 2.3 shows how the small metal plates called electrodes should be placed to record the mental activity from the scalp. Each electrode is labeled with a letter and number to indicate its position.

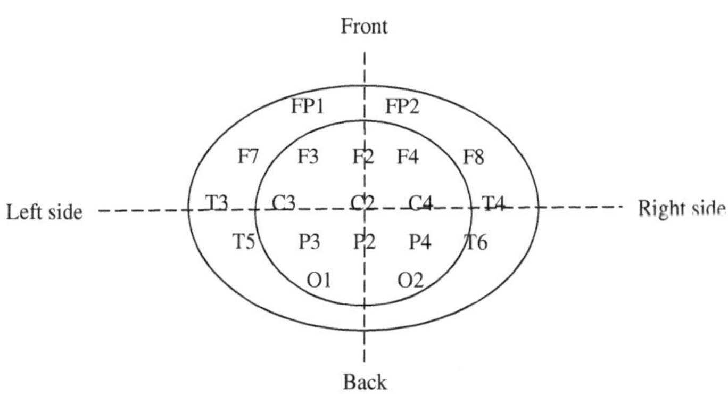

FIGURE 2.3 Placements of electrodes in the scalp for recording of mental activity.

2.3 Channel selection techniques

With the evolution of channel selection algorithms a large number of recording channels are used due to the of low-cost interfaces availability. The main motivation of channel selection is to improve the performance of manifold model, which provides dimensionality reduction and faster processing and identifies the area of brain that produces activity of class-event. The special steps that are followed by the channel selection are illustrated here. There are so many algorithms that are used to generate steps for heuristic search process; examples are complete search, sequential search, or random search. There are so many applications in which a trained expert selects channels based on his/her experience. These types of methods are used for evaluating of subsets. When a stopping criterion is satisfied, the process of channel evaluation is terminated (search is completed or a threshold is reached). At last, using prior knowledge, the selected subset channel subset is validated. Here, mainly five main candidate evaluation strategies are discussed, such as filtering, wrapping, embedded, hybrid, and human-based techniques.

2.3.1 Filtering

Using a search algorithm, the filtering method [16] uses an independent criterion of evaluation, such as measure of distance, information, dependency, consistency, and estimating the subset of channel. There are some advantages of this technique such as high speed, liberty from classifier, and scalability, but the drawback is that they have low accuracy.

2.3.2 Wrapper

To evaluate channel subset the wrapper method of classification algorithm in Ref. [17] is used, which is produced by a search algorithm. In wrapper, class A denotes a classifier and γ denotes the best value of the criterion evaluation. The evaluation of every candidate is obtained by training and testing a specific classification algorithm.

2.3.3 Embedded

During learning process the channels are selected based on criteria generated based on classifier construction using these embedded techniques [18]. Embedded techniques are used to achieve connection between the channel selection and the classification. They are not computationally much expensive and not prone to overfitting. With appreciated magnitude, they are used based on recursive way of channel elimination to keep only channels.

2.3.4 Hybrid

To avoid the prespecification of stopping technique a hybrid technique [19] is more useful than filtering technique and a wrapper technique. Mainly, an independent measure and mining algorithms are used for evaluating channel subset using two threshold values.

2.3.5 Human-based technique

Human-based techniques are used based on the previous knowledge of the observer, who evaluates the importance of channels for a specific application.

2.4 Different brain signals

There are three types of brain signals that are needed to be extracted from the acquired EEG signal for IoT applications:

- neural oscillation [20],
- event-related potential (ERP) [21], and
- somatosensory evoked potential (SSEP) [22].

Neural oscillation checks all levels of central nervous system. ERP is used to measure brain's electrical activity by employing electrodes on the scalp. SSEP is a brain reaction aroused by sensory stimulation having a constant frequency. The neuron of the brain generates electrical impulses that fluctuate with a rhythm of specific manner.

The activated neurons generate four kinds of waves in different states of neuron activity. These can be classified as alpha, beta, theta, and delta waves [16] as shown in Fig. 2.4. An awake but resting person produces rhythmic

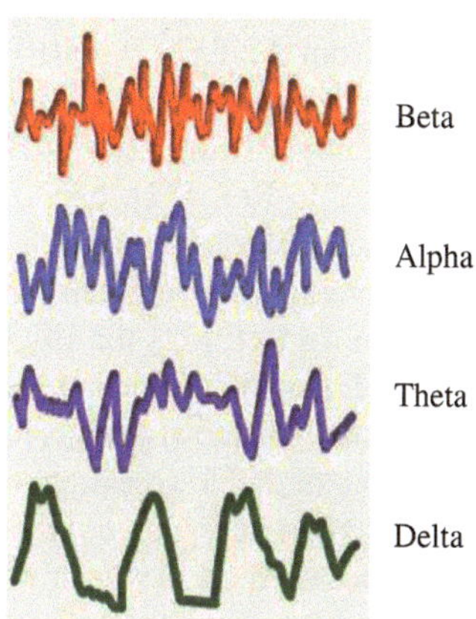

FIGURE 2.4 Pictorial representation of brain wave patterns.

alpha wave with a range of 8–13 Hz. Alpha waves originate from occipital region. An alerted person shows unsynchronized high-frequency EEG. During intense mental activity, beta wave originates from frontal and parietal region in the range of 14–30 Hz. Theta wave occurs in children as well as adult under stress and drowsiness in between 4 and 7 Hz. Higher amplitude and slower frequency waves, called delta waves, originate when a person is in deep sleep with low-frequency range 1–3 Hz. These brain waves can be detected by the electrodes placed on the scalp [23].

An ERP is used for used sensory, cognitive, and motor-related events [24]. For checking the response of brain for a certain stimulus, change in voltage is platted over a period of time. To verify the response to that stimulus the experimenter has to conduct many trails and then average the result. By averaging, random activity of the brain can be canceled out and the experimenter gets the remaining relevant waveform and this is known as ERP [25].

From touch stimulus the brain's electrical activation is known as SSEP [26]. This signal can easily be measured using noninvasive techniques. From upper or lower limb, this stimulation when reaches to the brain, then SSEP signal can be recorded from the scalp [27].

2.5 Preprocessing of electroencephalogram signals

Preprocessing of EEG signal [28] is an essential and important step in any brain–computer interface (BCI)–based applications. It aids to eliminate unwanted artifact from the EEG signal and makes it suitable for further processing. In Ref. [29] the preprocessing technique used is blind source separation (BSS). The EEG signal is initially filtered using a notch filter centered at 50 Hz, followed by BSS for EEG artifact removal. A finite impulse response filter has been used between 8 and 25 Hz to get the required EEG signal for feature extraction. In Ref. [30], basic filtering has been done to remove the unwanted artifacts from the EEG signal. The signal has been first high-pass filter with lower cut-off frequency 0.1 Hz followed by a low-pass filter with higher cut-off frequency 50 Hz. This band-pass filtering aided in the removal of power line noise. In Ref. [31], band-pass filtering is used to eliminate any signal that is not in the range of P300 frequency range of the EEG signal. Moreover, data averaging over many trials has been done to upgrade the signal-to-noise ratio. In Ref. [32] the EEG signal is filtered using a band filter between 8 and 12 Hz, which corresponds to the Mu rhythm frequency range. In Ref. [33] a new method, multichannel Wiener filter, has been proposed for enhancement of EEG signal. It is an advanced method than the BSS method. In Ref. [34] a new technique called joint approximate diagonalization of Eigen-matrices method has been implemented to calculate the independent components and hence remove unwanted artifacts from the EEG signal. In Ref. [35] an artifact removal technique has been proposed that incorporates Lifting Wavelet Transform, rather than wavelet transform (WT), with independent component analysis (ICA) technique for effective removal of artifacts from EEG signal. This method provides a better and efficient way to eliminate artifacts than traditional ICA method. In Ref. [36] the EEG signals were first band-pass filtered to get the desired band of frequency, followed by ICA technique to remove any unwanted artifacts from the signal. In Ref. [37], discrete WT (DWT) has been implemented to eliminate noise from the EEG signal. In Ref. [38] an adaptive filter through WT has been implemented for artifact removal from EEG signal. The signal is first decomposed up to eight levels using WT, and then it is subjected to adaptive filtering process. Wavelet reconstruction is used to get the artifact free EEG signal. In Ref. [39], improved WT technique, called the wavelet packet transform, has been implemented. It is an improvement over the traditional WT because in this case while fragmenting the signal into frequency subbands, it maintains the temporal form of the signal. In addition, the samples of the subbands after wavelet packet representation remain the same as that of the original signal.

In Ref. [40], various preprocessing techniques for EEG signal have been reviewed. The first technique described is the use of basic filtering to remove unwanted artifact from the EEG signal. A basic notch filter can be used to remove 50 Hz power supply signals. A band-pass filter can also be used to get the desired band of frequency. The second technique discussed is the adaptive filtering. Here, instead of a fixed filter, a filter that adapts to the spectrum of the recorded EEG is used for effective artifact removal. The last technique discussed is the BSS. In Ref. [41] also, other preprocessing techniques have been analyzed. Techniques, such as Wiener filters or adaptive filters, give better performance than conventional basic filtering of EEG signals. Another effective technique discussed in the chapter for artifact removal is the ICA, which is implemented for the removal of power line noise. Various preprocessing techniques have been discussed in Ref. [42] for effective artifact removal from EEG signals. Bipolar montage, Laplacian montage, and common spatial patterns are some of the techniques discussed in this chapter.

2.6 Feature extraction and classification from electroencephalogram signal

There are many feature extraction algorithms used to analyze EEG signal. Some of the well-known signal processing techniques used in EEG signals for IoT applications are described in this section. There are several signal feature extraction algorithms that are used currently. There are many research associates who sometimes combine multiple feature extraction methods for data analysis. But, this process often leads to feature dimension expansion and creation of certain features that are redundant in nature.

The reduction of feature space dimension can be done through feature selection, thereby increasing better results for IoT applications.

In general, the following three types of feature extraction method are used, as shown in Fig. 2.5.

- Time domain signal extraction [43]
- Spatial domain feature extraction [44]
- Feature extraction transformation model

Feature extraction of EEG signal is an important step in any BCI-based applications. It helps to extract the most relevant features from the EEG signal, thus giving a more precise description and hence making it suitable for further processing.

An EEG is arbitrary and unsteady signal; thus only fast Fourier transform (FFT) cannot efficiently differentiate EEG signals. Therefore other nonlinear methods are used to extract features, namely, sample entropy [45], Hurst exponent [46], Lyapunov exponent (LLE) [47], and multifractal detrended fluctuation analysis [48] which are popular for feature extraction.

An nondeterministic polynomial (NP) problem often arises out of optimal feature subset selection; hence, for optimal feature subset searching, genetic algorithm (GA) is often used [49]. A binary array having length equal to the number of features is used by the algorithm as individuals. The value comes to 1 for a feature if selected in the array, otherwise it is 0. The algorithm function is minimum of $(FPR - (1 - TPR))$, in which TPR and FPR are true positive rate (sensitivity) and false positive rate (fall-out), respectively.

This algorithm is used to conventional filtering and cannot be used in artifact. Because the EEG signal artifact has overlapping spectra. It is aligned to optimization algorithms. The adaptive feature is indicated by the error signal between the main signal and the output obtained from the filter. The least mean square algorithm [50] is the most commonly used criterion for optimization purpose. This algorithm is so very simple that this is used across the globe for its application.

The discrete Fourier transform (DFT) [51] is a digital signal that describes signal amplitude versus sampling time constant in time domain frequency. Comparing between the signal frequency and digital signal samples, it can be inferred that the former is more useful than the latter. There is a requirement to develop the digital signal based on its frequency and, that

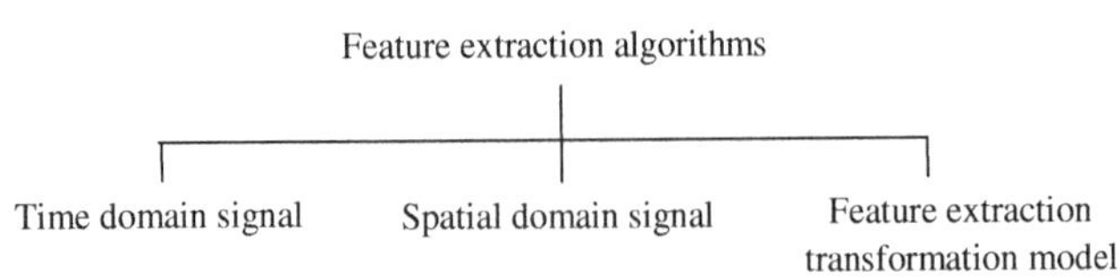

FIGURE 2.5 Different types of feature extraction methods.

too, in the domain of frequency, which is more commonly known as signal spectrum, is the algorithm that transforms time domain signal components to the frequency domain ones, thereby making correlation, or rather, connection between these two representations. So that, one can do analysis frequency by this DFT algorithms of a time domain sequence.

Discrete cosine transform (DCT) [52] is very useful for transforming the encoding video and audio tracks on computers. These algorithms actually apply in digital signal processing and particularly in coding transformation for data compression [53]/decompression. Correlated input data and concentrate of DFT are used in its energy in first few coefficients of transform.

Continuous WT (CWT) [54] method is used to represent the translation and scale parameter for the wavelet continuously. DWT [55] method is used to transform any wavelet to discretely sample. In using other WTs the main advantage is temporal resolution over Fourier transforms. DWT can capture both location information and frequency in respect of time. We can say that DWT is very useful for multiscale representation where it was the weakness of the CWT.

For analyzing a signal in both the frequency and time domains at the same time, time–frequency distribution analysis is used [56]. Instead of applying one-dimensional signal transform, a two-dimensional function is implemented in the two-dimensional real plane, where the time–frequency analysis is carried out via a time–frequency transform. In this case the output of the function can be real or complex-valued. The transformation is done by dividing the signal length into smaller intervals/segments (window-based method). For this, preprocessing is needed.

EEG data analysis by FFT [57] method involves mathematical tools or means. EEG signal characteristics are calculated by power spectral density (PSD) estimation [58]. An example of this method is Welch's method [59].

WT [55,60] performs a valuable role in the recognition and diagnostic field. It is used to compress the time-varying signal in biomedical engineering. As the EEG signal is a time-varying entity, this method is well suited for feature extraction from the raw data in time-frequency domain. This method is a spectral calculating method in which any general function of wavelet can be manifested as an infinite series. Since WT is applicable for variable sized windows, it provides a more adjustable way representing signal in time-frequency domain.

To estimate the frequency of signal and power from artifact prone measurement of EEG signals, eigenvector technique has been used in Ref. [61]. The parts of eigenvector methods are Pisarenko's method [62], Music method [63], and minimum-norm method (MNM) [64]. The Pisarenko technique is for the noise subspace eigenvector corresponding to be the minimum eigenvalue by finding out a linear combination of the whole noise subspace eigenvectors with the help of the minimum-norm technique. The Music method eliminates the false zeros by using the spectra's average equivalent to artifact subspace of the whole eigenvectors. MNM uses false zeros in the unit circle by separating them from real zeros. The PSD of the EEG signal can also be calculated by autoregressive (AR) method [65] using parametric approach. Hence, this method has no problem of leakage of spectral and thus yields better resolution of frequency not similar to nonparametric approach. Using estimating the coefficient that is the parameter of linear equation under consideration, we get the measurement of PSD. Coefficient and AR parameter are measured by the result of exploiting biased in the Yule–Walker method [66]. It approximates the data function using autocorrelation. Using this method, we find out the minimization of least squares of the forward prediction error. Burg's method [67] is an AR spectral estimation that reduces the error of the prediction of forward and backward errors for satisfying recursion of Levinson–Durbin. Burg's method calculates the coefficient of reflection directly not estimating the function of autocorrelation. This method has some advantages. Burg's method can measure the record of PSD looking exactly similar to original data. It can give closed paced sinusoids in signals when it has lowest level of noise.

The idea of sample entropy proposed by Richman and Moorman [45] results in better performance than approximate entropy proposed by Pincu et al. [68], which is a method to measure regularity for quantifying the complexity levels within a time series. Sample entropy is sometimes useful for extraction of signal features and defined by the given equation.

$$fse = sampleEntropy\,(x, sn, sm, sr) \tag{2.1}$$

where n is the signal length, m is the embedding dimensions, and sr is the similar tolerance.

Hurst exponent, proposed by H.E. Hurst [46], is utilized for chaos–fractal analysis of a time series data. With this exponent, whether that data are random walk or biased random walk can be predicted. The exponent is calculated as follows:

$$fhust = HursExponent(x) \tag{2.2}$$

Chaotic motions can be identified using Lyapunov exponent for fast computing. For EEG signals, features can be measured using this exponent by the following equation:

$$fue = LLE(Y) \tag{2.3}$$

GA is also called random search method [69]. It simulates the law of biological growth. Generally, GAs maintain probability optimization method rule. For searching the feature of domain of frequency, the standard *GA* is followed.

$$GA = \{C, E, P_0, M, \phi, \neg, \psi, T\} \tag{2.4}$$

where C is the chromosome coding, E is the fitness function of individual, P_0 is the initial population, M is the initial population's size, ϕ is selection operator, $\neg$ is the crossover operator, ψ is the mutation operator, and T is the given termination condition.

This method is used to follow GAs for searching a suitable classification features in the spectrum of frequency.

2.7 Classification of electroencephalogram signals based on features

With the rapid growing application of IoT, diversity in the process of classification is being observed for features extracted from EEG signal. It can be divided into two parts as shown in Fig. 2.6.

2.7.1 *k*-Nearest neighbor

k-Nearest neighbor (*k*-NN) is a classifier [70] which follows a nonparametric approach; it used to classify a given data point according to the neighbor's majority. Using (2.5), the neighbor is found, which uses distance metric (e.g., Euclidean distance).

$$\text{Distance}(x, y) = \sum_i (x_i - y_i)^2 \tag{2.5}$$

From training set it takes nearest k number of samples, after that it checks the vote of majority for their class in which to avoid ambiguity the value of k should be given an odd number. Fig. 2.7 describes the main architecture or

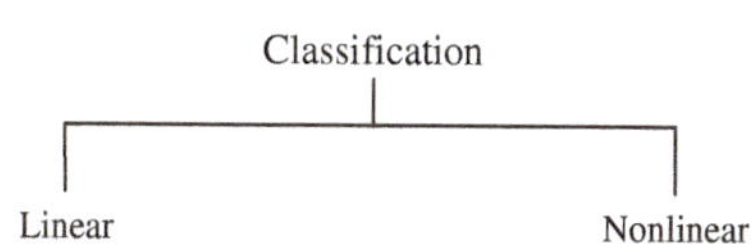

FIGURE 2.6 Types of classifiers for EEG signals. *EEG*, Electroencephalogram.

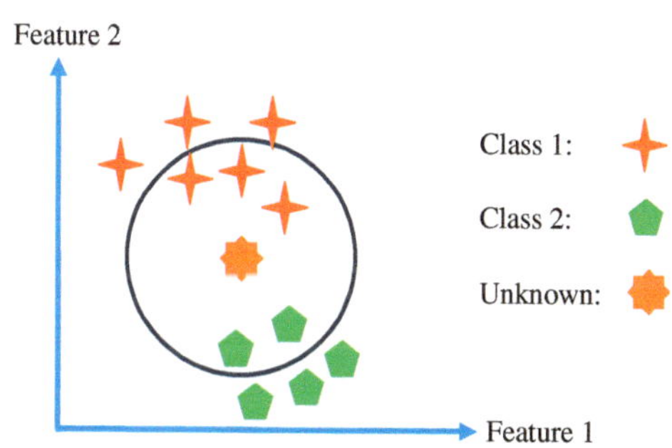

FIGURE 2.7 Illustration of *k*-NN model. *k-NN*, *k*-Nearest neighbor.

design of k-NN model. There are two types of classes, namely, classes 1 and 2. The stars indicate class 1, while pentagons indicate class 2. Unknown data point takes $k = 5$ classes among the nearest neighbors. As for class 1 and 2, 3 and 2 samples fall inside the circle, so the unknown data are labeled as class 1 by majority voting.

2.7.2 Linear discriminant analysis

In statistics, normal discriminant analysis or linear discriminant analysis (LDA) or discriminant function analysis are the methods used for generalization of Fisher's linear discriminant to find feature of linear combination [71]. The authors have used this pattern recognition algorithm to characterize or separate two or more no of events or classes of objects. A linear classifier using a hyperplane is used for classification of data points, as shown in Fig. 2.8.

For real-time BCI applications, this method is suitable for its lower time and space complexity. In addition, this algorithm is quite simple from the users' perspective. There are a number of IoT fields where LDA has proved its worth, for example, P300 speller, motor imagery-based BCI, asynchronous BCI, or multiclass. The main disadvantage of LDA is that it gives poor result, if complex nonlinear EEG samples need to be classified due to its linearity properties.

2.7.3 Decision tree

In prediction and classification, one of the most powerful and useful tools used for classification of data and prediction is the decision tree [72]. Fig. 2.9 shows an example of this tree used for classification of emotions from EEG signals. Here, each internal node of decision tree defines an attribute of test, each branch tree depicts a result of the test, and each leaf node (also called terminal node) shows a class label.

Decision trees are used to classify the instances by sorting them down, and the result of instance classification is provided from the root to some leaf node of the tree. An instance of the tree is classified from beginning at

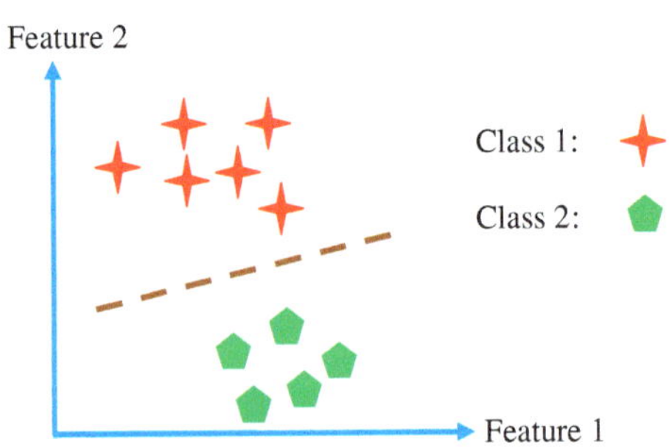

FIGURE 2.8 Illustration of LDA model. *LDA*, Linear discriminant analysis.

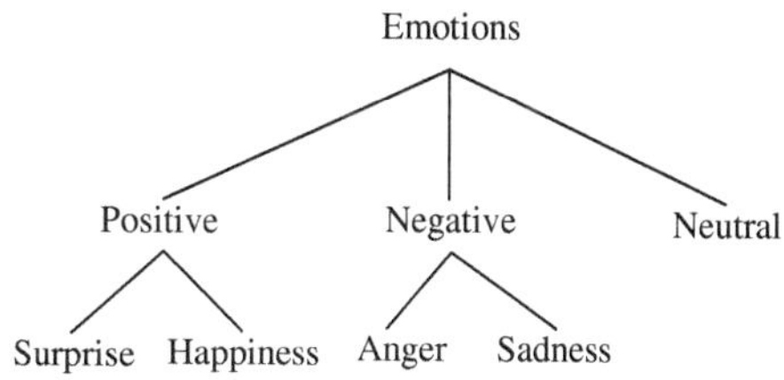

FIGURE 2.9 Illustration of decision tree.

the root node of the tree, test the given attribute by this node, and according to the value of the attribute move down the tree branch. Then, this process is repeatedly continued for the subtree rooted at the new node.

2.7.4 Adaptive Boosting

Adaptive Boosting (in short AdaBoost) follows metadata algorithms of machine learning. It is used for weak learner and is adaptive in the sense. It is weak for the instances that are misclassified by the previous classifiers. The properties of AdaBoost are sensitive for the data that are noisy and outliers. Overfitting problem is less susceptible in some problem than other algorithms of learning. The learners of individual are weak, but as long as each one's performance is slightly better than guessing randomly, after all the final model will prove the converge to a strong learner.

AdaBoost particularly is used to refer to a method of training of a boosted classifier. A boost classifier is a classifier which is represented as

$$F(x) = \sum_{T=1}^{T} \mathrm{ft}(x) \tag{2.6}$$

where each $\mathrm{ft}(x)$ is defined as a weak learner which takes x as input of an object and returns a value for indicating the object of a class. For example, the sign of the weak learner output in the two-class problem identifies the object class that is predicted and it will give the absolute value that confidence in that classification. The Tth classifier is positive if the sample in the class is positive or indicate negative otherwise.

2.7.5 Multilayer perceptron

A multilayer perceptron (MLP) follows feed forward method of artificial neural network. An MLP is a combination of the following three types of layers of nodes:

- An input layer
- A hidden layer
- An output layer

Apart from the neurons present in the input layer, other node of the neuron implements a nonlinear activation function in multilayer perception. In MLPs the idea of neurons using nonlinear activation function was developed based on the concept of biological neurons. In general, propagation method for training in a supervised learning is used for MLP. The method is used to distinguish multiple layers and nonlinear activation of MLP from a linear perception. It can be used for distinguishing data that are nonlinearly separable. Multilayer perceptions, that are sometimes called vanilla in neural networks, generally are used in a single hidden layer. A linear activation function of all neurons uses multilayer perception, that is, a linear function which is used to map the weighted inputs to the output for each neuron. Using linear algebra, any number of layers will be reduced to a two-layer input–output model.

To solve problems stochastically, MLPs are useful in research for their ability, which often prefer approximate solutions for extremely complex problems as example fitness approximation. In Cybenko's theorem [73] the MLP uses universal function approximation so that they can be used to generate mathematical models by using analysis of regression. As classification is a specific case where regression is very important issue, the use of MLP is proven to be fruitful as a classifier.

2.7.6 Naive Bayes

Naive Bayes classifiers [74] belong to the family of classifier where the concept of probability is used. This technique is based on the application of Bayes' theorem. An important application area of Naive Bayes is in the domain of automatic medical diagnosis. Naive Bayes classifiers require a number of parameters of linear variables that are highly scalable to a learning problem. Maximum-likelihood training is based on measuring a closed-form expression, which takes time of linear, in spite of iterative approximation, which is expensive as utilized for many other cases of classifiers. Bayes models have variety of names in the computer science literature: simple Bayes and independence Bayes.

Naive Bayes is used for creating classifier models, which allocate class labels to define problem instances, represented as vectors of feature values, where the class labels are taken from some finite set. Naive Bayes classifiers assume that the value of a particular feature is independent of the given value of any other feature in the given class. Parameter estimation for Naive Bayes models applies maximum likelihood in many practical applications, or the Naive Bayes model can be implemented without applying Bayesian probability or any type of Bayesian procedure.

2.8 Applications of electroencephalogram-based Internet-of-Thing applications

In the modern age of IoT and industry 4.0 the demand for big data, storage cloud network connectivity, edge computing, and high speed of processing is increasing in a rapid speed, unlike earlier days. In the International Patent Classification (IPC) analysis with category G06F19/00, equipment of digital computing or data processing is accepted for certain applications; it ranked fifth among the top 20 technical category of IPC with 130 patents and has been published between January 2006 and December 2015 [75]. Furthermore, for saving the energy of transmission, the use of battery-operated IoT devices decreases the volume of communicated data. For this purpose, some promising approaches are the following:

- For inclusion of comprehensive sensing, local in-network processing is used and data are compressed before transmission.
- The gathered data containing health information are processed using deep learning, which is a powerful tool than a chine learning and health informatics are for producing feature of optimized high level and semantic interpretation.

The high-dimensional vector of data has projected by multiplying linear embedded matrix with original signal [76], hence a low-dimensional subspace has projected by the high-dimensional data vector. This process has provided high-compression ratio that is used for the construction of challenging hardware. In real-time application, high-computational cost has the limit for reconstruction of signals. For example, the Orthogonal Matching Pursuit method that is used for recreation of the signal involving computation of heavy matrix. In addition, a trade-off is working between hardware energy-efficiency and accuracy of signal recovery. For example, more hardware consumption is expected for improving signal recovery accuracy when an optimization method is used (e.g., NuMax) [77]. High recovery accuracy cannot be guaranteed when nondata-driven random Boolean embedding is applied for enhancing the energy of hardware with better efficiency. The work is relevant to the research of e-Health where compression techniques have been used. In this case, different transformation techniques (such as vector quantization, Fourier or WTs, DCT), which are already stated in the previous section, is used. In short, most of the existing compression works are applied on layer of higher layers, ignoring feature of lower layers (e.g., wireless channels characteristics, signal-to-interference-plus-noise ratio, and rate of bit error rate). Battery-operated devices are costly for their computational complexity for implementing such schemes. On the contrary, transceivers that are designed for specific applications have recently gained interest. For enhancing future, transceiver architecture uses some efforts for coping with communication of long-range and RF transceiver of multistandard with high degree of reusability, scalability, and flexibility. Marine Systems granted radio transceiver (US patent 9473197 and European patent EP2951930), which has reversible time domain duplex transceiver technology [78]. This technology is used simultaneously for a specific architecture of single RF, for both receiving and transmitting through reversing electronics belonging to the RF chain between receiver and transmitter. This could be maintained through a complex combination of intelligent selection of intermediate frequencies and ultrafast switching. A patent for a radio transceiver enables multiple transceivers to share a single antenna; SRT Marine Technology Ltd., Bristol, obtained the patent in order to reduce the cost and installation effort (GB patent no. 2460012) [79].

Nowadays, many of the biomedical and robotic engineering applications are using IoT-based devices where robotic agents are maneuvered using EEG signals. Some of the applications concerning this domain are explained in this section. Today, Internet of Medical Thing (IoMT) technology is very advanced. In medical science, mobile health (m-health) system is used to get information about the delivery report and information of health care. Using IoMT technology, patient can observe their information of health without visiting the hospital or clinic. We can measure the signal from brain not only by using EEG but also by other technologies such as functional near-infrared spectroscopy and magnetoencephalography. The raw brain signals are passed to the cloud server with the help of the internet for IoT applications.

2.8.1 Seizure detection

Based on the report from the Epilepsy Foundation of America, around 200,000 cases of epilepsy are diagnosed each year [80]. Seizure is a sign or symptom of a brain abnormality, which is due to uncontrolled synchronous neuronal activities in the brain [81]. Features from EEG are suitable for seizure detection. To detect brain abnormality due to seizure from EEG, machine-learning approaches are commonly being used. Abdulhay et al. [82] utilize entropy spectral features for this purpose. Different approaches of using classifier can be observed in the works by Qaraqe et al. [83], Mutlu et al. [84], Diykh et al. [85], and Birjandtalab et al. [86]. In their methods, features are first extracted from the EEG signals, followed by employing these to train a classifier. The trained classifier is then used to discriminate normal and abnormal EEG.

2.8.2 Brain injury detection

To generate a model of a human brain, brain models from animals such as from rat are used in experiments, as implemented by Napoli et al. [87] and Fisher et al. [88]. To detect traumatic brain injury (TBI), machine-learning method is also used. McBride et al. [89] demonstrate the application of quantitative EEG (qEEG) analysis. Here, EEG signal is transformed using Fourier transform or WT. In the work by Mikola et al. [90], 186 qEEG features are extracted from the long-term EEG recording based on the power spectrum of EEG and then these features are analyzed by using the receiver operating characteristic curve. Albert et al. perform TBI diagnosis based on a model trained from clinically labeled EEG records [91]. Another approach for TBI detection makes use of activation maps. Variane et al. [92] implement correlated amplitude-integrated EEG to infants having a high risk of brain injury, while Franke et al. [93] explore the differentiation in EEG slow oscillations with mild TBI and posttraumatic stress disorder. Waveform analysis is also done by Weeke et al. [94] to preterm infants. Their approach classifies rhythmic EEG patterns and correlates the output with brain injury. Identically, for infants' brain injury, Nevalainen et al. [95] evaluate the usage of SSEPs and visual evoked potentials (VEPs) together with routine EEG to predict the occurrence of hypoxic ischemic encephalopathy (brain injury due to lack of oxygen).

2.8.3 Object controlling

One of the most common applications in BCI is gaming control, as in the works by Hawsawi and Semwal [96] and Hsieh et al. [97]. By implementing BCI, keyboard or game console can be replaced by EEG headset. Apart from that, BCI is also implemented in applications which are related to object control. Wheelchair control is one application that has been proposed in the literature. In the work by Velasco-Álvarez et al. [98] the user is required to have an imagery right-hand movement that corresponds to the audio cues. The imagery signal enables the user to control the wheelchair. On the other hand, Wang et al. [99] control the wheelchair by referring to eye blinks EEG features.

BCI can also be used for robot arm control application, which often requires user to have imagery movement [100]. Jiang et al. [101] give a Morse code-based BCI. Their method extracts codes of motor imagery task from recorded EEG and decodes them into special commands to control the robotic arm. Identically, Sunny et al. [102] use imagery action states and neutral states to control robotic arm. Action states move the robot arm, while neutral states enable the robot arm to remain idle.

2.8.4 Mental state recognition

BCI can be used to recognize imagery objects, taste, image familiarity, and movement intention. Classifier is used to classify the output into different categories. Other than the mentioned applications, BCI can be also used to classify attention stages. Mishchenko et al. [103] propose an attention stage detection implementing a virtual continuous attention vehicle control task. The system is able to differentiate operator attention states. In future improvement, this approach is able to provide attention state alert for drivers and reduce road accidents. Similar to attention state recognition that has been mentioned already, drowsiness detection system is also a useful BCI application that can avoid severe accidents from happen on operation line or on the road. Tripathy et al. [104] propose a detection method for drowsiness through continuous monitoring. A multivariate normal distribution is used to model the appropriation of the power spectrum in the alarm state. As the power spectrum of EEG reaches the alarm state, warning tone is given to alarm a drowsy state.

Furthermore, a gaze classification approach is proposed by Maleki et al. [105]. In their approach, EEG is recorded when subjects are staring at a rotating vane, both slow and fast speed. The recorded EEG is used to train a classifier, which subsequently used to classify slow and fast gazing. An approach of recognizing human responses is implemented by Park et al. [106]. Li et al. [107] propose a multichannel ERP lie detector. The equipment can recognize when the user is telling a lie, which can be differentiated by a trained classifier.

Furthermore, BCI is also applicable for emotion recognition [108] and mental task classification [109]. The BCI can be used for video content analysis system for perceiving emotions such as happiness and sadness. A classifier is commonly used for recognizing the emotions to categorize the recorded EEG from user into different emotions. Apart from emotion, BCI is also able to recognize voice familiarity, which is proposed by Smitha et al. [110]. In their work an union of features, such as mobility and complexity, are used and are shown to be feasible for detecting familiar and nonfamiliar voice signals, only by referring to the recorded EEG.

2.8.5 Rehabilitation and human assistance

BCI directly makes a connection between outside external device and human brain. Nowadays, the new trend in BCI research is transforming the thinking capability of human into physical actions such as wheelchair control. Other than the mentioned application, BCI is often used in providing aid and ease to one's life.

To aid patients with movement difficulties, Robinson et al. [111] proposed a hand movement trajectory reconstruction approach. The study proposes a method to remodel different attributes of hand movement trajectory from multichannel EEG. Subjects are asked to perform four hand gestural activities tasks at four different directions with two different speeds during recording of EEG. Multilinear regression is used to predict the parameters. In addition to rehabilitation purpose, Wairagkar et al. [112] use a new EEG analysis method to direct a virtual reality avatar and a software-based robotics rehabilitation tool. This BCI is capable of identifying and predicting the upper limb movement. In addition, Liu et al. [113] propose a method for gait training. Their approach decodes cerebral activity from EEG to dictate lower-limb gait training exoskeleton. Motor imagery of flexion and extension of both legs are estimated from the EEG. PSD is used as a measure to represent the motor imagery. For the purpose of rehabilitation, Luu et al. [114] come out with a virtual walking avatar control for engaging cortical adaption. In their approach, delta band EEG is used as main feature for prediction. Their work provides the feasibility by closed-loop EEG-based BCI-virtual reality to activate cortical adaption, encourage cortical entanglement, and observe cortical activity.

Apart from a steady-state VEP method, an amplitude-modulated stimulus is proposed by Chang et al. [115] to reduce eye fatigue. The proposed method succeeds to reduce eyes' fatigue efficiently with amplitude-modulated stimulus. Amplitude-modulated stimulus managed to give low-frequency information by a high-frequency carrier, which is able to reduce eye fatigue of user reading the delivered information. Verkijika et al. [116] introduce a BCI-based game to reduce students' anxiety when doing mathematics. Subjects are required to complete two sessions of mathematical games. The recorded EEG during the two sessions is analyzed for the changes of anxiety throughout the game session.

2.8.6 Neuro-marketing studies

Neuro-marketing is a study that implements neuroscience into marketing research. It studies consumers' responses and cognitive to marketing stimuli. EEG is one of the popularly used modalities due to its mobility and easy setup. The applications of EEG in neuro-marketing study deal with the response of subject's brain toward marketing-related stimuli. By analyzing the EEG recorded during stimulation, it is able to recognize the preferable marketing methods or products.

Khushaba et al. [117] use BCI to recognize physiological decision for subjects to decide preferred crackers described by shape, flavor, and toppings. Other than that, Yilmaz et al. [118] provide a technique that is able to forecast like/dislike response from a customer. Yadava et al. [119] present the similar approach, but their application is to learn consumer choice for e-commerce products rather than actual stock. The BCI is able to present the result of like or dislike when the participants are exposed to shoes' photographs while recording EEG. For product-based application, method proposed by Murugappan et al. [120] identifies the most preferred automotive brand in Malaysia. The participants are exposed to automotive advertisement while recording EEG. Similar as the method mentioned previously, EEG is analyzed for the preferable brand. Telpaz et al. [121] also propose a method that is able to predict customer's future choice. Their work predicts the customer decision after being exposed to several choices of consumer products by analyzing the recorded EEG during exposure to the choices. Other than the aspect mentioned previously, Gupta et al. also propose consumers' product preference recognition, targeting on preferred soap brand in New Delhi. Furthermore, a preference classification for bracelet is proposed by Teo et al. [122]. The study records the EEG and prefers when the participants are exposed to 60 bracelet-like objects as rotation visual stimuli on a computer display. In their approach, deep learning is used to classify the participants' preference. An approach has also been proposed by Kosters et al. [123] to predict the preference and choice of wine by recording the participant's EEG during wine tasting and smell procedures. To bring the impact of tourisms aspect, Bastiaansen et al. [124] came out with a vacation destination marketing evaluation approach. Participants are exposed to tourist destination in movies, and the method can evaluate their response toward the effectiveness of the stimulus.

2.9 Conclusion

In traditional systems, EEG signals are analyzed manually by medical practitioner, which may have some human errors and also the process of extracting relevant information is time consuming. To overcome this problem, automatic extraction of information from EEG signals to employ them for betterment of IoT applications, some recent states of the art literatures are explained in this chapter. The techniques are becoming very much popular in the field of IoT nowadays. In summary the chapter would be useful for the new researchers who are starting to work in this research domain.

References

[1] H. Berger, Über das elektrenkephalogramm des menschen, Eur. Arch. Psychiatry Clin. Neurosci. 87 (1) (1929) 527–570.

[2] C.Q. Lai, H. Ibrahim, M.Z. Abdullah, J.M. Abdullah, S.A. Suandi, A. Azman, Literature survey on applications of electroencephalography (EEG), AIP Conf. Proc. 2016 (2018).

[3] M. Modarreszadeh, R.N. Schmidt, Wireless, 32-channel, EEG and epilepsy monitoring system, in: Proceedings of the 19th Annual International Conference of the IEEE Engineering in Medicine and Biology Society. Magnificent Milestones and Emerging Opportunities in Medical Engineering (Cat. No. 97CH36136), vol. 3, 1997, pp. 1157–1160.

[4] S.E. Schmitt, K. Pargeon, E.S. Frechette, L.J. Hirsch, J. Dalmau, D. Friedman, Extreme delta brush: a unique EEG pattern in adults with anti-NMDA receptor encephalitis, Neurology 79 (11) (2012) 1094–1100.

[5] D.R. Stanski, Pharmacodynamic modeling of anesthetic EEG drug effects, Annu. Rev. Pharmacol. Toxicol. 32 (1) (1992) 423–447.

[6] J.C. Gillin, W. Duncan, K.D. Pettigrew, B.L. Frankel, F. Snyder, Successful separation of depressed, normal, and insomniac subjects by EEG sleep data, Arch. Gen. Psychiatry 36 (1) (1979) 85–90.

[7] G. Adler, S. Brassen, A. Jajcevic, EEG coherence in Alzheimer's dementia, J. Neural Transm. 110 (9) (2003) 1051–1058.

[8] L.A. Alvarez, S.L. Moshé, A.L. Belman, J. Maytal, T.J. Resnick, M. Keilson, EEG and brain death determination in children, Neurology 38 (2) (1988) 227.

[9] J. Friedberg, Shock treatment, brain damage, and memory loss: a neurological perspective, Am. J. Psychiatry 134 (9) (1977) 1010–1014.

[10] S. Waldert, Invasive vs. non-invasive neuronal signals for brain-machine interfaces: will one prevail? Front. Neurosci. 10 (June) (2016) 1–4.

[11] K.J. Burchiel, S. McCartney, A. Lee, A.M. Raslan, Accuracy of deep brain stimulation electrode placement using intraoperative computed tomography without microelectrode recording, J. Neurosurg. 119 (2) (2013) 301–306.

[12] M.C. Casdagli, L.D. Iasemidis, R.S. Savit, R.L. Gilmore, S.N. Roper, J.C. Sackellares, Non-linearity in invasive EEG recordings from patients with temporal lobe epilepsy, Electroencephalogr. Clin. Neurophysiol. 102 (2) (1997) 98–105.

[13] C. Onal, et al., Complications of invasive subdural grid monitoring in children with epilepsy, J. Neurosurg 98 (5) (2003) 1017–1026.

[14] T. Ball, M. Kern, I. Mutschler, A. Aertsen, A. Schulze-Bonhage, Signal quality of simultaneously recorded invasive and non-invasive EEG, Neuroimage 46 (3) (2009) 708–716.

[15] A. Pinegger, S.C. Wriessnegger, J. Faller, G.R. Müller-Putz, Evaluation of different EEG acquisition systems concerning their suitability for building a brain-computer interface: case studies, Front. Neurosci. 10 (SEP) (2016) 441.

[16] T. Alotaiby, F.E.A. El-Samie, S.A. Alshebeili, I. Ahmad, A review of channel selection algorithms for EEG signal processing, EURASIP J. Adv. Signal Process. 2015 (1) (2015) 66.

[17] A.R. Hidalgo-Muñoz, M.M. López, I.M. Santos, M. Vázquez-Marrufo, E.W. Lang, A.M. Tomé, Affective valence detection from EEG signals using wrapper methods, emotion and attention recognition based on biological signals and images 23 (2017).

[18] M. Dash, H. Liu, Feature selection for classification, Intelligent data analysis 1 (1–4) (1997) 131–156.

[19] H. Liu, L. Yu, Toward integrating feature selection algorithms for classification and clustering, IEEE Trans. Knowl. Data Eng. 17 (2005) 491–502.

[20] W. Klimesch, EEG alpha and theta oscillations reflect cognitive and memory performance: a review and analysis, Brain Res. Rev. 29 (2–3) (1999) 169–195.

[21] H. Woehrle, M.M. Krell, S. Straube, S.K. Kim, E.A. Kirchner, F. Kirchner, An adaptive spatial filter for user-independent single trial detection of event-related potentials, IEEE Trans. Biomed. Eng. 62 (7) (2015) 1696–1705.

[22] A.M. Norcia, L.G. Appelbaum, J.M. Ales, B.R. Cottereau, B. Rossion, The steady-state visual evoked potential in vision research: a review, J. Vis. 15 (2015) 1–46.

[23] K. Palani Thanaraj, K. Chitra, Multichannel feature extraction and classification of epileptic states using higher order statistics and complexity measures, Int. J. Eng. Technol 6 (1) (2014) 102–109.
[24] T.W. Picton, The P300 wave of the human event-related potential, J. Clin. Neurophysiol. 9 (4) (1992) 456–479.
[25] L. Mayaud, et al., A comparison of recording modalities of P300 event-related potentials (ERP) for brain-computer interface (BCI) paradigm, Neurophysiol. Clin. 43 (4) (2013) 217–227.
[26] M.R. Nuwer, E.G. Dawson, L.G. Carlson, L.E.A. Kanim, J.E. Sherman, Somatosensory evoked potential spinal cord monitoring reduces neurologic deficits after scoliosis surgery: results of a large multicenter survey, Electroencephalogr. Clin. Neurophysiol. Evoked Potentials 96 (1) (1995) 6–11.
[27] A. Turnip, K.S. Hong, Classifying mental activities from EEG-P300 signals using adaptive neural networks, Int. J. Innov. Comput. Inf. Control 8 (9) (2012) 6429–6443.
[28] P. Sarma, P. Tripathi, M.P. Sarma, K.K. Sarma, Pre-processing and feature extraction techniques for EEGBCI applications—a review of recent research, ADBU—J. Eng. Technol. 5 (2016) 2348–7305.
[29] K. Li, G. Sun, B. Zhang, S. Wu, G. Wu, Correlation between forehead EEG and sensorimotor area EEG in motor imagery task, in: Eighth IEEE Int. Symp. Dependable, Auton. Secur. Comput. DASC 2009, 2009, pp. 430–435.
[30] Y. Petrov, Analysis of EEG signals for EEG-based brain-computer interface, PLoS One 7 (10) (2012) e44439.
[31] A. Campbell, T. Choudhury, S. Hu, H. Lu, M.K. Mukerjee, M. Rabbi, et al., NeuroPhone: brain-mobile phone interface using a wireless EEG headset, Proceedings of the second ACM SIGCOMM workshop on Networking, systems, and applications on mobile handhelds, ACM, 2010, August, pp. 3–8.
[32] O. Diana, A. Mihaela, Comparison of classifiers and statistical analysis for EEG signals used in brain computer interface motor task paradigm, Int. J. Adv. Res. Artif. Intell. 4 (1) (2015) 8–12.
[33] H. Maki, T. Toda, S. Sakti, G. Neubig, S. Nakamura, EEG Signal Enhancement Using Multi-Channel Wiener Filter With a Spatial Correlation Prior, Graduate School of Information Science Nara Institute of Science and Technology (NAIST), Japan, 2015, pp. 2639–2643.
[34] C. Guerrero-Mosquera, A.N. Vazquez, New approach in features extraction for EEG signal detection, in: Proc. 31st Annu. Int. Conf. IEEE Eng. Med. Biol. Soc. Eng. Futur. Biomed. EMBC 2009, 2009, pp. 13–16.
[35] S. Jirayucharoensak, P. Israsena, Automatic removal of EEG artifacts using ICA and Lifting Wavelet Transform, in: 2013 Int. Comput. Sci. Eng. Conf. ICSEC 2013, 2013, pp. 136–139.
[36] Y. Wang, & S. Makeig, (2009, July). Predicting intended movement direction using EEG from human posterior parietal cortex. In *International Conference on Foundations of Augmented Cognition* (pp. 437–446). Springer, Berlin, Heidelberg.
[37] M. Kalaivani, V. Kalaivani, V.A. Devi, Analysis of EEG signal for the detection of brain abnormalities, Int. J. Comput. Appl 1 (2) (2014) 1–6.
[38] P.S. Kumar, R. Arumuganathan, K. Sivakumar, C. Vimal, Removal of artifacts from EEG signals using adaptive filter through wavelet transform, in: Int. Conf. Signal Process. Proceedings, ICSP, 2008, pp. 2138–2141.

[39] J.L. Cárdenas-Barrera, J.V. Lorenzo-Ginori, E. Rodríguez-Valdivia, A wavelet-packets based algorithm for EEG signal compression, Inf. Heal. Soc. Care 29 (1) (2004) 15–27.
[40] T. Kameswara, M. Rajyalakshmi, T.V. Prasad, An exploration on brain computer interface and its recent trends, Int. J. Adv. Res. Artif. Intell. 1 (8) (2013) 17–22.
[41] S. Motamedi-Fakhr, M. Moshrefi-Torbati, M. Hill, C.M. Hill, P.R. White, Signal processing techniques applied to human sleep EEG signals—a review, Biomed. Signal Process. Control 10 (1) (2014) 21–33.
[42] C. Vidaurre, N. Krämer, B. Blankertz, A. Schlögl, Time domain parameters as a feature for EEG-based brain-computer interfaces, Neural Networks 22 (9) (2009) 1313–1319.
[43] M. Yazıcı, M. Ulutaş, Classification of EEG signals using time domain features. In 2015 23nd Signal Processing and Communications Applications Conference (SIU), IEEE, 2015, pp. 2358–2361.
[44] J. Wang, Z. Feng, N. Lu, Feature extraction by common spatial pattern in frequency domain for motor imagery tasks classification. In 2017 29th Chinese Control and Decision Conference (CCDC), IEEE, 2017, pp. 5883–5888.
[45] J.S. Richman, J.R. Moorman, Physiological time-series analysis using approximate entropy and sample entropy, Am. J. Physiol. Heart Circ. Physiol 278 (6) (2000) H2039–H2049.
[46] B. Qian, K. Rasheed, Hurst exponent and financial market predictability, in: Proc. Second IASTED Int. Conf. Financ. Eng. Appl., 2004, pp. 203–209.
[47] J.H. Kang, C.H. Lee, S.P. Kim, EEG feature selection and the use of Lyapunov exponents for EEG-based biometrics, in: Third IEEE EMBS Int. Conf. Biomed. Heal. Informatics, BHI 2016, 2016, pp. 228–231.
[48] S. Chatterjee, S. Pratiher, R. Bose, Multifractal detrended fluctuation analysis based novel feature extraction technique for automated detection of focal and non-focal electroencephalogram signals, IET Sci. Meas. Technol. 11 (8) (2017) 1014–1021.
[49] I. Rejer, Genetic algorithms in EEG feature selection for the classification of movements of the left and right hand, Adv. Intell. Syst. Comput. 226 (2013) 9–11.
[50] A. Alın, S. Kurt, A.R. Mcintosh, M. Ozg, Partial least squares analysis in electrical brain activity, J. Data Sci. 7 (2009) 99–110.
[51] P. O'Brien, A primer on the discrete Fourier transform, Am. J. EEG Technol. 34 (4) (2018) 190–223.
[52] D. Birvinskas, V. Jusas, I. Martisius, R. Damasevicius, EEG dataset reduction and feature extraction using discrete cosine transform, in: Proc.—UKSim-AMSS Sixth Eur. Model. Symp. EMS 2012, 2012, pp. 199–204.
[53] D. Birvinskas, V. Jusas, I. Martišius, R. Damaševičius, Data compression of EEG signals for artificial neural network classification, Inf. Technol. Control 42 (3) (2013) 238–241.
[54] T. Ouyang, H.T. Lu, Vigilance analysis based on continuous wavelet transform of EEG signals, in: 2010 Int. Conf. Biomed. Eng. Comput. Sci. ICBECS 2010, no. 2008, 2010, pp. 0–3.
[55] D.D. Gautam, V.K. Giri, K.G. Upadhyay, Feature extraction of EEG signal using wavelet transform, in: 2017 Second Int. Conf. Converg. Technol. I2CT 2017, vol. 124, no. 2, January 2017, pp. 1030–1034.
[56] A. Ridouh, D. Boutana, S. Bourennane, EEG signals classification based on time frequency analysis, J. Circuits Syst. Comput. 26 (12) (2017) 1–26.
[57] S.K. Swee, L.Z. You, Fast Fourier analysis and EEG classification brainwave controlled wheelchair, in: Proc. 2016 Second Int. Conf. Control Sci. Syst. Eng. ICCSSE 2016, 2016, pp. 20–23.

[58] F. Cona, M. Zavaglia, L. Astolfi, F. Babiloni, M. Ursino, Changes in EEG power spectral density and cortical connectivity in healthy and tetraplegic patients during a motor imagery task, Comput. Intell. Neurosci 2009 (2009) 1–12.

[59] L. Zhao, Y. He, Power spectrum estimation of the Welch method based on imagery EEG, Appl. Mech. Mater. 278–280 (2013) 1260–1264.

[60] R.K. Chaurasiya, N.D. Londhe, S. Ghosh, Statistical wavelet features, PCA, and SVM based approach for EEG signals classification, World Acad. Sci. Eng. Technol. Int. J. Electr. Comput. Energ. Electron. Commun. Eng 9 (2) (2015) 182–186.

[61] T. Yamashita, K. Hikosaka, T. Nanba, Electroencephalogram analysis regarding visual information processing in a grapheme-color synesthete, Kawasaki J 16 (1) (2010) 13–15.

[62] E.D. Übeyli, Analysis of EEG signals by implementing eigenvector methods/recurrent neural networks, Digit. Signal Process. A Rev. J. 19 (1) (2009) 134–143.

[63] E. Derya Übeyli, Analysis of EEG signals by combining eigenvector methods and multiclass support vector machines, Comput. Biol. Med. 38 (1) (2008) 14–22.

[64] E.D. Übeyli, I. Güler, Features extracted by eigenvector methods for detecting variability of EEG signals, Pattern Recognit. Lett. 28 (5) (2007) 592–603.

[65] Y. Zhang, X. Ji, Y. Zhang, Classification of EEG signals based on AR model and approximate entropy, in: Proc. Int. Jt. Conf. Neural Networks, vol. 2015–September, 2015.

[66] A. Procházka, M. Mudrová, O. Vyšata, R. Háva, C.P.S. Araujo, Multi-channel EEG signal segmentation and feature extraction, in: INES 2010 – 14th Int. Conf. Intell. Eng. Syst. Proc., 2010, pp. 317–320.

[67] S. Aydin, Model order sensitivity of Burg Method for EEG diagnosis. In 2008 IEEE 16th Signal Processing, Communication and Applications Conference, IEEE, 2008, pp. 1–2.

[68] S.M. Pincus, Approximate entropy as a measure of system complexity, Proc. Natl. Acad. Sci. U.S.A. 88 (March) (2015) 2297–2301.

[69] T. Wen, Z. Zhang, Effective and extensible feature extraction method using genetic algorithm-based frequency-domain feature search for epileptic EEG multiclassification, Medicine (Baltimore) 96 (19) (2017) 1–11.

[70] A. Bablani, D.R. Edla, S. Dodia, Classification of EEG data using k-nearest neighbor approach for concealed information test, Procedia Comput. Sci. 143 (2018) 242–249.

[71] E. Neto, F. Biessmann, H. Aurlien, H. Nordby, T. Eichele, Regularized linear discriminant analysis of EEG features in dementia patients, Front. Aging Neurosci 8 (V) (2016) 1–10.

[72] M. Arvaneh, C. Guan, K.K. Ang, H.C. Quek, EEG channel selection using decision tree in brain-computer interface. In Proceedings of the Second APSIPA Annual Summit and Conference, 2010, pp. 225–230.

[73] G. Cybenko, Approximation by superpositions of a sigmoidal function, Math. Control Signal Syst. 2 (1989) 303–314.

[74] J. Machado, A. Balbinot, A. Schuck, A study of the Naive Bayes classifier for analyzing imaginary movement EEG signals using the Periodogram as spectral estimator. In 2013 ISSNIP Biosignals and Biorobotics Conference: Biosignals and Robotics for Better and Safer Living (BRC), IEEE, 2013, pp. 1–4.

[75] A.J. Trappey, C.V. Trappey, U.H. Govindarajan, J.J. Sun, A.C. Chuang, A review of technology standards and patent portfolios for enabling cyber-physical systems in advanced manufacturing, IEEE Access 4 (2016) 7356–7382.

[76] T. Strohmer, Measure what should be measured: progress and challenges in compressive sensing, IEEE Signal Process. Lett. 19 (12) (2012) 887–893.

[77] C. Hegde, A. Sankaranarayanan, NuMax: a convex approach for learning near-isometric linear embeddings, IEEE Trans. Signal Process. 63 (2012) 1–35.

[78] M. Kim, S. Chang, A consumer transceiver for long-range IoT communications in emergency environments, IEEE Trans. Consum. Electron. 62 (3) (2016) 226–234.

[79] A.A. Abdellatif, M.G. Khafagy, A. Mohamed, C.F. Chiasserini, EEG-based transceiver design with data decomposition for healthcare IoT applications, IEEE Internet Things J. 5 (5) (2018) 3569–3579.

[80] R.S. Fisher, et al., Epileptic Seizures and Epilepsy: Definitions Proposed by the International League Against Epilepsy (ILAE) and the International Bureau for Epilepsy (IBE), Epilepsia 46 (4) (2005) 1–3.

[81] M.Z. Parvez, M. Paul, M. Antolovich, Detection of pre-stage of epileptic seizure by exploiting temporal correlation of EMD decomposed EEG signals, J. Med. Bioeng. 4 (2) (2015) 110–116.

[82] E. Abdulhay, M. Alafeef, A. Abdelhay, A. Al-Bashir, Classification of normal, ictal and inter-ictal EEG via direct quadrature and random forest tree, J. Med. Biol. Eng. 37 (6) (2017) 843–857.

[83] M. Qaraqe, M. Ismail, E. Serpedin, Band-sensitive seizure onset detection via CSP-enhanced EEG features, Epilepsy Behav. 50 (2015) 77–87.

[84] A.Y. Mutlu, Detection of epileptic dysfunctions in EEG signals using Hilbert vibration decomposition, Biomed. Signal Process. Control 40 (2018) 33–40.

[85] M. Diykh, Y. Li, P. Wen, Classify epileptic EEG signals using weighted complex networks based community structure detection, Expert Syst. Appl. 90 (2017) 87–100.

[86] J. Birjandtalab, M. Baran Pouyan, D. Cogan, M. Nourani, J. Harvey, Automated seizure detection using limited-channel EEG and non-linear dimension reduction, Comput. Biol. Med. 82 (2017) 49–58.

[87] A. Napoli, M. Barbe, K. Darvish, I. Obeid, Assessing traumatic brain injuries using EEG power spectral analysis and instantaneous phase, in: Proc. Annu. Int. Conf. IEEE Eng. Med. Biol. Soc. EMBS, 2012, pp. 4692–4695.

[88] J.A.N. Fisher, et al., Real-time detection and monitoring of acute brain injury utilizing evoked electroencephalographic potentials, IEEE Trans. Neural Syst. Rehabil. Eng. 24 (9) (2016) 1003–1012.

[89] J.C. McBride, et al., Sugihara causality analysis of scalp EEG for detection of early Alzheimer's disease, NeuroImage Clin. 7 (2015) 258–265.

[90] A. Mikola, I. Ratsep, M. Sarkela, T. Lipping, Prediction of outcome in traumatic brain injury patients using long-term qEEG features, in: Proc. Annu. Int. Conf. IEEE Eng. Med. Biol. Soc. EMBS, November 2015, pp. 1532–1535.

[91] B. Albert, et al., Automatic EEG processing for the early diagnosis of traumatic brain injury, Procedia Comput. Sci 96 (2016) 703–712.

[92] G.F.T. Variane, et al., Early amplitude-integrated electroencephalography for monitoring neonates at high risk for brain injury, J. Pediatr. (Rio. J). 93 (5) (2017) 460–466.

[93] L.M. Franke, W.C. Walker, K.W. Hoke, J.R. Wares, Distinction in EEG slow oscillations between chronic mild traumatic brain injury and PTSD, Int. J. Psychophysiol. 106 (2016) 21–29.

[94] L.C. Weeke, et al., Role of EEG background activity, seizure burden and MRI in predicting neurodevelopmental outcome in full-term infants with hypoxic-ischaemic encephalopathy in the era of therapeutic hypothermia, Eur. J. Paediatr. Neurol. 20 (6) (2016) 855–864.

[95] P. Nevalainen, et al., Evoked potentials recorded during routine EEG predict outcome after perinatal asphyxia, Clin. Neurophysiol. 128 (7) (2017) 1337–1343.

[96] O. Hawsawi, S.K. Semwal, EEG headset supporting mobility impaired gamers with game accessibility, in: Conf. Proc.—IEEE Int. Conf. Syst. Man Cybern. 2014, January, 2014, pp. 837–841.

[97] C.H. Hsieh, H.P. Chu, Y.H. Huang, An HMM-based eye movement detection system using EEG brain-computer interface, in: Proc.—IEEE Int. Symp. Circuits Syst., 2014, pp. 662–665.

[98] F. Velasco-Álvarez, R. Ron-Angevin, L. da Silva-Sauer, S. Sancha-Ros, Audio-cued motor imagery-based brain-computer interface: navigation through virtual and real environments, Neurocomputing 121 (2013) 89–98.

[99] H. Wang, A. Bezerianos, Brain-controlled wheelchair controlled by sustained and brief motor imagery BCIs, Electron. Lett 53 (17) (2017) 1178–1180.

[100] S. Saha, A. Konar, A. Saha, A.K. Sadhu, B. Banerjee, A.K. Nagar, EEG based gesture mimicking by an artificial limb using cascade-correlation learning architecture, Proc. Int. Jt. Conf. Neural Networks (2016) 4680–4687. vol. 2016-October.

[101] J. Jiang, Z. Zhou, E. Yin, Y. Yu, Y. Liu, D. Hu, A novel Morse code-inspired method for multiclass motor imagery brain-computer interface (BCI) design, Comput. Biol. Med. 66 (2015) 11–19.

[102] T.D. Sunny, T. Aparna, P. Neethu, J. Venkateswaran, V. Vishnupriya, P.S. Vyas, Robotic arm with brain–computer interfacing, Procedia Technol. 24 (2016) 1089–1096.

[103] Y. Mishchenko, M. Kaya, Detecting the attention state of an operator in continuous attention task using EEG-based brain-computer interface. In 2015 23nd Signal Processing and Communications Applications Conference (SIU), IEEE, 2015, pp. 232–235.

[104] A.K. Tripathy, S. Chinara, M. Sarkar, An application of wireless brain-computer interface for drowsiness detection, Biocybern. Biomed. Eng. 36 (1) (2016) 276–284.

[105] M. Maleki, T. Kayikçioğlu, Classification of EEG signal during gaze on the different rotating vanes, in: 2016 24th Signal Process. Commun. Appl. Conf. SIU 2016—Proc., 2016, pp. 593–596.

[106] C. Park, D. Looney, D.P. Mandic, Estimating human response to taste using EEG, Conf. Proc. IEEE Eng. Med. Biol. Soc. (2011) 6331–6334.

[107] R. Li, H. Liu, J. Yong, Event—related potentials on the application of deception detection in the simulation of concealing cases, Am. J. Appl. Psychol. 6 (3) (2017) 42–50.

[108] R.M. Mehmood, H.J. Lee, EEG based emotion recognition from human brain using Hjorth parameters and SVM, Int. J. Bio-Science Bio-Technology 7 (3) (2015) 23–32.

[109] P.M. Lanke, P.R.K. Shastri, P.S.D. Biradar, EEG Signal Processing Techniques for Mental Task Classification, no. 1, 2015, pp. 66–74.

[110] K.G. Smitha, A.P. Vinod, K. Mahesh, Voice familiarity detection using EEG-based brain-computer interface, in: 2016 IEEE Int. Conf. Syst. Man, Cybern. SMC 2016—Conf. Proc., 2017, pp. 1626–1631.

[111] N. Robinson, A.P. Vinod, C. Guan, Hand movement trajectory reconstruction from EEG for brain-computer interface systems, in: Proc. – 2013 IEEE Int. Conf. Syst. Man, Cybern. SMC 2013, 2013, pp. 3127–3132.

[112] M. Wairagkar, I. Zoulias, V. Oguntosin, Y. Hayashi, S. Nasuto, Movement intention based Brain Computer Interface for Virtual Reality and Soft Robotics rehabilitation using novel autocorrelation analysis of EEG, in: Proc. IEEE RAS EMBS Int. Conf. Biomed. Robot. Biomechatronics, vol. 2016–July, 2016, p. 685.

[113] D. Liu, W. Chen, K. Lee, Z. Pei, J.D.R. Millan, An EEG-based brain-computer interface for gait training, in: Proc. 29th Chinese Control Decis. Conf. CCDC 2017, 2017, pp. 6755–6760.

[114] T.P. Luu, Y. He, S. Nakagome, J.L. Contreras-Vidal, EEG-based brain-computer interface to a virtual walking avatar engages cortical adaptation, 2017 IEEE Int. Conf. Syst. Man, Cybern. SMC 2017, January, 2017, pp. 3054–3057.
[115] M.H. Chang, H.J. Baek, S.M. Lee, K.S. Park, An amplitude-modulated visual stimulation for reducing eye fatigue in SSVEP-based brain-computer interfaces, Clin. Neurophysiol. 125 (7) (2014) 1380–1391.
[116] S.F. Verkijika, L. De Wet, Using a brain-computer interface (BCI) in reducing math anxiety: Evidence from South Africa, Comput. Educ. 81 (2015) 113–122.
[117] R.N. Khushaba, C. Wise, S. Kodagoda, J. Louviere, B.E. Kahn, C. Townsend, Consumer neuroscience: assessing the brain response to marketing stimuli using electroencephalogram (EEG) and eye tracking, Expert Syst. Appl. 40 (9) (2013) 3803–3812.
[118] B. Yilmaz, S. Korkmaz, D.B. Arslan, E. Güngör, M.H. Asyali, Like/dislike analysis using EEG: determination of most discriminative channels and frequencies, Comput. Methods Program Biomed. 113 (2) (2014) 705–713.
[119] M. Yadava, P. Kumar, R. Saini, P.P. Roy, D. Prosad Dogra, Analysis of EEG signals and its application to neuromarketing, Multimed. Tools Appl. 76 (18) (2017) 19087–19111.
[120] M.M. Subbulakshmi, M. Balaganapathy, C. Gerard, Wireless EEG Signals Based Neuromarketing System Using Fast Fourier Transform (FFT), 2014, pp. 7–9.
[121] A. Telpaz, R. Webb, D.J. Levy, Using EEG to predict consumers' future choices, J. Mark. Res 52 (4) (2015) 511–529.
[122] J. Teo, C.L. Hou, and J. Mountstephens, Deep learning for EEG-Based preference classification, AIP Conf. Proc., vol. 1891, 2017.
[123] J. Kosters, Prediction of Preference and Choice of Wines by EEG Derived Measures During Taste and Smell Procedures, 2017.
[124] M. Bastiaansen, S. Straatman, E. Driessen, O. Mitas, J. Stekelenburg, L. Wang, My destination in your brain: A novel neuromarketing approach for evaluating the effectiveness of destination marketing, Journal of destination marketing & management 7 (2018) 76–88.

Chapter 3

A case study: impact of Internet of Things devices and pharma on the improvements of a child in autism

Muhammad Javaid Afzal[1], **Shahzadi Tayyaba**[2], **Muhammad Waseem Ashraf**[3], **Farah Javaid**[4] **and Valentina Emilia Balas**[5]

[1]Government Islamia College Civil Lines, Lahore, Pakistan, [2]The University of Lahore, Lahore, Pakistan, [3]Government College University, Lahore, Pakistan, [4]Government APWA College for Women, Lahore, Pakistan, [5]"Aurel Vlaicu" University of Arad, Arad, Romania

3.1 Introduction

Internet of Things (IoT) is a new field for current technical, computing, and scientific world. This fast growing technology has always taken an emergent step. IoT can be applied in any field of life. The combination of IoT, pharma, and autism has proved a good research field for the betterment of children with autism. IoT devices can provide a mutual infrastructure with pharma for the betterment of these children.

3.1.1 Internet of Things devices

The IoT is an interconnected computer devices system. It is actually the networking of Internet-linked smart devices, which can intelligently collect and exchange any type of data [1]. It includes wireless and wired sensors, softwares, actuators, and all computer devices. It also includes the electrically/electronically embedded systems, Internet communication, and all systems of hardware. These devices can interconnect for communication and interrelate with other systems through Internet. All these devices can be controlled, observed, monitored, and organized remotely [2]. Any system with any type of sensor, which can send data to any computerized system with Internet communication, is in the field of IoT. There are many IoT devices, such as laptop,

Emergence of Pharmaceutical Industry Growth with Industrial IoT Approach.
DOI: https://doi.org/10.1016/B978-0-12-819593-2.00003-0

computer system, mobile, tablet, smart TV, security system, alarm system, refrigerator, android system in cars, temperature monitoring system, planes, radio, tape recorder system, washing machine, oven, sewing machine, camera, toy system, iron, coffee maker, pizza maker, juicer, hair dryer, smart glasses, Google Glasses, ear phone, embedded devices, cloud service system, weighing scale system, electric toothbrush, exercise equipment, microwave, VOIP phone, automobile, media player, smart bulbs, toaster, printer, remotes, wireless speaker, wearable infotainment devices, smart cities metering systems, e-health system, digital factories, e-mobility, smart grids, smart buildings, security cameras, and thousands more [3,4]. Some of them are shown in Fig. 3.1.

Some IoT devices are working well wirelessly, while some with wires. All of them are equipped with Internet facility. These devices can send and receive data with Wi-Fi systems and even without it. Internet is mainly divided in two categories, that is, Internet and Extranet. Internet is a communal, supportive, and self-sustaining capability. It is accessible to thousands of billions of people in this world. Actually, Internet utilizes a big portion of telecommunication and cable networks [5,6]. Extranet is a local and private networking system. It uses Internet communication systems with public telecommunication organizations to share information with contractors, sellers, associates, and consumers securely. It can be seen as part of any company's Internet, which can be extended to the external users of that particular company. IoT has many applications in every field of life [7]. Police car has IoT applications too. These cars are well equipped with many IoT systems such as cameras, sensors, tracking system, wireless, and computers. There is also

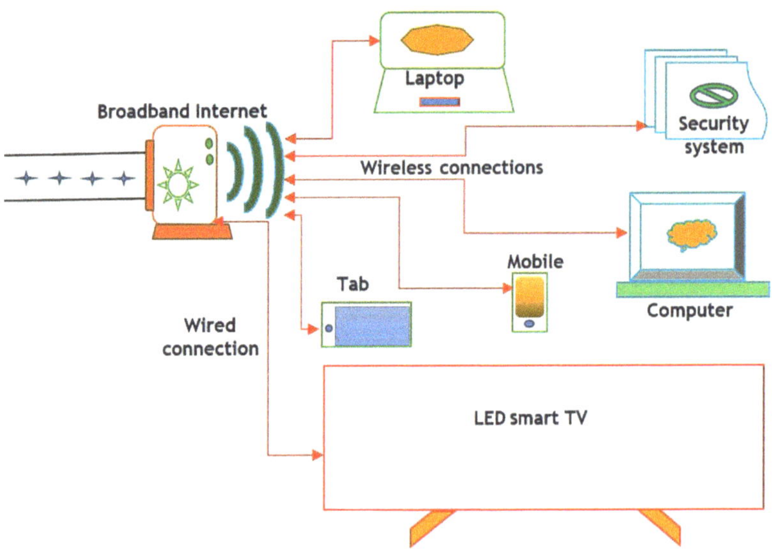

FIGURE 3.1 Internet of Things (IoT) smartest devices.

the requirement of bidirectional higher speed, safe, secure, steadfast, and reliable Internet communication [8]. Industrial sites depend on the widespread variety of electrical/electronical sensors and cameras to monitor and identify the manufacture developments, continuity in operations, and confirm their safety. These types of sensors can be found in odd locations frequently and need consistent, safe, and secure infrastructure for better communications [9]. The installation of surveillance cameras is a common thing these days due to security concerns. These cameras continuously require high-speed Internet for transmitting the videos on broadband to a central location [10]. Several hospitals trust on Internet-linked biomedical devices. They have latest equipment for saving lives of people. All types of sensors used in hospitals can receive and send data easily [11].

3.1.1.1 Internet of Things network requirements

The network requirements are always depending on particular smart devices and their applications [12]. The network may need

1. the capability of linking Internet with huge number of IoT components;
2. higher consistency;
3. real-time realization;
4. capability to protect and monitor the road traffic;
5. capability of programing for each and every custom applications;
6. road traffic observation and manage at local level; and
7. little cost of communication for huge number of smart sensors and devices.

Due to the widespread global interest, IoT has got numerous subareas and subnetworks. Several solutions of wireless heterogeneous communication coexist now. In these solutions the dominant solutions are Wi-Fi, cellular, Bluetooth, and ZigBee with MANET routing, and multihop, ad hoc, and rules of Internet protocols [13–16]. They work in regulating the networking properties and characteristics. All of them must be combined in order to have a unified communication organization [17].

3.1.2 Pharma in autism

According to scientific studies, medicine is an effective way to deal with autism, when it can combine with behavioral therapies. It has mood stabilizer medicines [18]. There is no medicinal cure for autism symptoms. Though autistic kids show distressing, repetitive, conventional, self-injurious behaviors. Mostly, when a kid hit himself or family members repeatedly, then medication involvement may be reasonable. The Food and Drug Administration (FDA) has permitted risperidone as the medicine for unstable aggressive behaviors of autistic child. There are other medicines for controlling these symptoms, but they are presently being studied. They have no approval from the FDA. Pharmacological involvements may enhance the capability of autistic kids in education with some other interventions [19]. They also remain in less restricting environments. Medicine can control anxiety, inattentiveness, aggression, hyperactivity, self-injurious, and stereotypic and obsessive behaviors with sleep disorders. Occasionally, Selective serotonin reuptake inhibitors (SSRIs) are used to control the autistic symptoms in kids and youngsters [20]. Different medications for autism spectrum disorders (ASDs) to control and enhance behaviors and other skills are shown in Fig. 3.2.

Mutations, adverse effects of drugs, environmental causes, and infections are the reason of mitochondrial dysfunctions [21]. A child can inherit or acquire this genetic disorder. Heart, brain, muscles, and lungs can be disturbed by mitochondrial dysfunctions. Omega 3, multivitamins, minerals, and supplements can control this type of disorder [22].

Chronic neuroinflammation is a brain disease in which irritation, irrelevant anger, frustration, swelling, enlargement of brain tissues, and inflammation in spinal cord occur. Antiinflammatory agents can play a role in its betterment [22,23]. Human immune system, heart, mental health, and metabolism can be affected by the long-standing exposure to stress and may distress the heart, metabolism, and mental health. Autistic kids take irrelevant stress on themselves. Antistress drugs and nonpharma treatments can be a better option for reducing the effects [24]. The imbalance between

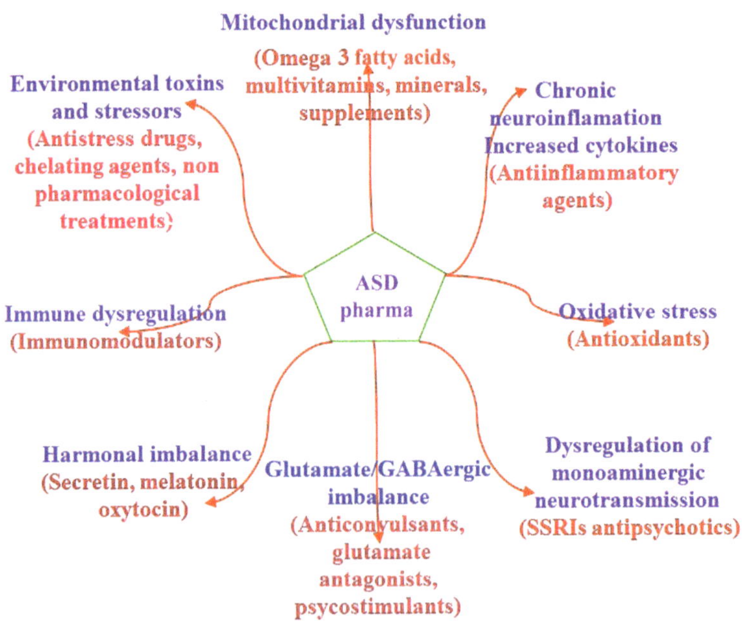

FIGURE 3.2 Pharma in autism spectrum disorder (ASD).

antioxidants and free radicals in human body is the reason of oxidative stress [25]. Free radicals have more number of electrons with oxygen-containing molecules. Therefore they can react certainly with other molecules. This is the reason of stabilizing the free radical, and hence, they become lazy and do not react. Antioxidants can be the better option to reduce this imbalance [26]. Immune disorder is also another type of disorder in human overactive/underactive of immune system. Immunomodulators (azathioprine and 6-mercaptopurine) are used to reduce these disorders. These two immunomodulators have chemical similarities [22]. Depression may be produced by the dysregulation of monoaminergic neurotransmission. SSRIs and antipsychotics are used in this disorder [27]. Glutamate is a very strong excitatory neurotransmitter [28]. Nerve cells release glutamate in brain. In normal persons, they are responsible for sending signals between nerve cells. They have a key role in understanding, memory, learning, and education. Their excess can produce hyperalgesia, irrelevant pain intensification, and inability to focus, nervousness, anxiety, impatience, irritation, restlessness, and ADHD symptoms. Anticonvulsants glutamate, antagonists, psychostimulants are used in this type of disorder. Little or excess hormones can create distress in human body. They have a key role in making our moods, behaviors, and every type of growth. Even smaller hormonal changes and imbalance can create side effects. There are some natural ways to balance our hormones, for example, by balanced intake of food—sufficient protein, healthy fats, green tea and fatty fish regularly among many—doing regular exercise, avoiding carbs, sugar, and stress, and refraining from overeating/undereating. In pharma, secretin, melatonin, and oxytocin are used for reducing these side effects. FDA has approved only "risperidone" as a drug for autistic children. It can be used to control the mental condition, irritability of an autistic child, bipolar disorder, and schizophrenia. With the use of this medicine a person or child can think clearly, and his or her day can be spent actively without these disorders. Risperidone is actually the atypical antipsychotics. Fifty percent of patients show sleepiness with the high dose of this medicine. In spite of these benefits, followings are some of the side effects:

1. headache;
2. dizziness, sleepiness, weariness;
3. tremors, muscle disorder;
4. agitation, anxiety, upset stomach, restless feeling;
5. depression, diarrhea;
6. constipation, dry mouth;
7. abrupt weight gain; and
8. runny and stuffy nose.

Table 3.1 presents ASD symptoms with their disorder and medication.

All of these medications cannot treat and control autistic spectrum completely. They are just time-consuming medicines. Some autistic kids can

TABLE 3.1 Pharma in autism spectrum disorder.

No.	Disorder	Symptoms	Medication
1.	Hormonal imbalance	Diarrhea, constipation, acne, irregular periods, brittle bones, indigestion, night sweats, heart rate, and mood	Secretin, melatonin, and oxytocin
2.	Glutamate imbalance	Hyperalgesia, irrelevant pain intensification, inability to focus, nervousness, anxiety, impatience, irritation, restlessness, and ADHD symptoms	Anticonvulsants glutamate, antagonists, and psychostimulants
3.	Immune system	Overactive/Underactive brain, mood swings, and snappiness	Azathioprine and 6-mercaptopurine
4.	Depression	Energy loss, loss of appetite, loss of sleep, recklessness, and loss of interest	SSRIs and antipsychotics
5.	Oxidative stress	Hypertension, heart disease, and neurodegenerative diseases	Antioxidants
6.	Chronic neuroinflammation	Irritation, irrelevant anger, frustration, swelling, enlargement of brain tissues, and inflammation in spinal cord	Antiinflammatory agents
7.	Environmental toxins and stressors	Human immune system, increased heart rate, mental health, and metabolism problems	Antistress drugs and nonpharma
8.	Mitochondrial dysfunctions	Mutations, adverse effects of drugs, environmental causes, and infections	Omega 3, multivitamins, minerals, and supplements

be sedated by these medicines because of their aggressiveness and violence. Some are controlled by therapies. Medical science has no cure for ASDs yet [29,30].

3.1.2.1 Food as pharma

There is a strong bond between eating, nutrition, and autism. Food therapies are also proposed in autism. There is widespread range of foods associated therapies for autistic persons. Gluten-, lactose-, and casein-free diets are proposed in autism [31]. Autistic kids may be sensitive for particular foods such as artificial sweeteners and dyes. Some kids may have picky habits of meal; therefore supplements are proposed for them. Food therapy is an important treatment in autistic kids. They have to follow strict elimination diet plan. Gluten found in wheat, and casein found in milk. All foods containing gluten and casein should be eliminated [32,33]. Food is very important in autism. It is either medicine or poison for an autistic kid. Autism can be worsening with following foods.

Dairy

Milk is present in each dairy product. Milk contains casein. It can be mixed in stomach and produced exorphin. This is poison for autism. It can produce brain fog, numbness to pain, no attention in any work or person, failure in concentration, and emotionlessness. Autistic kid started talking, communication, his/her hyperactivity is decreased and bowel problems solved with the removal of dairy from the diet [34].

Gluten

Gluten also acts as a poison in autism. It can be found in rye, barley, and wheat. Gluten can enhance the inflammation indigestion. Antibodies are created with gluten in the human body, which are not good for an autistic brain. It produces a negative effect on the function of cerebellum, which is already at lower working position [35,36].

Corn

Corn contains harmful fatty acid as it is not a vegetable. From corn, 22 different fungi are produced. For an autistic kid, it is not useful thing to eat. It produced aggression, anger, and depression in brain [37].

Sugar

Sugar produces inflammation in brain. There might be some insulin production in brain with sugar in autism. In this way the damage is magnified in brain. If sugar is avoided, then concentration is improved, decision power is increased, and thinking quality is increased [38].

Artificial ingredients

There should be no artificial ingredients in the diet of an autistic kid. They are not actually food. All additives, preservatives, and artificial color, flavor, and sweetener should be avoided in the diet of ASDs [39].

Now the question arises, what ASDs should eat? Honey, magnesium supplements, fish oil, melatonin, gut health and probiotics, sugar and gluten free, candida detox, vitamin D and C, and turmeric can be consumed by ASDs [40]. With the use of abovementioned diet rocking, teeth crushing, nervousness, anxiety, low concentration, low attention duration, reduction in hyperactivity, relaxing sleep, growth of useful gut bacteria and collagen, antioxidant movements, cognitive function, and cellular repair are the benefits to ASDs [41,42]. In autism, one can use foods that are given in Table 3.2.

TABLE 3.2 Good and bad foods in autism.

15 Good foods in autism	12 Bad foods in autism
Sweet corn	Strawberries
Avocados	Spinach
Pineapples	Nectarines
Cabbage	Apples
Onions	Peaches
Sweet peas (frozen)	Cherries
Papayas	Grapes
Asparagus	Celery
Mangos	Tomatoes
Eggplant	Sweet bell
Honeydew	Peppers
Melon	Potatoes
Kiwi	They have higher level of pesticide, therefore, they must be avoided
Cantaloupe	
Cauliflower	
Grapefruit	

3.1.3 Autism

Autism is a multifaceted neurodevelopmental disorder. It is generally identified in kids (18 months to 3 years). The cognizance, socialization, and communication of a kid can be affected by this disorder. Autistic kids can be improved well in time, if they receive early intervention. Hence, they can join the conventional schools [43]. There are no reliable data for the prevalence of autism in Pakistan. The statistics shows that there are around 350,000 autistic kids in Pakistan. Due to the lack of cognizance, knowledge, and expertise in identification and diagnosis of autistic kids, many of them remain undiagnosed unluckily. The poor parents suffer these facts without knowing, so the autistic kids have to bear their disorder for whole life. They mishandled by the parents, society, schools, and teachers. The visual and hearing impairment, physical disability, and mental retardation are the main concerns of ASD in Pakistan [44]. Unfortunately, autism is

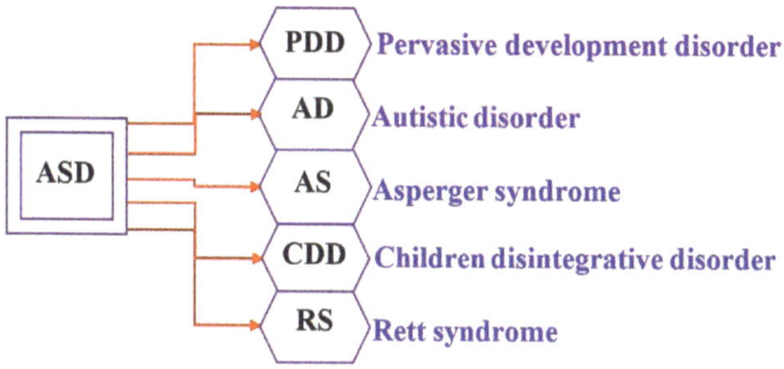

FIGURE 3.3 Types of autism spectrum disorder.

not considered a disability in Pakistan. The mental disability contains humiliation in Pakistan. Therefore parents cannot give proper treatment to their autistic kids. The spectrum of autism shows same insistence: lack of socializing, trouble in mingling with normal kids and resists variations in routine works, inappropriate giggling laughter, no fear of any dangers, little or no eye contact, constantly odd playing, apparent insensitivity to pain and discomfort, repeating irrelevant talks and words, loneliness, no cuddling, spinning different objects, zero response to parents, deaf acting, irrelevant affection for some objects, and feeling problem in communicating his/her needs [45]. Pollen allergy in autistic kids is also a main problem. It is severe in early change of seasons. There are more than 10,000 species of grass. Some autistic individuals may be allergic to multiple types because of different pollination cycles [46]. Autism has following disorders [47] shown in Fig. 3.3.

3.1.3.1 Pervasive developmental disorder or autism spectrum disorders

Pervasive developmental disorders (PDDs) include the lack of development delay in a child. It creates lack of socialization and communication. Due to this disorder, child feels uneasy when his/her routine changes and performs repetitive activities, and behavior problems arise. Children with PDD are nowadays termed autism spectrum disorder [48].

Symptoms

1. No social communication and interactions
2. Repetitive behavior
3. No speech
4. Zero or very little eye contact
5. No pain expressions
6. Cannot express their thoughts through language
7. High-/low-pitched voice
8. No communication

9. Cannot control their emotions
10. Twirling, rocking, hopping, and hand flapping [49].

Autistic children always repeat their playing habits. They are more concerned with only the parts of toys. These children need strict timetables, and they do not like changes to their routine works. Sometimes, they want to go to school on Saturdays and Sundays because of routine works. They cannot understand why the school remains closed on those days. This spectrum has a widespread range. Some ASD persons can live themselves, go to schools, and can do jobs. Some persons may have severe disabilities, while some lie in between them [50].

Causes

Science cannot find each and every cause of ASDs. Science knows that genetic is the important reason and do not have all the answers. There is no autism gene present in these kids. Apart from these, more things may be involved in genetics [51].

3.1.3.2 Autistic disorder

This type of disorder is the most common disorder in autism. Children with this type of disorder have differences in thinking, languages, behavior, and social abilities. This difference may appear before 3 years of age. It can be diagnosed in 1.5 years of age [52]. Their symptoms include

1. lack of socialization,
2. lack of communication,
3. language delay,
4. restricted range of behavior, and
5. stereotypic activities.

If your child have these symptoms then you need to discuss with your doctor and start the medication at the earliest.

3.1.3.3 Asperger syndrome

Children with Asperger syndrome (AS) have normal intelligence and language development, with some autistic traits. Sometimes, they have more intelligence but with language delay. They have trouble in socialization [53]. They have sensory issues in making transitions. They also need rigid routines for their activities. They can focus on one area at one time. It is a kind of obsession AS is actually a high-functioning autism [54]. Children with this syndrome may be very good at mathematics. If they understand the sequence of tables, then they can write all the tables at once. They may have very good puzzle-solving abilities; the normal kids may have the same pace. They have more intelligence than normal kids, but unfortunately they cannot express it. Their observation is also very strong than normal people. These children may have the ability to play games on two separate tablets with both hands simultaneously. They may have security issues, they sometimes even do not trust on their own parents. They are over sensitive. They may have dystonia (movement disorder). They may have unusual laughter and crying issues [55]. These children may have photogenic memory. It is the better kind of autism. These kids can improve a lot and can go to main stream schools [56]. They may live a happy life, but it all depends on their spectrum [57].

3.1.3.4 Childhood disintegrative disorder or Heller's syndrome

This kind of autism is very rare. A child with Heller's syndrome shows normal growth, and each area is developed normally. He/she attains appropriate verbal skills and sometimes shows nonverbal communication. They have social relationships. They are better in motor skills. They can play and show self-care skills. On the other hand, they show regression and almost lost completely between 2 and 10 years of age [58,59].

Symptoms after regression

1. They have delayed speech.
2. They have shown impairment in nonverbal behavior.
3. They have inabilities for a conversation.
4. They lack in playing abilities.
5. They lack in bladder control.
6. They lost their motor skills.
7. They lost their social skills.
8. They show sudden reversals in their improvements.

3.1.3.5 Rett syndrome

This type of disorder in autism is a genetic one. It is mostly found in girls. Its symptoms generally start from the age of 6 months to 3 years. These girls have normal developments before that age [60].

Symptoms

1. They have slowed physical growth.
2. They have slow brain development after birth.
3. They cannot move normally with no coordination.
4. They have lack of communication abilities.
5. Their eye movements are unusual.
6. They have breathing problems.
7. They have irritability and crying issues.

3.1.3.6 Difference between Asperger syndrome and autism spectrum disorder

AS is generally considered another kind of high-functioning autism. Mostly, kids with this syndrome are termed social and awkward. These kids want to have friends. Unfortunately, they lack social skills for the beginning and maintaining of friendship. They usually show less interest in friends. Moreover, high-functioning autistic persons lack in speech and communication. Asperger children tried very well not to show their lack of speech, but they are not successful in this. They cannot speak on different topics [61].

3.1.3.7 Diagnosis and treatment in autism spectrum disorder

The diagnosis is very important for therapies and services which can help parents to learn the treatments for the betterment of their kids. In order to diagnose this spectrum and the severity, doctors only can observe these kids and inquire some questions from parents. ASD has no blood test to diagnose. There is no cure to this disorder, only therapies can control their behavior. Every child has his/her own level of disorder [62]. First of all, diagnosis should be done as soon as possible. One can line up his/her resources then and help them to reach their full potential. The sooner we start, the better are the chances of having a healthy and strong later lives of these kids. To treat this or control their behavior, there is little help from pharma. Medical treatment works best when combined with therapies that can develop socialization and skills in these kids. The holders of this ASD experience this world differently. Their successes, encounters, and defeats might be different from the normal people. There is a need to appreciate them as they are. They have their own distinctive personalities, characters, and interests [63]. There is no cure for every type of autism. There are therapies for their development. Treatment may include the following:

1. They need supportive care in everyday activities.
2. They need nutritional care.
3. They need physical therapies and hydrotherapies to improve their flexibility and movements.
4. They need speech therapy sessions for better communication.
5. They need medication to treat and control seizure, breath, and motor skills.
6. They need some specialized equipment for problems in their orthopedic.

3.2 Parent history

The record of parental family medical history shows that they have no cardiovascular, gastrointestinal, diabetic, hypertension, and mental diseases. They are in good health at the time of their marriage. They are not even cousins. Both of them are aged 25 at the time of marriage.

3.2.1 Patient history

The patient's name is Muhammad Mujtaba Javaid. He is the third child of his parents with two elder sisters. He is 10.5 years old now. He was born on

August 22, 2008—after 10 years of his parents' marriage. His mother had no complications during the time of pregnancy and birth. He was a very good toddler. After 1 year of age, he was constipated all the time.

3.2.2 Herbal treatment

In order to treat his constipation his parents have given him herbal treatment. Parents had treated him by giving dates with milk, sugar, drinking more water, eating more soluble fiber. His constipation was not gone. Parents gave him Panadol and routine medicines in his fever and seasonal diseases.

3.3 General behavior up to 2.5 years

After 2 years of age, he developed some strange behavior. He could not imitate. He got interest in repetitive odd playing. He was not learning from his surroundings. He was not responsive. Parents called him by his name repeated times, but he gave no attention. He was very active and aggressive at that time. He shouted all the time with loud voice. He got an unusual laughter and crying issues. He did not like crowd. He had zero eye contact with no socialization and communication. He also got sleep problems.

3.4 Diagnosis

His parents lived in a small city (Gujarat, Punjab, Pakistan). Both his parents are professors in physics. Due to all these issues, parents took him to a local doctor. He advised them to take him to the Children Hospital Lahore. They arranged his hearing tests, but there was no problem with his hearing. After this test, doctors advised psychiatrist for him. Psychiatrist examined him, and a panel of doctors diagnosed ASD in him. They advised his parents to take regular six to eight sessions per day. Speech therapist was also advised for the beginning and development of speech. Peg boards are also advised for him for the development of his motor skills.

3.5 Autism spectrum disorder therapies, pharma, and Internet of Things

His parents were in shock when they heard ASD for the first time in their life. They were not ready for this kind of situation about their only son. Mujtaba was continuously being exposed by his ASD behavior. Soon his parents were ready to accept the autistic nature of their child. His therapies were started. In the beginning, his parents started these therapies by themselves. They took 8–10 sessions per day. The sessions were simple because of the age of child. Parents started following IoT devices. They learned about autism from Internet. They used tablets, different computers, and android mobiles for their own learning. They started reading new researches in this field from Internet. Net browsing proved beneficial in their learning. Mujtaba was small enough to use IoT devices. Soon they were able to download some programs for Mujtaba. They also used IoT for monitoring his conditions. They started small conversations by showing pictures of animals and fruits to Mujtaba. He also got his toilet training with the help of his tablet. The response was very good. Mujtaba started taking interest in IoT. A speech therapist in Gujarat was available at that time. She worked very hard in his speech and communication and motor skills. The results were very fine. She also used IoT devices in speech therapy and in other therapies. A special education teacher was also made available for Mujtaba. He taught him very well; in response, Mujtaba responded very well. He liked his routine very much. He had shown his aggression and violence in the absence of his teacher and speech therapist. A normal school was recommended for him by the Children Hospital Lahore. So he started going to school with a shadow teacher. Things were going in a good way in the small city. Mujtaba was busy in his morning-to-evening schedule: school—from 8.00 a.m. to 1.00 p.m.; speech therapy—2.00 p.m. to 3.00 p.m.; busy with his teacher in school homework and other social skills—from 4.00 p.m. to 5.00 p.m.; and after 5, his father took him to ground where he started learning cycling. Hence, a tough timetable was set for Mujtaba. He liked his routine very much. Mujtaba was not ready to take any sort of medicine. Even he was not taking his routine medicine for fever and throat infection. His parents treated him with some herbal and food treatment. They bore all his aggressive behaviors on their own. Therefore without medication, he was trained at home.

3.6 Assessment taken by Lahore Children Center at the age of 5.8 years

There were a lot of improvements at 5 years of age, so his parents wanted his assessment to be continued from a big city such as Lahore. The assessment was taken with following results.

3.6.1 Introduction

Muhammad Mujtaba Javaid, a 5.8-year-old boy, was seen for a consultation appointment looking at his development of speech and language skills. He was accompanied by his parents. The assessment took place using Preschool Language Scale, Fourth Edition UK, Renfrew Action Picture Test together with informal tasks. The results of abovementioned assessment tools along with discussion with his parents and observation have given the following profile. Parents reported that Mujtaba has received a diagnosis of ASD from the Lahore Children's Hospital.

3.6.2 Relevant medical and developmental history

There were no concerns or complications at time of Mujtaba's mother's pregnancy or during birth. No concerns regarding his early feeding and sucking patterns were reported. No significant pre- or postnatal complications have been reported. Mujtaba's motor milestones followed a normal developmental pattern. Moreover, no significant concern with regard to his vision and hearing was reported. Parents reported that Mujtaba settled into preschool easily and enjoys going there.

3.6.3 Language and family background

No significant family medical history was reported. Mujtaba lives with his parents and two sisters. He is exposed to English and Urdu languages.

3.6.4 Current functioning

Attention control:

Mujtaba showed fleeting attention in the clinical setting. He sat on the table, when provided with reminders for good sitting and with reinforcement. With regard to his attention span, it was observed that he kept focused on adult-directed tasks for less than 5 minutes and was able to sustain his attention on own choice of task for a reasonable length of time, that is, 10–15 minutes. Mujtaba was able to follow instruction containing single information-carrying element such as "put the dog," "put the ball" but demonstrated inconsistent listening and difficulty in following two information-carrying elements such as "put the dog on the train."

Social interaction:

Mujtaba showed reduced social interaction and inconsistent eye contact. He was unable to initiate and maintain a conversation with me. However, Mujtaba's parents reported that his eye contact has improved in comparison to his past performances. It was also mentioned by the parents that he does not interact much with his peers and adults and tends to play alone. However, his interaction with parents and other familiar people is better, and he shows empathy and affection toward them. Mujtaba shows preference for visually based materials. He likes writing numbers and alphabets. Parents reported that Mujtaba is able to write numbers up to 100 and alphabets from A to Z. It was also observed during the session that he can write his name both in English and Urdu languages.

Play skills:

He demonstrated reduced awareness of turn-taking, waiting, and sharing skills, and lack of symbolic play. It was observed that shared play needs to be developed.

Behavior:

Mujtaba exhibited some atypical behavior during the session including stomping feet and walking around. It was further added by the parents that

Mujtaba used to find difficulty in wearing shoes, getting a haircut, cutting nails, cleaning ears, and taking shower, but now he has improved a lot and does not show these behaviors.

Language skills:

Language refers to the way that an individual uses and understands words, sentences, and stories that are presented to him/her in both verbal and written form. Renfrew action Picture Test is a standardized test that assesses child's production in terms of information and the grammatical structures.

Receptive language:

Receptive language refers to the understanding of language. It involves the ability to listen, understand/process, and carry out the instructions and commands as well as the ability to follow the general thread of a conversation.

Mujtaba showed good understanding of everyday object labels, for example, ball, shoes, cat, cookies, balloons, glass, spoon, bird, banana, apple, cup, fish, duck frog, horse, car, bear, baby, ice cream apple, chair, book, and markers, some verbs, for example, sleeping, drinking, eating, washing, playing, and running. He also showed good understanding of colors and body parts but not early language concepts such as big/small, more/less, wet/dry, most, not, and half/full. It was observed during the session that Mujtaba can understand and follow simple instructions such as "sit down," "come here." Mujtaba was not able to sort objects into categories relating to food, animals, clothes, etc. Moreover, Mujtaba was unable to identify objects by their functions, for example, "which one do we cut with," "which one do we ride on," "which one do we cook with," "which one do we wear," and "which one do we drink with". It is important that underlying concepts are specifically taught to enable him to make use of them functionally in his daily life and/or at school. Furthermore, he presented weaknesses in his ability to process and retain complex auditory information of increasing length as well. As mentioned earlier, Mujtaba is able to follow instructions with one keyword but cannot consistently follow an instruction when this is increased to two keywords. This indicates significant difficulties in Mujtaba's understanding of language for a child of his age. Mujtaba's difficulty in receptive language has a significant impact on his ability to understand and communicate effectively. Moreover, these difficulties will impact significantly upon Mujtaba's ability to follow directions within a classroom setting and at home. Therefore it is important to develop his ability to follow instructions of increasing length on a consistent basis and generalize this across different tasks and in different settings.

Expressive language:

Expressive language refers to the ability to use language to interact with a variety of grammatically correct sentences and using language to convey meaning correctly. It also involves using vocabulary correctly. Parents reported that he has many words in his expressive vocabulary and is combining words into short phrases. His vocabulary includes everyday objects' names, colors, animals, food as well as names of some people. His vocabulary is in a mixture of English and Urdu, but his phrases are mainly in Urdu. Parents reported that he used to communicate using nonverbal means, which includes grabbing, pushing an adult to what he wants, using eye gaze, but now he communicates using words, and it was mentioned by the parents that at times he uses two words or phrases as well, for example, "chocolate do" (give chocolate), "shirt utaro" (take off shirt), "shirt pehnao" (put on shirt), and "ma'am roti do" (ma'am give food). Mujtaba has significant weaknesses in his expressive language skill. He showed difficulties in using language to convey information precisely due to significant gaps in his vocabulary. Mujtaba has limited expressive vocabulary and reduced length and range of phrase structure that he can produce. These difficulties with expressive language will have significant impact on Mujtaba's ability to convey his message effectively and will lead to irritation and frustration. So it is necessary to work upon developing Mujtaba's expressive vocabulary to enable him to express himself more clearly.

3.6.5 Summary and recommendations

Mujtaba presents with significant difficulties in aspects of his attention control, expressive and receptive language development along with weaknesses in play and social interaction. It is, however, important to note that Mujtaba is a good visual learner and presents with good imitation skills. Currently, he communicates using single words and some learned phrases consisting of two real words. I recommend Mujtaba to avail speech and language therapy sessions, which will focus on the following areas: enhancing Mujtaba's sitting behavior, attention, and listening skills through visual support system such as visual schedule, first/next/last board, "I am working for" board. Developing Mujtaba's eye contact, turn-taking skills, joint attention, and anticipation by engaging him in shared play tasks along with highly motivating toys accompanied with reinforcements. It would be good to extend his play skills in encouraging early imitative play and pretend play skills using large toys such as teddy or dolls carrying out different everyday activities.

To develop Mujtaba's ability to follow multistep instructions, two different task has been assigned through barrier boards, sheets and toy. For Mujtaba's better understanding basic language and linguistic concepts including position words (e.g., in/out, on/under, in front/behind, and top/bottom), size (big/small and tall/short), quantity (more/less, half/full, and most), and other concepts such as wet/dry and hot/cold has been taught. To improve Mujtaba's vocabulary within everyday topics using sorting, naming tasks (including names of people, animals, actions, objects, and verbs) through picture resources, categorization boards, etc. have been adpotied. These methods help Mujtaba's to improve a sentence structure comprised subject, verb, object, prepositional phrases, and adjective phrases. Color-coded system such as colorful semantics in combination with action picture cards will be useful in this regard. Mujtaba required fundamental vocabulary such as open/close, finish, I want, more, bye bye, please, give, gone, and hello. This should be done verbally along with the signs. Practice of all such things individually and in groups help him to improve his social behavior, skills and attitude.

3.7 Improvements up to 8 years of age

After 5 years of speech therapy and the efforts of special education teacher, Mujtaba still was not able to communicate well with his parents and fellows. He could barely tell his needs in a single word or a maximum of three Urdu or English words. "Mujhay pani do" (give me water), "Mujhay rooti do" (give me bread) are the kind of small sentences that he could speak. He understood very well the concept of positioning words (in/out, on/under, in front/behind, top/bottom), size (big/small and tall/short), quantity (more/less, half/full, and most), and other concepts such as wet/dry and hot/cold. He also gained the concept of names of people, animals, actions, objects, and verbs. Mujtaba developed his functional vocabulary such as open/close, finish, I want, more, bye bye, please, give, gone, and hello. These were the improvements after the assessment taken at the 5.8 years of age. Sleep problems have also gone. He learned to use IoT devices (tablet, Internet, mobile, and smart TV). By using IoT, he got some good changes in his behavior. His focusing span had increased. He got interest in mathematics. He learned some poems too with the help of IoT devices.

3.8 Changes in behavior after 8 years of age and use of Internet of Things

Everything was going well up to the 8 years of age. After this, Mujtaba got some more behavioral changes. But after that, speech therapist has moved to another city, resulting in lack of punctuality by his special education teacher. His visits were decreased to once or twice in a week, which continued for 2 years. Therefore his life suffered very much. His parents tried different teachers but all in vain. Only, the Quran teacher (Qari sb) was successful with him. Mujtaba waited all the time after school for his Quran teacher. Irregular speech therapy as well as visits of special education teacher is the key factor of disturbance in his life. Apart from that, his homework also got disturbed. Mujtaba was grown up and got some weight because of his eating habits. He came down from his level. With the excess use of tablet, he got eyesight problem. He left his little speech completely, and some ADHD behavior also added up. For 2 years, this routine had continued. The only things that he enjoyed was Quran teaching and cycling. He had lost his eye contact. His socialization was also on poor level. He could not sit on ground because of some muscular problems of legs. He could read English, Urdu and do some mathematics. His drawing pattern was good because of school. He learned by heart some surahs of the Quran. He learned some exercises on treadmill due to the use of IoT devices. He started searching nanotechnology on YouTube as he read this word from his science book. He often picked some words from science book and wrote them on YouTube and saw movie clips on that word. He played cycling, running, and different games on tablet. Some pictures of his tablet are shown in Fig. 3.4.

He started solving puzzles on IoT. Parents started using some playing techniques for him, which resulted in positive. This led him to take more interest in puzzles. Few pictures of puzzles are shown in Fig. 3.5.

Mujtaba played these by using IoT. He spent plenty of time, for which the balance got disturbed. He improved in IoT but with side effects due to its excess use. His speech and socialization went completely missing. His ADHD behavior has increased a lot. His parents had no idea about the importance of pharma in autism. They had not given him any medicines and bore all his behavior. They calm down Mujtaba by themselves in his meltdowns. Parents were very disturbed with the overall regression. They decided to move to a big city (Lahore, Pakistan) for more rehabilitation.

FIGURE 3.4 Pictures of games played using tablet as Internet of Things (IoT) device.

FIGURE 3.5 Different playing games.

3.9 Improvements up to 10.5 years of age

Mujtaba's parents enrolled him in "Global Institute for Autism and Special Needs" near their new house in Lahore. This autism school helped him a lot. The institute has every type of equipment and facilities for kids like Mujtaba. They have a special room for sensory needs of these kids. Surah Al-Rehman is also recited in a room to calm down such kids. Mujtaba started talking again with the therapies given in school, that too with much improvement in his behavior than before. The following assessment is taken by "Global Institute for Autism and Special Needs" after 4 months of his admission.

3.10 Applied behavior analysis therapy assessment report

Report Date: April 15, 2019

Name: Muhammad Mujtaba Javaid

Gender: Male

Date of Birth: August 22, 2008

Muhammad Mujtaba Javaid is a 10.8-year-old boy. He is residing with his parents and two sisters in Lahore. There was no family history of autism spectrum disorder in his extended family. As per his mom's pregnancy history, Mujtaba was born full term. There were no complications during the pregnancy. Mujtaba can tolerate peers very well; he would experience anxiety sometimes, especially when he cannot complete the task given to him. Mujtaba would be sad sometimes because he is having a hard time to say or express himself. No food allergies have been reported by the parents. He requests for the items that he wants, by words, or holding an adults hand toward the item. He can speak names of his preferred items.

Developmental milestones of a child by 10–11 years of age:

According to the Center for Disease Control and Prevention, the milestones for a child by the age of 9 years are:

Emotional/social changes:

1. Show more independence from parents and family.
2. Start to think about the future.
3. Understand more about his or her place in the world.
4. Pay more attention to friendships and teamwork.
5. Want to be liked and accepted by friends.

Thinking and learning:

1. Show rapid development of mental skills.
2. Learn better ways to describe experiences and talk about thoughts and feelings.
3. Have less focus on one's self and more concern for others.

ABLLS-R assessment:

The ABLLS-R assessment tool was used. The ABLLS-R provides a comprehensive review of 544 skills from 25 skill areas including language, social interaction, and self-help, academic and motor skills that are acquired most typically by the developing children. The task items within each skill area are arranged from simpler to more complex tasks. Expressive language skills are assessed based upon the behavioral analysis of language as presented by Dr. B.F. Skinner in his book, *Verbal Behavior* [64].

The skills are listed as follows:

Letter	Title	Explanation/remarks
A	Cooperation and reinforcer effectiveness	How well a child responds to motivation and others
B	Visual performance	The ability to interpret things visually, such as pictures and puzzles
C	Receptive language	The ability to understand language
D	Motor imitation	Being able to mimic the physical actions of others
E	Vocal imitation	Being able to mimic the sounds and words others make, also, called echoic in ABA
F	Requests	Also called manding in ABA

(*Continued*)

(Continued)

Letter	Title	Explanation/remarks
G	Labeling	Naming objects, or their features, functions, or classes
H	Intraverbals	Responding to only the stimulus of words. Objects/ motivators not present
I	Spontaneous vocalizations	Using language without being prompted
J	Syntax and grammar	How well words and sentences are put together
K	Play and leisure	Solitary and group play skills
L	Social interaction	Abilities regarding interaction with peers and adults
M	Group instruction	Ability to learn in a group setting (not just one-on-one)
N	Classroom routines	Ability to follow rules and common school routines
O	N/A	
P	Generalized responding	The ability to generalize material learned and use it in real-life or novel situations
Q	Reading	Alphabet, prereading, and reading skills
R	Math	Numbers, counting, less-more-equal, basic addition and subtraction
S	Writing	Coloring, drawing, copying, and writing skills
T	Spelling	
U	Dressing	Ability to dress or undress self independently
V	Eating	Basic self-help skills regarding eating and preparing of food
W	Grooming	Basic self-help skills regarding grooming and hygiene
X	Toileting	Basic self-help skills regarding toileting
Y	Gross motor skills	Large motor activities such as playing ball, swinging, crawling, running, and skipping
Z	Fine motor skills	Fine motor activities such as writing, peg board, turn pages in a book, cutting, and pasting

The results of the assessment are as follows:

Skill area: cooperation and reinforce

Present level of achievement: During assessment, Mujtaba can take the reinforce when offered. He takes the item within 3 seconds; he can take a common object when offered; he can look at the common objects when asked; he can respond on controlled reinforce; he can look at teacher for instruction, but it is not consistent; he cannot response on instruction quickly; he can scan the items; he can take the reinforce out of two choices; and he can respond on the social reinforcements.

Goals:

1. Look at a nonreinforcing item.
2. Respond to instructor controlled reinforces.
3. Wait without touching stimuli.
4. Look to instructor for instruction.

Skill area: visual performance

Present level of achievement: When Mujtaba was given a puzzle with a single-piece type of inset, he can place eight pieces independently. He can sort the shapes in a form box. When he was given identical objects to match, he was able to match at least four objects in a display of five items in single-item-per-task presentation. Mujtaba can match 10 pictures of 11 pictures on display independently; he was not able to make the block design coping with the pattern from card and blocks. He was not able to sort nonidentical items, in the array of two samples. Mujtaba was able to do a board type of puzzle independently up to eight pieces.

Goals:

1. Match identical objects to sample.
2. Match objects to picture.
3. Match pictures to objects.
4. Sort nonidentical items.
5. Jigsaw puzzles.

Skill area: receptive language (the ability to understand or comprehend language heard or read)

Present level of achievement: Mujtaba can respond to his name. He follows instructions to look at his reinforcing item but requires minimal assistance. He is able to an enjoyable action in context. He followed instructions that required him to give a named, nonreinforcing item. He follows instruction to touch a positional reinforcing/nonreinforcing item. He can follow an instruction in a routine situation. Mujtaba can give a nonreinforcing item when asked him to give the named item. He can do a simple motor action when instructed him to do. He is able to select one reinforce out of two choices, he can give, touch, and take the common object out of two choices. He is able to touch his own body parts and cloths on request. Mujtaba is able to select one picture of common object in an array of six picture cards. He is able to follow the hand signs. He can go to the person on request and give the named item.

Goals:

1. Acquire new selections skills without intensive training.
2. Varied instructions to select using a response.
3. Point to a body parts on others or pictures.
4. Touch parts of items.
5. Follow an instruction to walk to someone and get a named item.

Skill area: motor imitation (imitation becomes the foundation upon with other important skills are based, e.g., verbalization, play, social, and self-help)

Present level of achievement: Mujtaba was not able to imitate a motor activity with an object upon request. He imitates 10 actions when verbal prompts were provided. He was not able to imitate the motor action with lower body. He was able to do 10 motor actions with upper body on request.

Goals:

1. Motor imitation using objects
2. Motor imitation using objects in a discrimination
3. Imitation of leg and foot movements
4. Imitation of gross motor actions modeled in a mirror
5. Imitation of head movements
6. Motor imitation of fine motor movements.

Skill area: vocal imitation

Present level of achievement: Mujtaba was able to imitate sounds upon request. He is able to imitate initial sound of words.

Goals:

1. Imitation of multiple separate sound combinations
2. Imitation of short and fast versus elongated/slow sounds
3. Imitation of the number of repetitions of a sound
4. Imitation of held a sound to a second sound

Skill area: request

Present level of achievement: Mujtaba is able to request for his needs independently, he can request for what he wants by using words or gesture. He is able to ask for help.

Goals:

1. Acquire novel request without intensive training.
2. Request using and adjectives.
3. Request using prepositions.
4. Request information using "what."
5. Request information using "where."

Skill area: labeling

Present level of achievement: Mujtaba is able to label his reinforces and some of the common items.

Goals:

1. Label common objects.
2. Label pictures of common items.
3. Label body parts.
4. Label piece of clothing.
5. Label common ongoing actions.
6. Label pictures of common actions.

Skill area: intraverbals

Present level of achievement: Mujtaba is able to fill in words from songs. He can tell about his personal information only such as his name and he is fine. He is able to make animal sounds on request.

Goals:

1. Fill in the blanks regarding fun items and activities.
2. Answer questions regarding personal information.
3. Fill in the describing common activities.

Skill area: spontaneous vocalization

Present level of achievement: Mujtaba can vocalize identifiable speech sounds. He can say words spontaneously; he can sing a song and model the actions of song.

Goal:

1. Spontaneous conversation

Skill area: play and leisure skills

Present level of achievement: Mujtaba can explore the toys in the environment, he allows peers to touch his toys.

Goals:

1. Independent outdoor activities.
2. Independent indoor leisure activities.
3. Play with toys manipulates toys as designed.
4. Independently play with toys and engages in verbal behavior.

Skill area: social interaction skills

Present level of achievement: Mujtaba can sit appropriately with peers, he tolerates the touch of other kids and is able to take an item offered by the peer or adult in a group sitting. Mujtaba is able to return the greetings.

Goals:

1. Physical approaches and engage others.
2. Look at others in anticipation of completing a reinforcing action.
3. Imitate peers.
4. Respond to approaches and attempts to interact from peers.
5. Label items for others.
6. Eye contact.

Skill area: group instructions

Present level of achievement: Mujtaba can sit appropriately in a small group of kids and is able to follow the instruction of teacher in a group. He cooperates with teacher and other kids in group.

Goals:

1. Follow group instructions with a discrimination.
2. Raise hand to get teachers attention to do an activity.
3. Raise hand to answer a question.
4. Raise hand and names items.

Skill area: reading

Present level of achievement: Mujtaba can identify random letters of the alphabet, receptively. He can receptively select the corresponding letter when given the sound associated to the letter. He can label letters and sounds of letters on request. He is able to match word to picture on request and word to word on request.

Goals:

1. Name letters in words reading left to right.
2. Match individual letters to letters on word card.
3. Fill in the missing letters of words.

Skill area: maths

Present level of achievement: Mujtaba is able to do rote count independently; he can add, subtract, and multiply.

Goals:

1. More
2. Less
3. Some
4. All
5. Greater
6. Walk and get specified numbers objects from a larger set
7. Time telling
8. Identify coins by name.

Skill area: writing skills

Present level of achievement: Mujtaba is able to hold pencil, crayon, or marker but not with the pencil grip. He is able to trace, color, and copy the words he can write independently.

Goal:

1. Hold pencil with pencil grip.

Skill area: spelling

Present level of achievement: Mujtaba is able to spell three-letter words.

Goals:

1. Match individual letters to letters on word card.
2. Fill in the missing letter of words.
3. Write in missing letters of words.
4. Spell words vocally.

Skill area: dressing skills

Present level of achievement: Mujtaba can pull his pants up/down, independently. He can take his shoes off by himself.

Goals:

1. Use snaps.
2. Buttoning shirts on and off.
3. Use buckles.
4. Adjust clothing when needed.
5. Tie shoes.

Skill area: eating

Present level of achievement: Mujtaba is able to eat by himself with the use of fork and spoon; he can drink from the glass and straw independently.

Goals:

1. Spread with a knife.
2. Pour liquid into a cup.
3. Clean up table after meals.
4. Keep eating areas clean.

Recommendation

As the report indicates, there is a gap between a typically developing 10.8 years old and Mujtaba. He is recommended to do 5–10 hours of ABA therapy 5 days a week to address the skills where he scores low.

3.11 Autism schools in Pakistan

There are more than 20 autism schools in Pakistan. All schools are working at their own level. Some are very expensive, and some are cheap. Some are private and some are on different funding from government or others. There are more schools, but still awareness, facilities, and treatment for autism are limited in Pakistan.

1. Global Institute for Autism and Special Needs
2. Roots and Wings Lahore
3. The Trust School
4. Autism School Lahore
5. Autism Care Centre
6. Help Autism in Pakistan
7. Autism Resource Centre Lahore
8. Oasis School
9. Amin Maktab
10. The Rising Sun
11. The Fountain House
12. Centre for Autism Karachi Pakistan
13. Pakistan Centre for Autism
14. Pakistan Centre for Autism—PECHS Br
15. Autism Society of Pakistan

16. Apples The Grooming School
17. Autism Resource Centre Islamabad
18. Headstart School—Kuri Campus
19. Hope Inn Islamabad
20. Ghaliya's Montessori House for Autism
21. Autism Resource Centre, Islamabad
22. Institute for Special Children Quetta
23. Zobia School for Special Children Mirpur
24. APS Bright Horizons Mirpur

3.12 Cost analysis of some schools with autism

Here is a comparison of fees of some schools (Table 3.3).

3.13 Recommendations for autistic child discipline with Internet of Things and pharma

It is very difficult for parents to decide and apply the best method to control the behavior of their autistic child. Therefore it is very important to modify the bad behavior into positive behavior in autism. There are six different recommendations (methods), which can bring discipline in the life of an autistic kid, which are as follows:

Recommendation 1 (child-centered discipline)

1. Please do not forget that an autistic child is a child.
2. Be patient with him/her.
3. Please stay positively attentive with him/her.
4. Please handle his/her meltdowns with great care.
5. Please keep a peaceful voice and demeanor with him/her.
6. Please use recommended medicine for him/her.
7. Please get the proper knowledge about autism with IoT devices.

Recommendation 2 (timetable)

1. Please make a predictable timetable and apply it with care.
2. Please use picture timetables to produce order in routine works.
3. Please be regular with the timetable.
4. Adjust the timetable excellently with his/her growth.
5. Organize lot of time with him/her relaxation.
6. Schedule lot of time for fun with IoT devices that can make him/her happy.
7. Schedule some energetic moments because of his/her hyperactive tendencies.
8. Resolve sleep problems with good food and pharma.

TABLE 3.3 Fee comparison.

School	Fees in PKR per month	Funding
Global Institute for Autism and Special Needs	60,000	No funding
Oasis School	80,000	Sharif Trust
APS Bright Horizons Mirpur	Free	Pakistan Army
Roots and Wings Lahore	25,000	No funding from anywhere
The Trust School Wapda Town	25,000	No funding from anywhere
Amin Maktab	Free	Donations
The Rising Sun	7500	Government funding and others
The Fountain House	Free	Government funding
Autism Resource Centre Islamabad	33,000	No funding from anywhere
Pakistan Centre for Autism Karachi	12,000 + 22,000	No funding from anywhere
Pakistan Centre for Autism—PECHS Br. Karachi	12,000 + 22,000	No funding from anywhere
Apples The Grooming School	8760	No funding from anywhere
Centre of Clinical Psychology PU	25,000	No funding from anywhere
Help Autism in Pakistan	30,000	No funding from anywhere

Recommendation 3 (behavior problems)

1. Please set an excellent example before him/her.
2. Communicate with him/her about handling emotions.
3. Please help him/her from a traumatic condition by using pharma or IoT devices.
4. Set redirection for the stressed child.
5. Do not pressurize him/her about smaller things.
6. Keep your expectations low.
7. Admiration of an autistic child is very fruitful for his/her good behavior
8. Try to explain him/her the positive outcomes of good behavior, use IoT devices for this purpose.

Recommendation 4 (exact discipline strategies)

1. Learn how to calm down the child.
2. Give positive reminders for him/her mistakes.
3. Give a pleasant warning if they acting out.
4. Give instant consequences if he/she refuses to change their bad behavior.
5. Modify the punishment style.
6. Please stay constant.
7. Avoid punishments such as spanking, slapping, and hitting.
8. Do not criticize child's behavior.
9. Use proper IoT devices and pharma as discipline strategies.

Recommendation 5 (reward system)

1. Give the rewards again and again for positive and good behavior.
2. Admiration is very important for your child.
3. Set sensory rewards for him/her.
4. Practice modification in the reward scheme to make him/her happy.

Recommendation 6 (learn and accept the cause)

1. Please believe that autistic children can think correctly.
2. Learn and understand the purpose of the autistic behavior.
3. Work out for the cause of the bad behavior.
4. Use IoT devices to understand the cause.

3.13.1 Recommendation of nutritional interventions in autism spectrum disorder

The link between nutrition and ASD, which is a complex developmental and neurological disorder manifesting itself in significant delay or deviation in interaction and communication, has provided a fresh point of view and signals that nutrition may have a role in the etiology of ASD as well as play an active role in the treatment by alleviating symptoms. People with ASD often repeat behaviors and have narrow obsessive interests, these type of behaviors can effect eating habits and food choices which can lead to following health concerns:

- Limited food selection or strong food dislikes
- Not eating enough food
- Constipation or diarrhea
- Leaky gut

Diet order for autism (GFCF)

You may have heard that a gluten- or casein-free diet can improve symptoms of ASD.

Gluten is a type of protein found in wheat, rye, and barley. Casein is a protein found in milk. Proponents of diet believe people with autism have a leaky gut or intestine, which allows part of gluten or casein to seep into the blood stream and affects the brain and central nervous system.

The GFCF diet is very important because of the following:

1. The GFCF diet seeks to heal the gut and calm down the immune system.
2. It improves digestion and cognition.
3. It decreases aggression and anxiety.
4. It improves sleep quality.

The vitamin B6 helps (fish, chickpeas, liver, potato, and fortified food) in the following areas:

1. It improves speech.
2. It decreases aggression.
3. It improves eye contact.
4. It improves social responsiveness.

The omega 3 fatty acid helps (soybean oil, fish oil, flex seeds, fish, canola oil) in following areas:

1. It decreases hyperactivity.
2. It decreases aggression.
3. It improves language.
4. It improves learning skills.

The probiotic yogurt helps in the improvement of intestinal microflora profile. In autism, some foods (gluten, casein, sugar, simple carbohydrates, artificial ingredients, dyes, and colors) must be avoided for the betterment of the child.

3.14 Conclusion

In this decade, there is a big development in the innovative field of IoT. Health-care treatment systems get advancements from IoT. IoT can give an authentic system that can monitor the daily activities of mature and immature, disabled, autistic persons and patients in order to support their healthy, fit, and safe livings. This chapter discusses about how IoT-led health care in autism involves the convergence of pharma to facilitate the autistic persons in a better way.

References

[1] L. Da Xu, W. He, S. Li, Internet of Things in industries: a survey, IEEE Trans. Ind. Inf. 10 (2014) 2233–2243.

[2] S. Mukherjee, G. Biswas, Networking for IoT and applications using existing communication technology, Egypt. Inform. J. 19 (2018) 107–127.

[3] R. Khan, S.U. Khan, R. Zaheer, S. Khan, Future Internet: the Internet of Things architecture, possible applications and key challenges, 2012 10th International Conference on Frontiers of Information Technology, IEEE, 2012, pp. 257–260.

[4] Y.-B. Lin, Y.-W. Lin, C.-Y. Chih, T.-Y. Li, C.-C. Tai, Y.-C. Wang, et al., EasyConnect: a management system for IoT devices and its applications for interactive design and art, IEEE Internet Things J. 2 (2015) 551–561.

[5] I. Lee, K. Lee, The Internet of Things (IoT): applications, investments, and challenges for enterprises, Bus. Horiz. 58 (2015) 431–440.

[6] S.C. Mukhopadhyay, N.K. Suryadevara, Internet of Things: challenges and opportunities, Internet of Things, Springer, 2014.

[7] S. Madakam, R. Ramaswamy, S. Tripathi, Internet of Things (IoT): a literature review, J. Comput. Commun. 3 (2015) 164.

[8] B.K. Gandhi, M.K. Rao, A prototype for IoT based car parking management system for smart cities, Indian J. Sci. Technol. 9 (2016).

[9] J.M. Talavera, L.E. Tobón, J.A. Gómez, M.A. Culman, J.M. Aranda, D.T. Parra, et al., Review of IoT applications in agro-industrial and environmental fields, Comput. Electron. Agric. 142 (2017) 283–297.

[10] E. Bertino, N. Islam, Botnets and Internet of Things security, Computer 50 (2017) 76–79.

[11] C.S. Nandyala, H.-K. Kim, From cloud to fog and IoT-based real-time U-healthcare monitoring for smart homes and hospitals, Int. J. Smart Home 10 (2016) 187–196.

[12] P. Schulz, M. Matthe, H. Klessig, M. Simsek, G. Fettweis, J. Ansari, et al., Latency critical IoT applications in 5G: perspective on the design of radio interface and network architecture, IEEE Commun. Mag. 55 (2017) 70–78.

[13] T. Chi, M. Chen, A frequency hopping method for spatial RFID/WiFi/Bluetooth scheduling in agricultural IoT, Wireless Netw. 25 (2019) 805–817.

[14] V. Dattana, A. Kumar, A. Kush, S.I.A. Kazmi, Manet for stable data flow in smart home and smart city, 2019 4th MEC International Conference on Big Data and Smart City (ICBDSC), IEEE, 2019, pp. 1–4.

[15] S.S. Salwe, K.K. Naik, Heterogeneous wireless network for IoT applications, IETE Tech. Rev. 36 (2019) 61–68.

[16] R. Wankhade, S. Karhade, P. Mohite, K. Dhole, A. Ganvir, B. Khedkar, and S.S. Sangekar, Home Automation System Based on IoT Using Cellular Devices, International Journal of Scientific Research in Science and Technology, 6, 2019, 480–484

[17] Z. Qin, G. Denker, C. Giannelli, P. Bellavista, N. Venkatasubramanian, A software defined networking architecture for the Internet-of-Things, IEEE Network Operations and Management Symposium (NOMS), 2014, pp. 1–9.

[18] B.L. Hoffman, E.M. Felter, K.-H. Chu, A. Shensa, C. Hermann, T. Wolynn, D. Williams, B.A. Primack, It's not all about autism: the emerging landscape of anti-vaccination sentiment on Facebook, Vaccine 37 (2019) 2216–2223.

[19] J.A. Rankin, C.A. Paisley, T.S. Tomeny, S.W. Eldred, Fathers of youth with autism spectrum disorder: a systematic review of the impact of fathers' involvement on youth, families, and intervention, Clin. Child Fam. Psychol. Rev. (2019) 1–20.

[20] C. Croteau, L. Ben amor, D. Ilies, L. Mottron, J.-E. Tarride, M. Dorais and S. Perreault, Impact of psychoactive drug use on developing obesity among children and adolescents with autism spectrum diagnosis: a nested case–control study, Child. Obes. 15 (2019) 131–141.

[21] A.L. Lopresti, Mitochondrial dysfunction and oxidative stress: relevance to the pathophysiology and treatment of depression, Neurobiology of Depression, 2019, pp. 159–168.

[22] B. Kumar, A. Prakash, R.K. Sewal, B. Medhi, M. Modi, Drug therapy in autism: a present and future perspective, Pharmacol. Rep. 64 (2012) 1291–1304.

[23] I. López-González, R. Pinacho, È. Vila, A. Escanilla, I. Ferrer, B. Ramos, Neuroinflammation in the dorsolateral prefrontal cortex in elderly chronic schizophrenia, Eur. Neuropsychopharmacol. 29 (2019) 384–396.

[24] M. Wang, J. Zhou, F. He, C. Cai, H. Wang, Y. Wang, et al., Alteration of gut microbiota-associated epitopes in children with autism spectrum disorders, Brain Behav. Immun. 75 (2019) 192–199.

[25] A. Chauhan, V. Chauhan, Oxidative stress in autism, Pathophysiology 13 (2006) 171–181.

[26] M. Hamidpour, R. Hamidpour, S. Hamidpour, M. Shahlari, Chemistry, pharmacology, and medicinal property of sage (Salvia) to prevent and cure illnesses such as obesity, diabetes, depression, dementia, lupus, autism, heart disease, and cancer, J. Trad. Complement. Med. 4 (2014) 82–88.

[27] S.S. Valvassori, R.B. Varela, J. Quevedo, Animal models of mood disorders: focus on bipolar disorder and depression, Animal Models for the Study of Human Disease, 2017, pp. 991–10001.

[28] R.T. Fremeau, J. Burman, T. Qureshi, C.H. Tran, J. Proctor, J. Johnson, et al., The identification of vesicular glutamate transporter 3 suggests novel modes of signaling by glutamate, Proc. Natl. Acad. Sci. U.S.A. 99 (2002) 14488–14493.

[29] A. Speaks, What is autism, 2011, https://www.autismspeaks.org/what-autism, Retrieved on November 17, 2011.

[30] A.A. Subramanyam, A. Mukherjee, M. Dave, K. Chavda, Clinical practice guidelines for autism spectrum disorders, Indian J. Psychiatry 61 (2019) 254.

[31] A. Rao, K. Minakshi, G. Sabyasachi, V. Kumar, Food allergy investigations and its significance in autism spectrum disorders, Int. J. Pharma Bio Sci. 1 (2010).

[32] J.H. Elder, The gluten-free, casein-free diet in autism: an overview with clinical implications, Nutr. Clin. Pract. 23 (2008) 583–588.

[33] K.A. Schreck, K. Williams, Food preferences and factors influencing food selectivity for children with autism spectrum disorders, Res. Dev. Disabil. 27 (2006) 353–363.

[34] F. Navarro, D.A. Pearson, N. Fatheree, R. Mansour, S.S. Hashmi, J.M. Rhoads, Are 'leaky gut'and behavior associated with gluten and dairy containing diet in children with autism spectrum disorders? Nutr. Neurosci. 18 (2015) 177–185.

[35] C. Millward, M. Ferriter, S. Calver, G. Connell-Jones, WITHDRAWN: gluten-and casein-free diets for autistic spectrum disorder, Cochrane Database Syst. Rev. 4 (2019) CD003498.

[36] K. Reichelt, J. Ekrem, H. Scott, Gluten, milk proteins and autism: dietary intervention effects on behavior and peptide secretion, J. Appl. Nutr. 42 (1990) 1–11.

[37] P.C. Calder, Fatty acids and inflammation: the cutting edge between food and pharma, Eur. J. Pharmacol. 668 (2011) S50–S58.

[38] B. Torrico, AD. Shaw, R. Mosca, N. Vivó-Luque, A. Hervás, N. Fernàndez-Castillo, P. Aloy, M. Bayés, JM. Fullerton, B. Cormand, C. Toma, Truncating variant burden in high-functioning autism and pleiotropic effects of LRP1 across psychiatric phenotypes, J. Psychiatry Neurosci. 16; 44 (4), 2019, 1–10.

[39] S. French, A.P. Inamdar, I.A. Macdonald, S.T. Francis, Methods and Compositions for Improving Cognitive Function, Google Patents, 2019.

[40] P.J. Gillies, W.S. Harris, P.M. Kris-Etherton, Omega-3 fatty acids in food and pharma: the enabling role of biotechnology, Curr. Atheroscler. Rep. 13 (2011) 467–473.

[41] S. Mccullough, Autism symptoms reversed in new study, Genetics 3 (2019) 14.
[42] K. Srivastava, Autism and diet: an insight approach, Emerging Trends in the Diagnosis and Intervention of Neurodevelopmental Disorders, IGI Global, 2019.
[43] L.G. Klinger, G. Dawson, K. Burner, M. Crisler, Autism spectrum disorder, Child Psychopathol. (2019).
[44] F. Suhaib, A. Saeed, H. Gul, M. Kaleem, Oral assessment of children with autism spectrum disorder in Rawalpindi, Pakistan, Autism 23 (2019) 81–86.
[45] D.-W. Kang, J.B. Adams, D.M. Coleman, E.L. Pollard, J. Maldonado, S. Mcdonough-Means, et al., Long-term benefit of Microbiota Transfer Therapy on autism symptoms and gut microbiota, Sci. Rep. 9 (2019) 5821.
[46] L.M. Walter, K. Tamanyan, L. Nisbet, A.J. Weichard, M.J. Davey, G.M. Nixon, et al., Pollen levels on the day of polysomnography influence sleep disordered breathing severity in children with allergic rhinitis, Sleep Breath. 23 (2019) 651–657.
[47] N. Francoeur, M. Gandal, X. Xu, K. Sarpong, J. Johnson, P. Sklar, et al., Assessing the role of long noncoding RNAs (LncRNAs) in autism spectrum disorders, Eur. Neuropsychopharmacol. 29 (2019) S960.
[48] K.R. Stanek, E.M. Youngkin, L.L. Pyle, J.K. Raymond, K.A. Driscoll, S. Majidi, Prevalence, characteristics and diabetes management in children with comorbid autism spectrum disorder and type 1 diabetes, Pediatr. Diabetes (2019).
[49] P. Kennedy, P. Sinfield, L. Tweedlie, C. Nixon, A. Martin, K. Edwards, Brief report: using the social communication questionnaire to identify young people residing in secure children's homes with symptom complexes compatible with autistic spectrum disorder, J. Autism Dev. Disord. 49 (2019) 391–396.
[50] G.M. Miller, V. Kheifets, Treatment of Pervasive Developmental Disorders With Redox-Active Therapeutics, Google Patents, 2019.
[51] G.S. Fischer, H. Su, L. Dickstein-Fischer, K. Harrington, E.V. Alexander, System and Method of Pervasive Developmental Disorder Interventions, Google Patents, 2019.
[52] S. Akhondzadeh, Microbiota and autism spectrum disorder, Avicenna J. Med. Biotechnol. 11 (2019) 129.
[53] C.-T. Liu, L.-M. Chen, Comprehending conjunctive entailment of disjunction among individuals with Asperger syndrome, Clin. Ling. Phon. (2019) 1–15.
[54] S.L. Calhoun, A.M. Pearl, J. Fernandez-Mendoza, K.C. Durica, S.D. Mayes, M.J. Murray, Sleep disturbances increase the impact of working memory deficits on learning problems in adolescents with high-functioning autism spectrum disorder, J. Autism Dev. Disord. (2019) 1–13.
[55] W.-J. Chou, R.C. Hsiao, H.-C. Ni, S.H.-Y. Liang, C.-F. Lin, H.-L. Chan, et al., Self-reported and parent-reported school bullying in adolescents with high functioning autism spectrum disorder: the roles of autistic social impairment, attention-deficit/hyperactivity and oppositional defiant disorder symptoms, Int. J. Environ. Res. Public Health 16 (2019) 1117.
[56] B. Mirkovic, P. Gérardin, Asperger's syndrome: what to consider? Encéphale 45 (2019) 169–174.
[57] C.Y. Alverson, L.E. Lindstrom, K.A. Hirano, High school to college: transition experiences of young adults with autism, Focus Autism Other Dev. Disabil. 34 (2019) 52–64.
[58] S. Ozonoff, A.-M. Iosif, Changing conceptualizations of regression: what prospective studies reveal about the onset of autism spectrum disorder, Neurosci. Biobehav. Rev. 100 (2019) 296–304.
[59] N.P. Rosman, Childhood disintegrative disorder: part of the autism spectrum? Dev. Med. Child Neurol. 61 (2019) 503.

[60] A. Banerjee, M.T. Miller, K. Li, M. Sur, W.E. Kaufmann, Towards a Better Diagnosis and Treatment of Rett Syndrome: A Model Synaptic Disorder, Oxford University Press, 2019.
[61] A.T. Wieckowski, D.M. Swain, A.L. Abbott, S.W. White, Task dependency when evaluating association between facial emotion recognition and facial emotion expression in children with ASD, J. Autism Dev. Disord. 49 (2019) 460–467.
[62] H. Brentani, C.S.D. Paula, D. Bordini, D. Rolim, F. Sato, J. Portolese, et al., Autism spectrum disorders: an overview on diagnosis and treatment, Braz. J. Psychiatry 35 (2013) S62–S72.
[63] T. Buie, D.B. Campbell, G.J. Fuchs, G.T. Furuta, J. Levy, J. Vandewater, et al., Evaluation, diagnosis, and treatment of gastrointestinal disorders in individuals with ASDs: a consensus report, Pediatrics 125 (2010) S1–S18.
[64] B.F. Skinner, J. Vargas, Verbal behavior, J. Behav. Educ. 12 (1957) 185–206.

Chapter 4

Internet of Things–based pharmaceutics data analysis

Pranshu Dhingra[1], N. Gayathri[2], S. Rakesh Kumar[2], Vijayakumar Singanamalla[1], C. Ramesh[3] and B. Balamurugan[4]

[1]*Galgotias University, Greater Noida, India,* [2]*Anna University, Chennai, India,* [3]*Bannari Amman Institute of Technology, Sathyamangalam, India,* [4]*School of Computing Science and Engineering, Galgotias University, Greater Noida, India*

4.1 Introduction

In this era of technology, Internet is ruling the world. Everything that we can see around us is going to be connected. Smart technologies are being introduced in every domain of the industry, and the pharmaceutical industry is one of them although being at its infancy stage. The introduction of Internet of Things (IoT) and big data in the pharmaceutical industries opens its way to a plethora of opportunities for it to grow and progress by not only optimizing the cost of the manufacturing processes of drugs or their logistics but also in providing patient-centric care and customizable drug development for effective care of patients.

The pharmaceutical industry has not been too much progressive in adopting these technologies; and hence, its effects cannot be noticed so strongly. This chapter introduces the concept of applying IoT and big data in various fields of pharmaceutical industry and shows how all of these work in synchronicity. It also introduces the reader with all the basic concepts required to understand and apply analytics in the data procured. As most of the data obtained from the pharmaceutical industry is highly unstructured, various methods for dealing with this kind of unstructured data are also explained in this chapter.

Section 4.1 deals with the introduction to IoT, big data, and the pharmaceutical industry. Section 4.2 deals with the related work in these domains, and Section 4.3 deals with the proposed work.

4.1.1 The era of Internet of Things

Internet is going to rule in the coming times as all the physical objects around us are going to be connected and worked through the Internet. So,

Emergence of Pharmaceutical Industry Growth with Industrial IoT Approach.
DOI: https://doi.org/10.1016/B978-0-12-819593-2.00004-2

what this IoT says is that the scope of the Internet is not going to be limited to just computing and the connection of the computer devices, rather all the physical objects that can be seen, such as lights, refrigerators, fans, air conditioners, microwave ovens, and anything and everything is going to be interconnected. Not only in our homes but also in various business organizations, all the workplaces, and hospitals, the Internet of various things is being implemented effectively.

These physical objects are going to be fitted with embedded systems, embedded electronics and information technology so that there is some basic computing platform to make them communicate with each other. Besides, communication without any human intervention, gives rise to a new concept of M2M communication, that is, machine-to-machine communication, which will be discussed later. Until now, over 9 billion "Things" have access to connection with the Internet, and these "Things" have been anticipated to cross 20 billion in the near future.

The coalescence of the embedded systems, cloud computing, various communication protocols, big data, machine learning, and networking embodies IoT. There are various IoT enablers including radio frequency identifications (RFIDs), nanotechnologies, sensors, and smart networks. A study reveals that almost 5 quintillion bytes of data are produced every day by these IoT devices out of which less than half of the unstructured data is used for analysis and in business intelligence, and almost less than 1% of unstructured data is used for the same. So, what is basically needed is a substantial process of storing and analyzing this huge amount of data so that it can be used for providing such useful results, so that putting them into action makes better informed and profitable decisions for the company and good health care, services, and facilities for the people. This process is depicted in Fig. 4.1.

The number of things being connected to the Internet is growing every day, and the integration of the existing devices and the naming of each of these devices with unique address and constrained number of nodes in single framework will soon lead to address crunch. To address these problems, Internet protocol version 6 (IPV6) will be used, which is still being worked upon.

In terms of application domains of IoT as in Fig. 4.2, it plays a very important role in several sectors, the most prominent domain spheres being manufacturing and business on top with 40.2%, then comes the pharma industry and health-care domain with 30.3%, on third comes retail with 8.3%, and finally the security domain with 7.7% of the total devices present. Sticking to the pharmaceutical domain, as per the chapter, the most important spheres of pharma in which IoT has a vital role revolves around portable health-care monitoring, impeccable production, electronic record keeping, quality control measures, pharmaceutical safeguards, keeping track of sterile environmental conditions, behavioral human patterns, and so on. All these will be discussed in much more detail in the following sections.

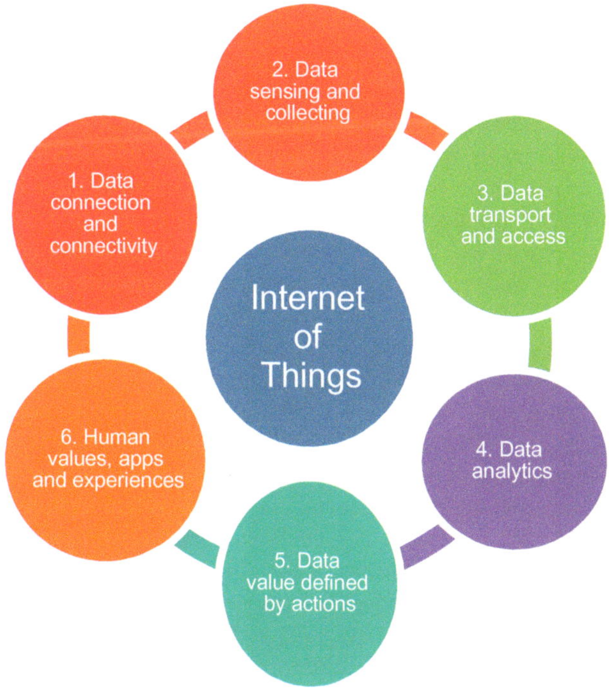

FIGURE 4.1 Utility of IoT and its devices in real life. *IoT*, Internet of Things.

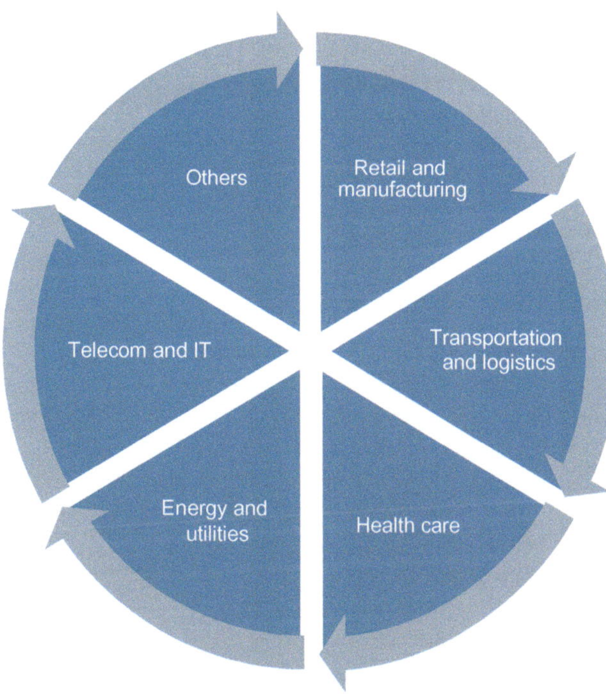

FIGURE 4.2 Contribution of IoT in real-life applications. *IoT*, Internet of Things.

4.1.2 Intermittent connectivity

This is required because it might happen sometimes that a particular activity may get covered up by two or more than two nodes, thus sending the same information to the server twice or, more than that, creating unnecessary trafficking of network. So, the covering up of a particular area or activity should be done such that if one node is covering it, the other nodes covering the same area should become inactive. This is what intermittent connectivity is. Processing of the data sent to the servers by WSNs (wireless sensor networks) is yet another important thing. Some data might be needed to be stored in a well-structured format, and some data might be required in its full form, without any losses. For the prior one, Structured Query Language (SQL) is used and for the latter one, NOSql platform, such as Cassandra, is used. Let us understand this with an example. If there is a need to keep track and remotely control the environmental conditions of temperature and lighting conditions of the room in which the "sick" patient, who gets affected by temperature and lighting conditions of the room, is residing, the data in the structured format on SQL would suffice our needs, whereas if there is a need to keep track of all the little details, such as any kind of movement happening in the room along with temperature, pressure, smoke, or infrared radiations, it is needed to get data in NOSQL as there are chances of loss of data if this kind of information is considered in SQL.

By now, it is clear how IoT works superficially but how everything is connected and communicated is still concealed and yet to be revealed. For building up IoT, Arduino chips or Raspberry Pi needs to be programmed and then integrated with sensors and actuators. The information thus sensed and collected by the devices are then sent to the servers from where they can be processed to give useful insights based on analyzing the patterns in the data. This thing involves cloud computing and fog computing, fog computing being the subset of cloud computing.

4.1.3 Connectivity technologies

The devices, once embedded [1] with the programmed chips, need reliable connectivity technologies to form networks and share data. There are certain communication protocols, each with different characteristic, built up to serve different purposes. These protocols [2] include Zigbee, Bluetooth, 6LoWPAN, HART, ZWAVE, and RFID near-field communication (NFC). The selection of appropriate connectivity technology is critical to the good performance of the objects being implemented through IoT, for reduced costs and higher profitability and to avoid hindering of scalability and efficiency of the devices in the long run. The factors on which the critical selection of the technology might depend include coverage of the area, energy efficiency, quality of service, cost, security, scalability, and interoperability.

4.1.4 Quirky machine-to-machine communication

Suppose there are two cars, both of them fitted with emergency sensors and they collide. As soon as the collision occurs, an alert is generated, and the information is sent to the servers. From the servers the data is transferred to the nearest hospitals and the emergency wards so that the ambulance can be dispatched accordingly besides putting the doctor and the nurses on alert. This is M2M communication does not need any third person to communicate the information to some other person or device. There is minimal human interaction, and the devices are self-sufficient to transfer and work in a synchronized manner. These systems may include certain embedded devices such as sensors or RFIDs and certain computing platforms to interpret the data and take automated actions accordingly.

As the things being connected to the Internet are growing at a fast rate, the sensors, RFIDs, and other embedded devices responsible for sensing and transferring data are also growing; and hence, the data being collected is growing at a really fast rate. This data is of no importance if it cannot be used and processed to make better informed decisions. Thus a centralized place is needed where all these data, from heterogeneous platforms, can be effectively collected, combined, and processed for its analysis and made the actuators work accordingly. This centralized space is "cloud" where all the data collected can be analyzed efficaciously for improving various decision-driven processes as well as profit the organizations.

But, unfortunately, the requirements and designing of the platform for IoT [3] makes "cloud" quite infeasible, especially when our goal is typically to serve a wide variety of IoT applications. If all of the data collected were to be sent from the devices to the "cloud," then it will have some network and communication issues as the bandwidth required in the latter case will be too high to undisturbed and rapid transfer of data. This is the first obstacle in the way of depending totally on "cloud computing" for data storage.

Many devices responsible for sensing and collecting data might be characterized by low-power communication or lossy signals because of short reaching frequency, which becomes the second obstacle. So, because of these restrictive constraints the communication or transfer of data from these devices to the endpoints (cloud) is infeasible.

The possible solution to this problem will be "fog computing" [4]. "Fog computing" was first discovered by CISCO, and it aims to bring cloud services near to the IoT devices. It serves as a middle layer between cloud and the IoT devices, which owes to processing and analyzing of the data by several applications within the network rather than communicating it through the centralized "cloud." So, with the help of fog computing, latency is being reduced, and the rate at which the actions are taking place is being increased by getting response from the severs in less time than when compared to cloud.

Let us consider this with the help of an example. Suppose that the situation is related to medical health care. The biosensors worn by the patient can be used to continuously monitor health condition of the patient. Even a little change in oxygen level in the bloodstream can be detected by it, and the information can be sent to the cloud processing. The response time being more, due to reduced latency, destroys the real timeliness of the situation and the patient might as well be dead by then. So, the whole idea behind "fog computing" is to reduce latency providing real-time data processing and immediate response as well as reducing the network traffic leading to fast and feasible transfer of data.

Fog computing does not serve as a competitor of cloud, rather they serve as good companions, and it depends on the platform designer, depending on the use cases and situations, whether the endpoints should be looked after by the cloud or the fog or a combination of both of them.

Besides this, the usage of cloud or fog or its combination depends very largely on the cases involved, as in, nontime-sensitive data can be sent to the cloud, whereas time-sensitive data can be analyzed and processed through fog.

4.1.5 Revolutionization of Internet of Things in pharma industry

IoT is not only accountable for building smart cities, smart agriculture, connected industries, smart energy, smart retail, and so on but is also unrolling its applications in the domain of pharmaceutical industry at a very rapid pace. It has the capability to enhance several processes in pharmaceutical industry [5], which includes the manufacturing of drugs, aiding in pharmaceutical logistics, medical equipment and drug tracking, clinical trial optimization, maintaining digital hospitals, remote control of patients, medical emergency management, drug storage, fighting pharmaceutical errors, regulatory compliance consistency or quality of drugs, complete real-time monitoring, smart pills and implanted devices, pharmaceuticals anticounterfeiting, pharmaceutical packaging, drug interaction checking, and control and monitoring of additive manufacturing processes. The digitization of the pharma industry is depicted in Fig. 4.3.

Each of these applications will be discussed in detail in the following sections.

Now, once these embedded devices and WSNs [6] have been used to collect data, our next step is to make the data homogeneous. As the data is received from different platforms and devices owing to different geographical locations, heterogeneity will prevail and before working on that data, the data needs to have some common platform and homogeneity needs to be brought to it. This data that is collected, because of its huge volume and veracity, can no longer be handled by the traditional ways of handling data and will require some revolutionized ways of dealing with it. The firm will need to adopt cloud-based services along with blockchain and similar

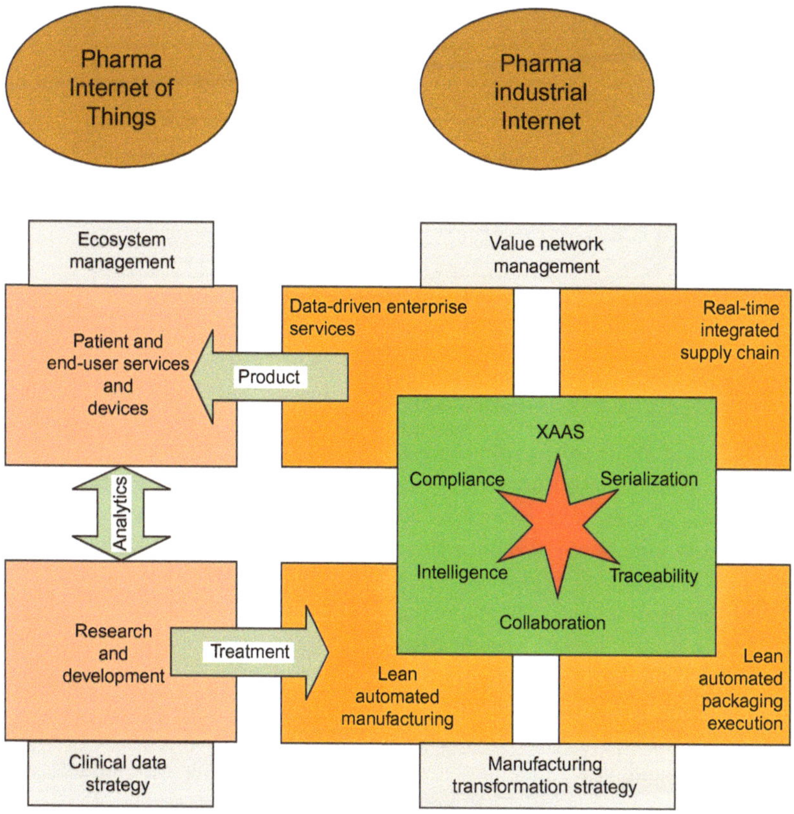

FIGURE 4.3 Framework for digitalization of pharma.

technologies to handle the data, because digitalization of the processes will lead to the throwing of data at a much faster pace; and hence, its handling needs to conform to that pace. Furthermore, this data that is collected, once stored, needs to be processed and analyzed to have complete control over real-time monitoring scenarios or to make better informed decisions improving the quality, efficacy, and optimization of the processes. Many manufacturers and suppliers are using advanced analytic techniques to scrutinize this big data that is collected by the smart machines and embedded devices. The maximum utilization of this data is important, and the trends and patterns can be recognized using data mining and machine-learning algorithms.

The two major challenges [6] lying in its way are as follows:

1. Building of such big data centers that can handle this much amount of big data.
2. Protecting this data from attackers and building up of strong cybersecurity to avoid any kind of data breaches. These are shown in Fig. 4.4.

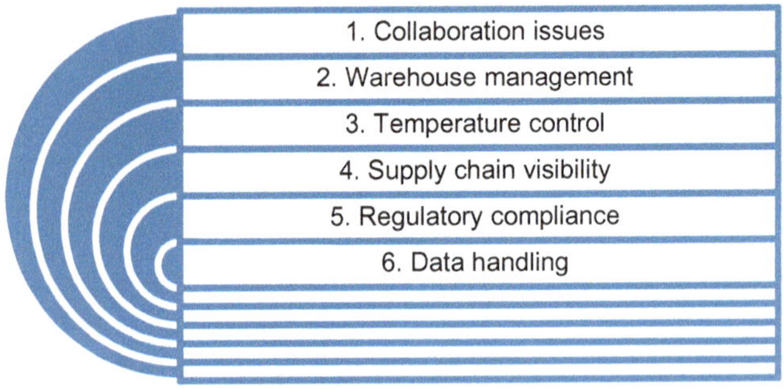

FIGURE 4.4 Difficulties in flow of pharma supply chain.

4.1.6 Revolution of Industry 4.0

After entering the fourth revolutionization [7] of the industry ensures to digitalize every domain and progression on its way is very much visible by the millions of sensors and nanochips and other IoT devices that are being embedded in almost every device to monitor and control the processes digitally. IoT promises to bridge all the gaps that aim to minimize the profits due to high operational costs and enhances the efficiency of managing warehouses for drugs. Today, the pharmaceutical industry has in its hands a golden opportunity of growing and progressing toward a more profitable and digitalized scenario with the help of this game-changing technology "IoT." The manufacturing, regulation, and distribution of drugs had been unchanged for decades, and now, this new technology is disrupting old models and bringing in new innovations that are worth appreciating.

4.1.7 Big data in pharmaceutical industry

Size is the very first term that comes into the mind when talking of "big data." The term "big data" [8] is applied to those datasets that are too large to handle by traditional software tools or systems to store, manage, and process data well within an unobjectionable time limit. Big data is constantly increasing in its size because of the mere fact that the number of things from which data is being collected is growing at a rapid rate. The obvious fact is that there are several challenges involved in the storing, managing, processing, analyzing, and visualizing of this data that is produced in quintillions each day, but the advantage that lay hidden besides this is that the patterns and facts that are discovered after the scrutinization of the highly detailed data give surprising results and limitless advantages.

Even the modern information technology industries are built on IoT, data warehousing, and cloud computing, where the huge datasets are stored in the data warehouses. But the main challenge lies in extracting the precise information from these warehouses, which uses techniques of data mining or statistical analysis. Another issue is that even the data mining techniques are not able to handle such large datasets successfully and that is probably because of dearth of coordination between database systems and the analytic tools [9].

The complexity theory of big data very well eases out the problems aforementioned. It helps to find the complex patterns in the data, gives knowledge abstractions in a better way and easily understandable representations along with the guidance of computing models on big data. Furthermore, it is to be noted that not all the data available is useful for analyzing or decision-making; and hence, industries are rather more interested in disseminating the data before initiating the analyzing process.

The automation of any tasks in the pharmaceutical industry be it manufacturing of drugs, patient health monitoring, clinical development and trials, supply chain management, medical product fault monitoring, data-driven preventive care and health interventions, sterile environment checker, predictive maintenance of medical equipment, or any kind of pharmacy services leads to terabytes and petabytes of data generation. This can help us to make "data-driven" decisions to improve the efficacy of the processes and the control systems, thereby reducing the costs and facilitating rapid postapproval changes. With increasing complexity of the molecules and the processes, there are increasing chances of generation of larger volume of supporting data. Biologicals and, to be particular, biosimilars are good examples of this.

Despite the increased focus on the quality of the drugs manufacturing, there has been an increase in the number and severity of quality defects as researched by the FDA. So, the manufacturing department of drugs is under high pressure to comply with the quality standards and thus needs to focus on improving their processes. Thus the implementation of IoT and embedding of big data in their decision-making processes could be the best possible solution. But it is not as simple as it looks!

Data integrity is an important concept. And "metadata" is an important part of data integrity. Data is often of no importance when the context in which it is retrieved or processed is unknown. This is where metadata comes into play. Metadata is "data about data," and this can include user id/person who generated the information, date or time stamp so as to know when the data was generated, the unique instrument identification that was used in retrieving the data. Data contextualization is a term that brings life to the data and is easily comprehensible owing to the knowledge it aims to provide. It should be performed well in advance of any execution of the plan so as to verify how effective it will turn out.

Pharmaceutics is a critical part of pharmacy that is very much concerned with pharmaceutical formulations and drug delivery, which is responsible for preparing an active pharmaceutical ingredient (API) into a safe and efficient medication for use by the patients. The emergence of big data for pharmaceutical industry is no doubt, quite insightful, but the challenge lies in upgrading the skill sets from that which were sufficient to analyze relatively small datasets to those which can analyze quite big ones, be it of clinical data trials or that of unstructured data such as physician's notes, pathology reports, scans, and images. A lot still needs to be done and to create standardized methods for sharing and making sense of the anonymized EHR. Although now, it is possible to link different data sources and thus helps to address complex research questions [10].

Suppose, the EHR data of a patient collected during real-time monitoring of the patient is analyzed. This analysis report gives detailed information about the treatment patterns and clinical trials along with an opportunity to better understand the disease of the patient. This information serves as valuable insight complimentary to that gained by clinically observing the patient.

Real-world data comes from different sources and in different formats making it quite messy with lots of missing data, potential biases, and inconsistencies making it more difficult to handle and analyze. This pressurizes the data scientists to devise ways to answer critical research questions. Hence, there is an emerging need for analysts and data scientists who can handle such ambiguous and inconsistent data and present the full scenario in a comprehensible way.

Various factors contributing to the lodging of big data and IoT in pharmaceutical industry includes:

- failure of R&D production;
- medical equipment cost optimization (basically, reducing global pharma cost);
- necessity of value-based drug delivery models;
- movement of pharma industry to a more digitized approach;
- declining health of people;
- reduction in operational margins; and
- side effects caused by imprecise medicines.

Thus despite many complexities in the run of adopting big data technology and connectivity in pharmaceutical companies, the infinite benefits it has overpower it all.

4.1.8 Linking Internet of Things with big data

The scope of Internet has risen to such an extent that machines are autonomously handling innumerable activities via Internet thus creating this "IoT." These machines and gadgets are becoming the users of Internet just like the humans are users of web browsers. Knowledge acquisition from the innumerable IoT devices is the biggest challenge being faced by the data handlers and professional analysts. IoT devices [9] are continuously sending streams of data to the cloud, and continuous research is going in developing an

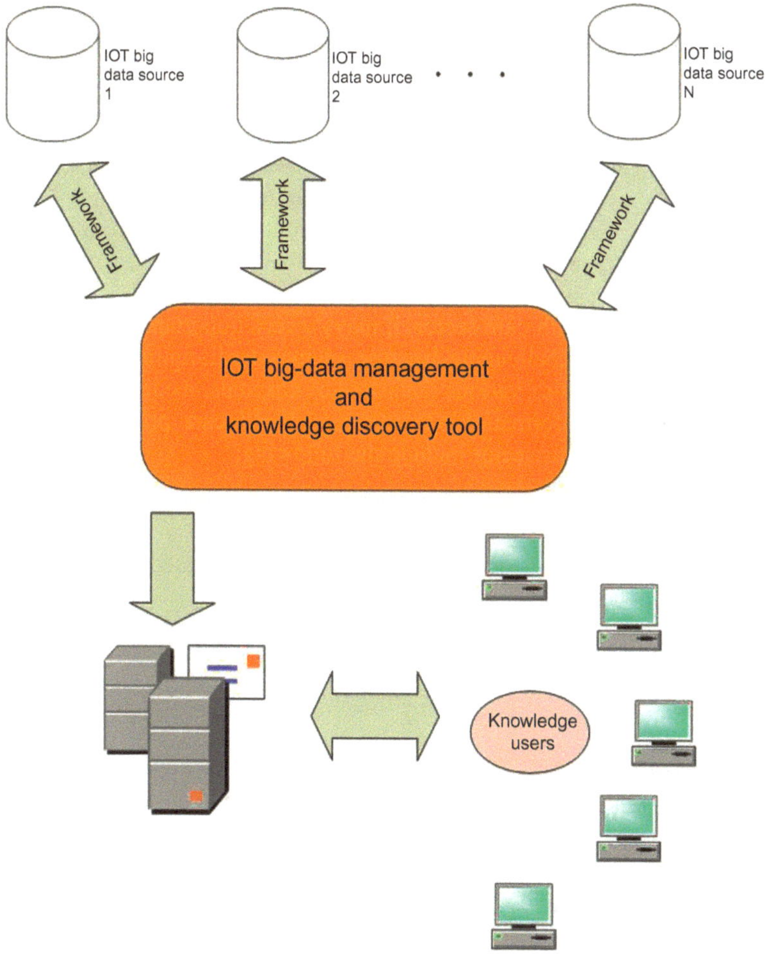

FIGURE 4.5 Relation of big data and IoT. *IoT*, Internet of Things.

infrastructure that can effectively handle this data to give useful insights. The current technology being used for the same is machine-learning algorithms and computational intelligence solutions from the perspective of IoT as depicted in Fig. 4.5.

There are several big data platforms, and each of them has specific functions. Some of the platforms include Apache Hadoop and MapReduce, Apache Mahout, Apache Spark, Dryad, Storm, and Apache Drill. Each of the platforms has its own way of functioning and cooperating with the task of analyzing. Some are good for real-time monitoring of big data, whereas some are designed for batch processing. Then, different techniques used for the analysis include cloud computing, data computing, data stream computing, data mining, machine-learning, and intelligent analysis.

4.1.9 Analysis of pharma data

Pharma intelligence leads to a high scope of improvement in the current ways of controlling the processes of pharmaceutical industry, and it gives the manufacturers the access to benchmark their products in terms of quality and increase the competitiveness in the market. This pertains to the manufacturing processes in the industry.

Coming to the health-care sector, it plays a very important role in pharma domain but is quite challenging because of its interdisciplinary nature that includes combinations of various databases of hospitals, medical researchers, visualization tools, information retrieval, and health-care practitioners. The challenges faced by this sector include the following:

- the domain specific analysis, as in, understanding of the medical terminologies and concepts according to the domain;
- the inadequate knowledge of programming by the medical researchers and practitioners; and
- lack of adequate exposure to mathematical and statistical concepts and techniques.

Data is mainly classified under two main categories: structured and unstructured.

Structured data is that which is organized and formatted in the database such that the information can be used in an addressable way to make effective analysis of the data, whereas unstructured data is highly unorganized without any specific format or predefined style of storing data and may contain heterogeneous data.

Most of the data in pharmaceutical industry is unstructured such as doctor's notes, patient's prescriptions, electronic health record, medical image reports, pathology reports, number of days the patient admitted, new or refilling of medicines, supply of materials to clinic, date of admit and discharge of patients, patients communication patterns and history, and drug bar codes. To handle such data, modern analytical tools are required to handle vast amount of real-world unstructured as well as structured data [11].

For handling pharma data, many open-source tools are available, each used for a different purpose and situations. For new drug discovery, machine-learning algorithms have been proved as the best. Classification and prediction algorithms are preferred by most of the researchers for drug discovery based on predictive analytics. Other algorithms that may be used are decision trees, artificial neural networks, Naive Bayesian classifier, support vector machines (SVMs), etc. Different techniques are used at different stages of drug discovery, which uses various machine-learning techniques

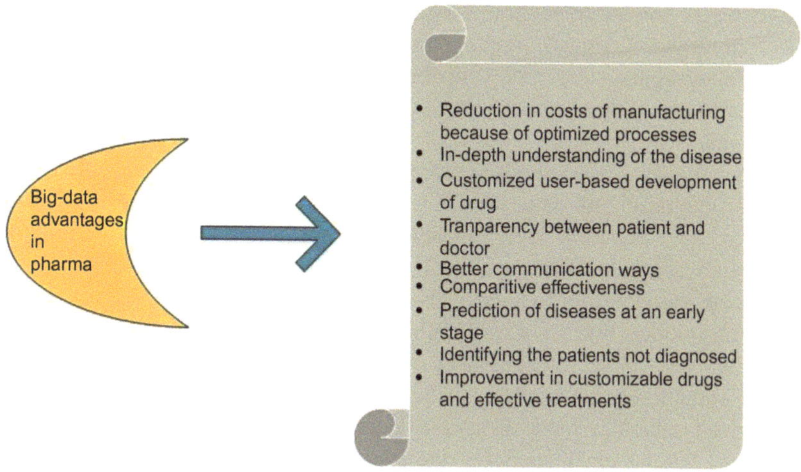

FIGURE 4.6 Advantages of big data in pharma.

such as partial least squares, multiple linear regression, *K*-means, decision trees, and ensemble methods [12].

Even for designing clinical trials and targeting patients for the same, data analysis is important. It also proves vital in monitoring and mining the results of clinical data and patient records to identify the side effects or negative effects as well as the benefits from the use of a drug. Analyzing the data can also help the pharmaceutical industries to develop personalized medicines taking into account the genetic variation and response of specific drugs toward individuals so that medicines can be tailored according to individuals. The major advantages of big data in pharma industry are depicted in Fig. 4.6.

4.2 Related works

4.2.1 Interoperability of Internet of Things devices

Since IoT is spanning a variety of domains, there has been diversity in the devices that are collecting the data, the standards, the payload data semantics, and the communication protocols being used. The devices are fitted with embedded devices that consist of microcontrollers, sensors, actuators, energy source, and a wireless transceiver for performing their actions. Because of their cheap costs, they are constrained in terms of power/energy, bandwidth and communication capability. The data from the sensors is collected from heterogeneous devices and platforms, which use different variety of network connectivity options, different communication standards and protocols. Sometimes, the export of data is avoided because of the security issues or bandwidth limitations. This limits the leverage of the value-added services that uses these economical IoT devices. So, to enable the sensors to interoperate the data and communicate with each other and to overcome certain limitations that might prevent the transfer and communication of data, different models have been proposed each of which has its own benefits and challenges.

This chapter [13] proposes a cloud-based software architecture that enables integration of data from heterogeneous devices and platforms by means of an adaptation layer, which offers uniform device abstraction to hide the diversity in devices, network connectivity methods, protocols, and application models from IoT users. The provision of CoAP interface over the device abstraction layer makes possible for highly heterogeneous as well as legacy IoT devices to integrate. For each of the device with different protocols or communication models, the cloud-based platform creates its virtual counterpart, which is made available as an IPV6 endpoint (virtual device). The cloud takes care of mapping the interactions to the specified/particular constrained device while the client interacts only with the virtual environment (device). This is basically the sensor as a service paradigm that has been followed in this chapter, and the main advantage behind using this is that there are significantly more resources available as service when a virtual device is considered as opposed to the constraints of physical real devices, which hinder the processing and communication of data. This design not only helps in integration of devices with heterogeneous platforms but also great service development of constrained IoT devices while neither burdening the service developers nor the constrained IoT devices.

This chapter [14] proposes a Hadoop-based design of principal component analysis to leverage efficiently the distributed embedded heterogeneous systems and processing the collected data on distributed computing system to efficiently manage and process fine grained data chunks. The Hadoop system assumes high portability by assuming each node as a computing CPU-based system with localized memory and processing. MapReduce, which is a base paradigm that serves as a platform for problems that need to handle large datasets, decomposes the problem into many processing tasks that are key/value pairs, which gets mapped to a large number of computing nodes. The inputs in addition to this chapter adopt cuBLAS to run the parallel computing capability on GPGPU device; cuBLAS library being implemented on top of the CUDA driver. The key/value pairs are dynamically collected in a preallocated data array that is denoted by PreDataArray to achieve high data parallelism. A native JNI function is invoked, after the data has been collected, to transfer it to the device GPGPU through CUDA API. Thus high computation parallelism can be attained through this way of data collection, and even the data transferring overhead can be very well avoided. The heterogeneous embedded computing along with GPGPU leads to high-performance technology as the performance of single core processor hits its limits; and hence, multicore system for parallel processing using computer clusters and graphic processing units can accelerate our process.

This chapter [15] presents an approach that can integrate devices based on IoT context-aware infrastructure. This infrastructure named ContQuest has a well-defined architecture and a reflective description model, which offers a set of middleware services for context-aware application and adopts an inversion of control principle to offer customizable definitions and behavior via method overwriting. The architecture of ContQuest is based on SOA (service oriented), which means that some of its functionalities are defined within specialized services with well-defined interfaces. In this a device is considered as a resource and is represented by resource agent, and each of the resource agents is structured on the basis of an architecture that points to ease the integration of various IoT devices and their deployment. This model also includes design solutions that can adapt devices having different communication protocols.

This chapter proposes text and data mining that can build structured databases from the unstructured information to yield the most relevant information for not only biomedical research but also patient's health-care monitoring. The unstructured information includes doctor's notes, the prescriptions, electronic health-care reports, pathology reports, and medical image reports. There are two major roles of text mining; the first one is devising patterns and trends to complement traditional ways of working, and another one is to build structured databases proficient enough to provide valuable information. In this model, unstructured data is used to build knowledge base automatically and offers search capabilities for full text, curated data or abstracts. It leans on rule-based engines and machine-learning engines, and thus by combining the information from patient care, their health and management by dealing with the diagnoses and treatment, health-care industry can experience an enormous shift in its digitization.

This chapter [16] proposes architectures that are able to integrate data from heterogeneous sources, and the frameworks proposed are often self-contained, which tends to create a cluster of devices that are framework-compatible and hence can interoperate. Some of these kinds of architectures include the following:

- Cumulocity, a platform that provides a unified and service-oriented HTTP REST interface to devices.
- AllJoyn, a framework that enables devices to implement an attachment to a software bus among various applications or connecting to AllJoyn router via a thin library, which, in either way, enables the integration of even constrained devices.
- Xively, which offers an API to implement custom message bus thus enabling the devices to be interoperable among different application protocols such as CoAP, XMPP, MQTT, and HTTP.

- OMA-LwM2M, an ecosystem implemented by Open Mobile Alliance, which defines a custom layer over CoAP and hence focuses on exchanging instances, that is, objects and operates them via custom interfaces.

The chapter [17] proposes a new framework called MODALITi (FraMewOrk for EmbeDded and CollAborative data anaLysIs with HeTerogeneous DevIces), which not only provides interdevice communication protocol for deploying machine-learning models reliably and efficiently but also takes into account the diversity in data collected from heterogeneous devices deploying predictive models into devices. This enables communication protocol in such a way that the parts of the models are sent only to those devices that can process them, and they only communicate with those devices that can run the machine leaning model further. Thus MODALITi serves as a hybrid framework that supports predictive model training and runs models on a network of low-energy devices.

4.2.2 Pharma logistics: helping hand from Internet of Things

The waste products generated during the diagnosis, treatment, or immunization of human beings or animals or while performing research activities or while testing and production of biological are biomedical waste that needs to be handled properly or otherwise would lead to hazardous effects such as mass infection. Thus smart ways of handling this data have been proposed, which aims to reduce human interaction besides completely automating the waste management system for pathological labs, hospitals, and laboratories. For the automation purposes, IoT devices can be easily deployed as they are not only cheap but also easy to setup. Generally, Chain of Things of IoT components is used. According to this chapter [18], the proposed system would be implemented at the sources itself, and the color-coded bags will be marked with RFID tags, which are automatically indexed by a system. Besides this, the bins in which these color-coded bags are put would have weighing sensors that would trigger the van that carries the waste material to common biomedical waste treatment facility. This quantified weight will be sent to government authorities directly through IoT-based microcomputers connected to Internet and can prove helpful in analyzing and fetching data every month or even every second. This system is centralized and based on real-time data and hence can overcome the flaws of decentralized system. This type of system can also prevent any kind of frauds in the data and fool the authorities as at any point of time the data is present on many nodes before being synchronized into the main server as backup.

Another application of IoT in the field of pharma is designing of digital drug administration interface [19], which introduces syringe infusion pump with IoT for its dual control. It enables to have controlled discharge of drugs, which works effectively in the fields of neonatal medicines and in ICUs as well. Even a small deviation from the required quantity can be fatal to the patient. This type of administration of drug not only proved blissful for patients as of diabetes, who need round-the-clock injections, but also saves a lot of time as well as human effort. It also proves to be beneficial in keeping track of the patients and the syringe status remotely from the doctor's cabin itself. In cities and towns and villages with acute shortages of doctors and maintenance of OPD and ICUs, the doctors have to keep the patients waiting, which is against the morality of the profession. But this advancement in technology helps to keep track of the syringe status remotely. Two of these pumps can be easily combined and drove through an electronic system that provides gradient-programming operation and mixed mobile phase synchronously arranged so that one of the pumps can be filled while the other is being used to obtain continuous elocution by the patient. Thus with the introduction of IoT in syringe infusion pump has even made it more secure as it checks and prevents even a small error in the dose. Not only this, it also gave patients the freedom to move while receiving treatment and hence could also receive their treatment on an outpatient basis giving the opportunity for serious inpatient cases to be treated.

Anticounterfeit technology in pharmaceutical industry perspective [20]: inadequate drug regulation and growth of international free trade have led to the expansion of trade in counterfeit drugs worldwide, and the best way to avoid this problem is technological protection. A medical product is claimed as counterfeit when it has false representation in relation to its identity or source, and this may apply to the product, its packaging or labeling information or its container. Counterfeiting may exist in both branded and generic drugs with correct or wrong ingredients, without APIs, with incorrect amount of API or even with fake packaging. The Indian Government is getting tactful in dealing with this problem and has formulated rules to mandate bar codes, and even the pharmaceutical companies have been lately employing these technologies to avoid illegitimate drugs in their supply chain.

Some anticounterfeiting technologies may include the following:

- Holograms, which can provide overt first-line authentication while second-line authentication for trained examiners can be provided by covert features as scrambled images or microtext.
- Track and trace technology, a process of assigning a unique identity to each unit of stock, which remains with it from the manufacturing to its final consumption through the supply chain. This technology includes Pedigree, mass serialization, Global Trade Item Number (GTIN), Serialized GTIN, data carriers that include 2D bar codes and RFID tags.

These technologies can be used to verify the supply chain and also the composition of the drugs, analytical methods such as chromatography, isotopic characterization, and optical spectroscopy.

The concept of Industry 4.0 aims to revolutionize the industries and factories in a smart way that meet and exceeds the challenges of shorter life cycle of products, stiff global competition, and highly customizable products according to the needs. In the light of Industry 4.0, machines and various products act as cyber physical systems, which exchange information autonomously without any human intervention and trigger and control each other's actions independently. And thus, smart manufacturing environments are being developed where the manufacturing processes are being digitized and automated through various programming-oriented platforms and IoT. Thus manufacturing systems are integrating intelligent processes and providing it flexibility, monitoring of interruptions, and generating indications for production and its management. Also, the integration of the beacon technology in the production environment to manage additive manufacturing processes allows for communication of this data extracted from the machines through devices such as mobile phones allowing for real-time monitoring of the production processes [21].

The manufacturing of pharmaceuticals on a large scale revolves around connected data, which not only provides better understanding of performance of equipment but also predicts maintenance and hence prevents breakdown of saving costs and resources. The scope of IoT spans the entire pharmaceutical processes from the manufacturing to the pharmacist. The sensors being deployed to keep a record at every step of the manufacturing, connects to a central network from where it is transmitted to a central database, which is provided with a dashboard to access real-time data and hence give scope to any amendments in the equipment or the processes that might be needed. IoT devices can be deployed in various steps of manufacturing, including warehouses, local hospitals, pharmacies, and delivery vans, the main goal being to deliver information related to product quality, temperature information, and real-time alerts if the safety limit is being crossed and hence conforming to the standards. This will not only help in making informed decisions but also improving overall effectiveness of equipment such as its cleaning, maintenance, and scheduling batches. And because all the processes taking place and equipment being used need to conform with good manufacturing practices, the equipment needs to maintained and hence safety of drugs needs to be assured. IoT backs up and helps in achieving this level of maintenance with high accuracy but collecting the data from the deployed IoT devices and converting the raw data to get meaningful information by informing about its performance and conditions.

Using the information from the sensors and hence controlling the operations from the back end even helps the technicians to develop self-learning algorithms and thus generates automatic request to look into the process or at the step where it is not working conforming to the rules. This helps in minimal disruption in the supply chain and increases its reliability. The ability to leverage the use of real-time data obtained can help manufacturing teams to optimize the performance of the equipment, improve resources allocation, and reduce machine downtime, hence lowering the manufacturing costs and shorten cycle times.

Access to real-time visibility of warehouse can also be gained by relaying metrics as well as real-time data to the warehouse managers without which it would have been difficult to track products in the warehouse and leverage the use of operators and transport equipment. Then, again, there is the active and passive temperature loggers attached to the refrigerators in warehouses to continuously record the temperature and IoT could connect these devices and compare their measurements to thermostability tables and hence prompt them to generate alerts if there are any temperature variances.

Due to very high costs of maintaining inventory and hence the warehousing costs, it is futile to keep the stock in bulk, hence comes the need for analyzing the inventory required to ensure continuous and timely supply of medicines in a cost-effective way and to avoid any extra amount being spent on its storage. Sensors are placed on the inventory items to interpret vital information such as inventory details, product location and to report any misplaced products or inconsistencies [6].

The chief challenge for pharmaceutical industry lies in keeping track of logistics, which requires timely, accurate transfer of consistent information during transportation of pharma goods to allow for real-time traceability of the goods in the supply chain. Recently, the logistics are being incorporated with RFID tags, wireless sensing devices, Global Positioning Systems (GPS), temperature monitoring systems and so on, which aims to overcome the risks associated with supply chain of pharmaceutical network. The difficulty lies in that the data and information procured from these technologies are not integrated and functions independently making the information control ineffective. But since the emergence of IoT in pharma field, it is possible to overcome this issue. According to this chapter [22], an IoT-based smart logistic system has been designed in which two-layer network architecture of IoT platform is introduced; the first layer being RFIDs and the second one being WSNs layer. The RFID layer acts as an asymmetric tag reader link and, WSN acts as a temporary network between reader nodes. By this, live monitoring of goods during shipment or transportation becomes easy. Moreover, there is the scope for immediate action if any inconsistencies not conforming to the standards are found in the logistics chain.

There are certain pharmaceutical products such as biologics having high value of active ingredients with shorter shelf lives that are highly sensitive to temperature and requires conforming to the temperature requirements even during the transportation. This is cold chain logistics, and they must be kept in temperature controlled containers during its transportation. The existing products in cold chain logistics approximates about $260 billion, and approximately 20% of these goods are wasted during transportation and shipping [6]. But due to coverage of this sector through IoT devices, the pharmaceutical manufacturers can monitor remotely the real-time cold chain environments with autostart and autoshut mechanisms during transportation in shipments or warehouses or vehicles using smartphones.

Another application of IoT in pharmaceutics includes health-care monitoring and emergency situation alerts of patients. Telemedicine monitoring is another idea that is patient centric and is focused on improving the diagnosis and medications of the patients by collecting data from patient's lifestyle as well as past medical records in order to reduce health-care costs and provide continuous improvement in the medical care. It also has promising implementation in the research work by developing new drugs within reduced time frame along with adhering to the regulations, speeding up the production, and making the drugs available to the consumers at low cost. RFID tags and sensors can be used for collecting data for discovery of new drugs along with machine-learning algorithms [5].

IoT has helped researchers to design innovative technologies to provide high-quality health services to patients with lower costs and reliable care. The application of IoT in health care is enormous with an estimation of approximately 30% of its application in all fields. With rapid advancement of wearable IoT (WIoT) devices, cloud computing and mobile applications, there is a huge role of IoT in transforming and facilitating the traditional health-care services to smart, personalized, and patient-centric ones. The IoT-enabled health-care systems helps in monitoring several medical parameters such as blood pressure, glucose levels, body temperature using smart wireless sensors, computer networks, and remote servers along with providing suggestions for treatment based on machine-learning algorithms or analysis of the previously recorded data of the patients. IoT-based health care is basically an interdisciplinary research area that deploys different methodologies ranging from those of engineering and computer science to those of pharmaceutics for designing of innovative methods for application to practical medical issues. Presently, many manufactured products are employed with unique identifiers such as RFID, Quick Response codes, bar codes, and intelligent sensors in order to help keep a track of these goods.

The medical IoT system is a sophisticated setup containing a variety of systems and mechanisms such as medical equipment, network gateways, smart sensors, cloud computing, big data, and clinical information systems to cooperate in determining and controlling the health-care environment. The medical devices, such as smart watches or mobile phones, can be used for keeping in control the health parameters that are remotely recorded by the back-end systems. After analyzing the recorded data, it provides appropriate feedback to the clinical staff, which helps them to determine current health condition of the patients besides alerting them of any action they should take in the case of critical cases [23].

Sensors form the heart of the IoT-based technologies as they form the critical part of monitoring and collecting the data from every possible field you can probably think of. For example, the pulse oximeter helps a physician in monitoring a patient's heart rate and blood oxygen saturation. Other types of sensors include pressure, temperature, water quality, and smoke sensors. Then there are also instruments embedded with various sensors in order to analyze motion such as accelerometers, surface electrodes, and gyroscopes. There is a huge scope of transforming all the received data from the sensors to digital form immediately transmitting it over the network for its analysis and results. Wireless sensors have made it possible for people to wear portable sensors that collect the data in an automated way.

Glucose level monitoring—Due to the need of continuous monitoring of glucose levels in diabetic patients, wearable sensors (medical IoT device) capable of tracking the health parameters can be used and the data collected can be transferred via Internet protocol (IPV6) to health-care providers. The tracking device consists of a blood glucose collector, an IoT-based medical acquisition detector and a mobile phone to monitor the glucose levels. This type of monitoring can help to detect the individual modification in patterns in glucose level to decide about the meals, medication times, and physical activities of the patients.

Electrocardiogram monitoring—In ECG monitor consists of a wireless transmitter and a receiver along with an automated application that can identify an abnormal heart activity. The IoT system uses algorithms for the continuous monitoring of ECG. The system keeps track of the heart rate and the basic rhythm besides myocardial ischemia and prolonged QT intervals by recording electrical activity of the heart.

Blood pressure monitoring—The machine used to record the blood pressure of the patient consists of an apparatus with network-based communication technology. A wearable sensor device such as Blipcare, which uses a home Wi-Fi network can be used to continuously monitor the blood pressure of the patient.

Wheelchair management—Smart wheelchairs for persons with disabilities, which use IoT application, have been developed. It uses wireless body area networks technology to control and coordinate with different sensors. Besides this, it also keeps track of the status of the person using the wheelchair by sensing and providing information about his surrounding as well as sitting position of the patient.

Body temperature monitoring—The change in body temperature forms an essential part of health-care service as it helps in identifying homeostasis. A medical IoT device, TelosB mote, has embedded sensor to record body temperature [24].

This chapter [25] proposes the use of IoT devices, that is, Arduino Uno and sensors such as temperature sensor, heartbeat sensor, ECG sensor, heartbeat sensor, and timer. Human temperature depends heavily on the metabolism rate and is directly related to it. Here is an algorithm that detects the abnormal functioning of the patient taking into account temperature rate, heart beat rate, and eye blink rate of the patient using MEMS (microelectromechanical systems) sensor and Arduino board.

```
TR-> Temperature rate
HR-> Heart beat rate
ER-> Eye blink rate
mem- >MEMS sensor
if(TR> =45 || TR< =30)
{
raise alarm
}
else if(HR> =100 || HR< =60)
{
raise alarm
}
else if
{
for(int i =20;i< =40;i++)
{
if(ER==null)
{
Patient has slept
}
}
}
else if(mem is dislocated)
{
Patient has fallen
raise the alarm
}
END
```

In today's society the busy life of people has made them to forget many things in day-to-day activities, the elderly people being the victims of chronicle diseases, and hence reduced life span due to missing of their daily doses. This is because they suffer from dementia, forgetfulness of their daily activities. Diseases are increasing in large amount and sooner or later, people will come in contact with these diseases some of which are temporary and some are permanent life-threatening diseases in such a way that once they get mixed up with the human body, they cannot leave the body ever and increases in rapid time. To overcome or live a better life, medicines need to be taken regularly in large amounts. Very many times, the prescription changes in every few days with respect to the effectiveness of the treatment being given and hence creates a problem, especially for the elderly people as it becomes difficult to keep in mind the medicines they need to take with every changed prescription. To overcome these problems, PIC microcontroller based smart medicine box is made, which uses real-time clock, so the patient cannot delay the time on which he needs to take the pills. It becomes compulsory for the patient to take out that medicine from the box at the right time else it starts to make large sound. The smart box is a combination of electronic and mechanical pill boxes or dispensers that has a pill dispenser having different prescribed administration schedules. This device also has a pill detector that is responsible for generating a signal to make the patient aware to take the prescribed medicine. Thus the introduction of such devices in IoT could help us to get important information about patients anytime and anywhere and hence avoid any calamities besides giving proper treatment to the patients. IoT also aims to keep the smart box connected to the Internet in order to help manage the patients' treatments, especially, of the elderly people, in a better way [26].

This chapter [27] proposes iMedBox, an intelligent medicine box that serves as a home health-care gateway, which expands the coverage of traditional health-care systems from a confined hospital environment to a patient's comfortable home zone. IoT devices such as wearable sensors and iMedPack (intelligent medicine packaging) get connected to iMedBox through heterogeneous networks. These heterogeneous networks are compatible with existing wireless network standards. The Biopatch worn by the patients can detect and transmit, in real time, the users' biosignals to the iMedBox and all the information collected are displayed and stored locally on the iMedBox that can further be forwarded for clinical diagnosis or analysis. The iMedPack is linked with iMedBox via RFID link that helps the users to go according to their prescribed medicines.

This chapter [28] proposes a platform that involves intelligent medicine box with enhanced connectivity and aids in integration of devices and services enabled by Zigbee and actuation capability that is flexible and wearable biomedical sensor–enabled device. The values from the sensors [heartbeat, temperature, MEMS, and light-dependent resistor (LDR)] are read and the messages are sent accordingly through Zigbee. After receiving the values from Zigbee, it is displayed in LCD and transmitted through Wi-Fi.

MEMS allows the development of smart products aiding in the computational ability of microelectronics with control capabilities and perception of micro sensors. LDR is light-dependent resistor which is light-controlled variable resistor.

This chapter [29] proposes a system to avoid infant abduction threat to the newborn babies in hospitals and birth centers by using the RFID tags attached to the ankle bracelets that are made to wear to infants shortly after their birth. These RFID tags can prevent the abduction of babies by verifying their location and sending warning signals. This proposed system also keeps a track of the in time and out time of every child and only allows entry to authenticated persons inside the nursery. The system also uses Universal Product Code scanning technique that includes a tag having sound sensor and bar code; and if anyone tries to abduct the child or cut the band, the sensor produces an alarm to alert the parents and administration. Only after releasing the band, it is possible to take the child out of the hospital premises.

With such an advanced technology, certain chronic diseases can be controlled as well as spread of those diseases can be prevented with the help of "IoT." The hospitals are there to ensure proper treatment as well as reduce suffering of the sick people; and hence, proper sanitation needs to be maintained in the hospitals to avoid prone of already sick people to a plethora of diseases. The overflowing bins itself is a major cause of spread of various diseases because of the presence of recuperating patients. Hence, it is very important to get the bins cleared of the garbage as and when they get filled but the manual checking is quite strenuous and disturbing for the residing patients as well. Thus this chapter [30] proposes a model that uses sensors such as ultrasonic sensors to detect the garbage levels, displays them, and intimidates the central server of the hospital for immediate cleaning of the trash in each ward or room. The amount of waste produced in the bin is determined by two factors: first, the population present in any given hospital and second, the patterns of consumption in it.

Its working is presented in the form of a flowchart as shown in Fig. 4.7.

Another scope of IoT in pharmaceutical industry is the distribution of medication to patients at a pharmacy. The patients have to wait in a long queue to receive their medications due to inefficient way of distribution of medicines. This chapter [31] introduces a system to solve the current system in certain pharmacies, which is purely paperwork based. The proposed work makes use of electronic sensors, that is, the touch and temperature sensors, which minimizes the scope of any mistakes. This approach makes use of Intel Galileo Board to which the electronic sensors are connected and hence defines a system made of hardware and software components featuring a simple graphical user interface.

The revolution of Industry 4.0 has led to the digitalization of almost every field, and IoT can help in the digitalization of the hospitals leading to the development of smart hospitals, which can overcome certain disadvantages such as fixed information point, fixed networking mode, and manual input of medical information. On the other hand, IoT helps to connect almost every item with the Internet to share, transfer, and communicate information hence aiding to implement intelligent recognition, tracking, monitoring, and management via RFID, GPS, infrared sensors, and other sensing equipment. This chapter [32] gives an in-depth study of the model introduced in this chapter to help in the development of smart hospitals. It aims to combine the existing three structure architecture of IoT (Application Layer–Network Layer–Perception Layer) with the characteristics of hospital scenes which,

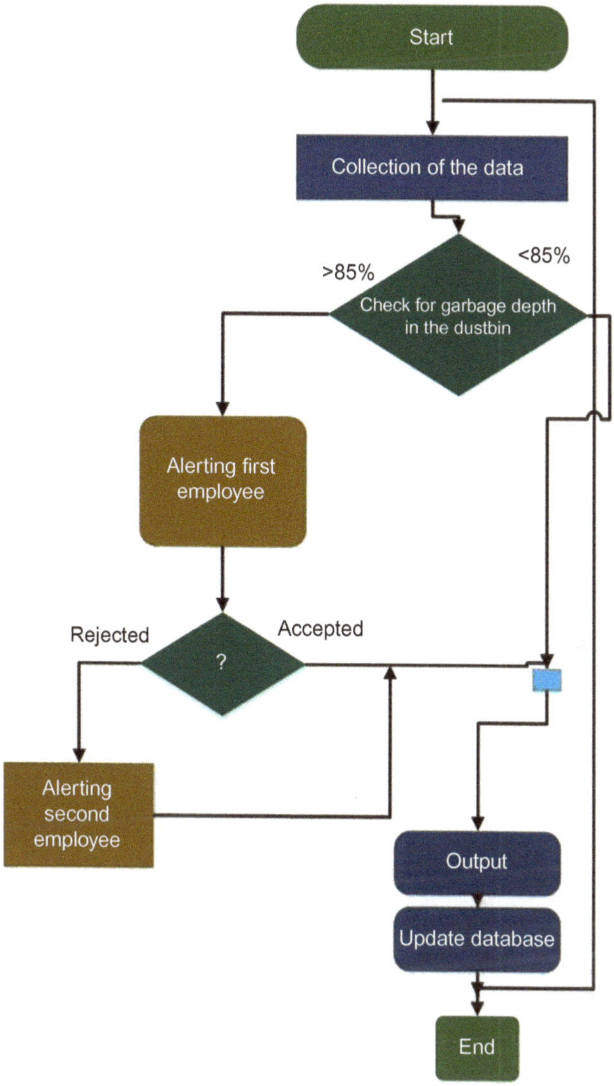

FIGURE 4.7 Flowchart for treatment of biological waste in hospitals.

in turn, compiles up the information specifications and standards and helps in the construction of embedded mobile electronic medical records application platform and hence unified network platform. This, basically, is the key technology in creating smart hospitals.

Coming to the drug stores, which is the go-to place for every person suffering from any acute or chronic disease or illness, needs to have proper management of the drugs so that they can be dispensed as and when required without the patients having to wait for too long. There are also chances of the drugs getting expired or getting misplaced or stolen. In order to ensure safe and quality dispensing of drugs to the patients and avoiding dispensing of any expired medicines to them, which will proof to be harmful to the patients, this chapter [33] proposes drug and medicine monitoring model, which aims to make smart drug stores with the help of Raspberry Pi that uses IoT and RFID technology. The medicines and drugs brought in the store are labeled with RFID tags that deliver and store complete information (manufacturing date, expiry date, storage conditions, and their location) in the database. Moreover, many sensors such as light intensity sensor, temperature sensor, and humidity sensor are placed in the drug store so that the owner of the store is completely aware of the environment of his store. The sensors are arranged in the form of bus or mesh technology so that failure of one node does not affect the working of and information collection of other nodes. These are the most reliable topologies. Then the data collected by the sensors is passed on to the Raspberry Pi via ADC, which is finally uploaded on cloud using IEEE 802.15.4 for the purpose of transparency in the system. Then it follows 6LoWPAN for network module and then TCP or UDP as data transfer protocol and finally the data is accessed using CoAP or MQTT protocols. This information can be accessed by the users anytime from the cloud database. This "Smart Drug Store" Model will also help in complying with the rules and regulations made by the government for storage conditions and quality of the drugs. It will also help in maintaining and getting better outcomes related to patient safety.

The rising cost of health care and the chronic diseases prevalent in increasing number of people demands a patient-centric environment, diagnosis, and treatment of the disease rather than hospital centric. The era of smartphones has led to the establishment of mHealth that focuses on person-centric care and can even detect a user's location, movement patterns, location through various sensors embedded in the smartphones, still it lacks the ability to collect information regarding the bodily health of the patient. This extends the scope to "WIoT" that introduces us to intelligent fabrics worn by the body having sensors that can communicate to each other. Wearable devices also encircle a variety of functions that includes collection of data from on-body sensors, preprocessing of the data and its momentary storage, and finally transfer of the data to Internet-connected devices such as smartphones or to remote servers. The concept of IoT basically provides a concrete foundation for interconnecting wearable sensors and the smart devices to cloud computing platforms for easy interactions. Hence, WIoT aims to transform current health practices to more patient centric with low costs and effective treatment which also helps in the early detection of diseases [34].

Patient safety is one of the most important things that need to be taken care of and drug compliance as well as adverse drug reactions (ADRs) stand out as the most important issues regarding patient safety. Prevalence of ADR is around 6.7% in the hospitals throughout the world, and the death rate due to it is 0.32% of the total population. This occurs due to patient's noncompliance due to drug dosage and their intake schedule besides suffering from polypharmacy. However, with a strict follow-up of drug treatment, the incorrect drug complications can be reduced. And for this, an innovative system based on IoT for the identification of drug and monitoring of medication has been proposed in this chapter [35]. The main technologies used in IoT use traditional bar codes or modern RFIDs for the identification of drugs together with its version for smartphones NFC. These combined with communication protocols of IPV6 extends the feasibility to identify as well as locate and connect all the people to technological devices. After the drug has been identified, the compatibility is verified with the patient's profile through Pharmaceutical Intelligent Information System via Internet or embedded knowledge-based system in devices such as smartphones or smart tabs when there is no Internet. Thus these systems check the drug suitability according to patient medical history or personal health cards based on RFID and the allergy profile.

Thus the concept of IoT being applied to the pharmaceutical industry will certainly shake things up and take it to whole new levels contributing to the revolution of Industry 4.0. The concept of IoT is not just measuring one parameter of an asset, rather it aims to connect assets so as to bring to the forefront every vital parameter related to them. The use of technology and its components will surely open up many more opportunities and possibilities with IoT in future.

4.2.3 Data and its analysis—a way to optimize pharmaceutical processes

The pharma industry is generating enormous amount of data on a daily basis and due its advancement in technologies of sensors, image capturing devices, mobile phones, the pharma industry is not only limited to just the development of drugs or monitoring health care of patients, but it also deals with merging of diverse disciplines such as chemical people, academics, mathematicians, and business intelligence for improving its processes from discovery of a new drug and its logistics to highly advanced analysis of drug interactions and creating smart hospitals and drug stores. The most challenging task that is encountered while dealing with pharmaceutical industry data is the large volume of highly unstructured data that includes medical image reports or physician's prescriptions or pathology reports.

Although structured data gets itself arranged in a predefined format and is easy to be handled, it does not reveal much information for the pharma company to make any significant decisions based on that data. On the other hand, the unstructured data is difficult to deal with but reveals even the finest of details and hence enables significant and more precise decision-making.

This unstructured data might as well include the disease, genome structure, drug descriptions, and mechanism of action of drugs and the data in measured in qualitative as well as quantitative terms (i.e., the structured data) might include the symptoms of a disease, laboratory results of patients, pains and discomforts of patients, which are generally utilized for proper qualitative analysis [11].

As mentioned earlier, data, without its analysis, is useless, and hence, if the right tools for data mining and machine learning are implemented, it can lead to better results in the logistics as well as health care. Thus "Pharma Data Analytics" (PDA) is the process of inspecting large chunks of data to fetch information that can be used in business intelligence processes in the pharma industry with the help of specialized systems and software.

Now, as it is very much clear of all the pharmaceutical fields from where the data is collected, stored, and accessed, it is time to analyze this data to give useful insights in the respective fields, be it early diagnosis of diseases, using machine-learning techniques to help in the discovery of the drugs, postmarket vigilance or pharmacovigilance, medical product fault monitoring, care coordination and delivery management, personalized health care and targeted treatments, predictive maintenance of medical equipment or pharmacy services.

Pharmaceutical industry is one of those industries that can benefit maximum from the use of big data and use of predictive analytics. The big data can create more than $100 billion value for the medical as well as pharmaceutical sector as it can use analytics to aggregate research data to for quick prediction of drugs reducing the costly paths to introduce new drugs to the market. Analyzing the real-time collection of the reports having adverse cases will help to enable pharmacovigilance and safety signals to hint at the clinical trials itself. This data analysis is seen vital as it could even help companies avoid drug withdrawals and can even help develop personalized medicine in which specific drug responses and genetic variation is taken into account.

This chapter [11] gives various machine-learning techniques for pharma data analysis. Taking from the scratch, machine learning is that branch of artificial intelligence in which the computer starts learning and identifying by itself the hidden patterns in the data using certain algorithms, and this learning is basically classified as supervised (predictive), unsupervised (descriptive), and reinforcement learning. This type of technique is useful where the solutions are not available in the form of algorithms or there is lack of knowledge about application domain and formal models.

- Artificial neural networks: This technology models patterns and has the capability of recognizing neural networks of the brain. It adapts analogues of biological neurons on stimulating the biological nervous system and they do not even require rigidly structured designs and can also work well with incomplete data making it a powerful tool for simulation of nonlinear systems. They are made up of large number of layers that are connected by nodes.

- Bayesian networks: These networks are kind of probabilistic model that uses graphs and Bayesian inference for the computations of probability aiming to model conditional dependency and causation by representing with the help of edges the conditional dependence in a directed graph. It is structured and represents graphically the probabilistic relations among several random variables.
- Inductive Logic Programming: This is another area belonging to artificial intelligence and uses logical programming used to construct first order clausal theories from background knowledge. This Inductive Logic Programming is considered unique from other machine-learning techniques in the sense that it uses expressive representation language and its ability to logically encode background knowledge.
- SVMs: These are the supervised machine-learning algorithms belong to kernel method that is used extensively in the field of pharma data analysis, clinical data analysis and drug discovery. It is used in quantitative structure–activity relationship, it being a part of drug discovery process as well as its powerful feature being a good classification and regression tool.
- Multivariate data analysis method: It is a set of data analysis methods established in various sectors of the industry. By this type of analysis, we mean the type of analysis having more than one independent or dependent variable of interest which usually is the case in process control where there are many concerns such as yield, cost, and purity.

4.2.4 Big data handling of pharma data

A huge repository of data is generated in terabytes and petabytes from modern digital technologies, cloud computing, and IoT, and the analysis of this data is required for decision-making at multiple levels. The main aim of big data analysis is to process high volume, velocity, veracity, and variety of data using various computational techniques. Generally, data warehouses are used to maintain the large sets of data, but the extraction of information for precise knowledge gaining and understanding is difficult and not easily available through the present approaches used in mining of the data from these large chunks. Thus the main problem that lies in the way of gaining insightful decision-making ideas is the lack of coordination between the existing database systems as well as the analysis tools.

A large number of tools are available for processing big data although every tool has its own advantages and disadvantages. This chapter [9] emphasizes on the mot emerging and effective tools, some of which are MapReduce, Apache Spark, and Storm. Apache Hadoop infrastructure such as Mahout and Dryad are used as batch processing tools, whereas stream data applications are used for analysis required in real time. Large-scale streaming platform example includes Strom and Splunk. Another platform for interactive analysis includes Dremel and Apache Drill that are used by users for their own analysis in real time.

4.2.4.1 Tools for analytics

1. Apache Hadoop and MapReduce:

   This is the most widely used software platform consisting of MapReduce, Hadoop kernel, Hadoop Distributed File System (HDFS), Apache Hive, etc. Hadoop works on master node and the worker node; the master node divides the received input into smaller subgroups. These subgroups are then distributed to the worker nodes. After the completion of the task the master node combines the outputs generated from all the subgroup problems in reduce step. Coming to MapReduce, it is a programming language that works on the basis of divide and conquer, which is further implemented in two steps: map step and reduce step. Combining the work of Hadoop and MapReduce creates a powerful software framework capable of tolerating fault storage and solving big data problems.

2. Apache Mahout:

   It is a distributed linear algebra framework aimed to provide scalable machine-learning techniques for large-scale analysis applications. The chief algorithms of this framework include pattern mining, dimensionality reduction, regression, classification, clustering, evolutionary algorithms running on top of Hadoop platform through a specialized framework of MapReduce. The objective of Mahout is to provide a tool for overcoming big data challenges by building a responsive and diverse community platform to give way to discussions on potential use cases. Google, Yahoo, IBM, Twitter, and Facebook are some of the companies that have implemented these scalable machine-learning algorithms.

3. Apache Spark:

   Apache Spark is a big data processing framework, which is open source, easy to use and helps to write applications in Java, Python, or Scala. It supports SQL queries, machine learning, and graph data processing besides map-reducing operations. It runs on top of HDFS infrastructure for speed processing and sophisticated analytics. Its primary feature is Resilient Distributed Datasets, which helps in providing fault tolerance and stores data in memory without replication. It consists of components such as cluster manager, worker nodes, and driver program.

4. Dryad:

   It is a general-purpose repository of data that makes it freely usable and discoverable consisting of a cluster of computing nodes. It is a kind of programming model that handles large context based on dataflow graph. A user uses its resources of computer cluster so that they can run their program in a distributed way. Its main advantage is that the user needs not be aware of the concurrent programming, and it provides various functionalities such as scheduling of machines for available processes, generation of job graphs, transition failure handling, and collection of performance metrics.

5. Apache Drill:

   It is an open-source software framework and a distributed system for data-intensive applications for interactive analysis of big data having more flexibility for supporting certain types of query languages and data formats. It is specially designed to exploit nested data and scales up on more than 10,000 servers to process trillions of data in seconds. In Apache Drill, map reduce is used for performing batch processing and HDFS for storage.

6. Jaspersoft:

   It is a commercial open-source software for reporting and analytics to produce reports from database columns, which has fast analyzing and visualizing capability on storage platforms such as Cassandra, Redis, and MangoDB with its most important property being its capability to handle and explore big data without extraction, transformation, and loading. It can also build HTML reports and dashboards directly from big data store and can be shared by anyone who is a part of the user's organization.

7. Splunk:

   By combining the cloud technologies up to this moment and big data, it helps to exploit the big data generated through machine learning. It is a real-time and intelligent platform that helps users to analyze their machine-generated data through web interface exhibiting the results in the form of graphs and reports. The objective of this platform is to diagnose problems for systems and information technology infrastructures as well as provide metrics for various applications.

8. Storm:

   Storm is a real-time processing computational system for processing large chunks of data, in contrast to Hadoop, which is used for batch processing. It is a fault tolerant and distributed platform that is scalable and easily operable. Storm clusters as well as Hadoop cluster are quite similar. For different storm task, different topologies are run on the storm clusters, whereas map-reduce jobs are implemented in Hadoop for corresponding applications; the basic difference between map-reduce jobs and topologies being topologies run and process messages until terminated by the user, whereas map-reduce jobs eventually finish. Storm cluster consists of two kinds of nodes: the master node and the worker node implementing two kinds of roles as the nimbus and supervisor role respectively.

Revolution of medicine, treatment, and diagnoses in an IoT hospital: The patient will have an ID card, which when displayed will be read by RFID reader and all the information stored relating to that ID card/number such as electronic health records of lab results, medical and prescription histories will be easily accessible by the doctors and nurses. The driving force behind all the wearable sensors is the data that is generated. In Holland the Philips and Salesforce on *HereIsMyData*, a database where the health data of patients can be stored, and access to selective people can be provided, collaborated with the Radboud University Medical Centre. The Salesforce platform support Veeva, a customer relationship management platform that is widely used in the pharma industry. Thus Salesforce effectively bridges the gap between the medical data of the patient and pharma. The vast storage of billing data in certain centers of medical services can get their data mined in order to promote high-value care. Moreover, the rate of readmission of patients has been attempted to be reduced by predictive artificial intelligence algorithms based on the analysis of historical data of patients to indicate people at highest risk of suffering from a disease [36].

Another way of handling big data has been proposed by this chapter [37], which consists of two main subarchitectures: Meta-Fog-Redirection (MF-R) and Grouping and Choosing (GC) architecture. Apache Pig and Apache HBase are the big data technologies used by MF-R for collection and storage of the data generated by various sensors and devices. Then the GC architecture is used for the secured integration of fog computing with cloud computing and also uses data categorization function for providing high-level security services. The given framework also makes use of MapReduce-based prediction model for predicting heart diseases.

The quantification and understanding of medical data can be beneficial for the patients, physicians, insurance companies, investors, pharmaceutical manufacturers, drug testing companies, etc., and hence, data mining techniques leverage the information from the data. These techniques give a new face to the pharma industry and prove as an innovative application in information technology field in the health-care industry.

Sometimes, the data generated from various fields may be structured or unstructured. This chapter [38,39] presents certain tools available for big data analytics.

- Text analytics:

  This technique is basically text mining which refers to extraction of information from the textual data like from that of prescription data. This type of analytics involves statistical analysis, machine learning and is useful for analyzing the unstructured data to support decision-making. *Information extraction* technique helps to extract structured data from the unstructured data such as drug name, dosage, and frequency can be extracted in a structured format from medical prescriptions. *Question answering* techniques provide answers to questions as if a normal conversation is going between humans in natural language, such as the conversation between the doctor and the patient. This could help us get more information out of the data.

- Audio analytics:

  This kind of analytics analyzes and extracts information from the audio data like, when applied to human language. It is unstructured data, and this type of analytics is also called speech analytics. In health care, diagnosis and treatment of certain medical conditions that affects the patient's communication patterns is supported by audio analytics. Speech analytics/audio analytics follows two common technological approaches: the transcript based approach and the phonetic based approach.
- Video analytics:

  It is also known as video content analysis and involves a various techniques to monitor, analyze, and extract meaningful information from the videos. The two leading contributors of giving rise to video analytics are the increasing prevalence of CCTV cameras and popularity of video sharing websites. The primary application of video analytics is in security and surveillance systems in spite of their high cost. This is because labor-based surveillance is, in any way, more prone to risk and hence less effective than the automatic systems. Video analytics can help in detecting breaches of restricted zones, detection of loitering in a specific area, identifying objects removed or left unattended, etc. The data that is generated from these CCTV cameras can be extracted out for further processes in business intelligence, operations management, and marketing being the primary area for that.
- Predictive analytics:

  This technique comprises a variety of techniques that can predict the future outcome of anything based on its historical and current data and it can be applied to almost every discipline for its analysis. It seeks to find the hidden patterns and uncover the relationships in data. Some techniques such as moving average and linear regression focus on discovering historical patterns in the outcome variables and find the interdependency between outcome as well as explanatory variables respectively. This technique is based on statistical analysis. It can help detect the patterns that could inform us of a machine breakdown during drugs manufacturing processes well in advance so that it could be worked on. This type of analytics can also help in the early detection of diseases in human beings so that they can be treated at their infancy stage. This can also help in identification of the patterns of human beings with an intention for infant abduction and can generate an alert preventing any mishappening. The techniques can even be subdivided into two groups based on methodology: regression and machine-learning.

4.2.5 Visual analytics in pharma industry

Visualization is the key aspect in achieving the goal in presenting and understanding large chunks of data. But what matters is the type of visualization used. For extreme volumes of data, traditional charts or graphs are not able to present the complete and satisfactory graph in order to clearly understand it. Hence, more effective visualizations, which are fully interactive, and have effective visualizations to communicate the information the data wants to reveal must be used. This is the very first aspect to be kept in mind. Another thing is, when progressing toward the pharmaceutical pipeline, a proper understanding of various types of data such as numeric, categorical, text documents, genome sequences, is required. Visualizations provide high value to the data.

4.2.5.1 Gene expression

There has been a development of a number of microarray techniques which makes it possible for us to study the level of expression of all genes within a cell at once which helps in identification of genes in various diseases or under certain set of conditions, such as the treatment through drug over time.

The heat map visualization shown in Fig. 4.8, provides an overview of the data values with their distribution range in the dataset; in which, each gene is represented by a row in the map and each column corresponds to the condition tested. In this map, color scale is presented instead of text representation of individual data values for a visible structure to the map. In the above map, dendrograms from the OmniViz TreeScape visualization is used, which shows hierarchical relationship between rows and columns.

FIGURE 4.8 Heat map visualization of gene expression data.

4.2.5.2 *Target discovery*

Target discovery is basically focused on what cellular components should be augmented by a particular drug and one of the important steps to stick to this process is to identify the genes functioning and the proteins related to them respectively. This thing can be achieved by Basic Local Alignment Search Tool—Blast or to be more particular with the recent variations, Blast2 is used. The approach that it commonly follows is comparing the genes or protein sequences of one set to another set of objects (Fig. 4.9).

This is a good presentation to similarity in sequences and their alignment but still is insufficient for representing large volumes of data, for it can compares one sequence against others and many of the comparisons are required when industrial scale sequencing is required. Thus to compare many sequences against each other, Blast analysis for each protein against all others is performed, which requires a different approach for effective visualization. Hence, OmniViz Galaxy visualization is used, which shows the relation of each protein with every other protein.

With the flow of data at such a rapid rate across all domains of pharmaceutical industry, information visualization has become a critical component, for it aids in discovery, development, and business by providing a framework for understanding immense volumes of data as well as revealing unexpected relationships. It is the new class of visualizations that has a great impact on analyzing, and these visualizations cover all domains and data types helping in integrating analyses and supporting fast, effective decisions.

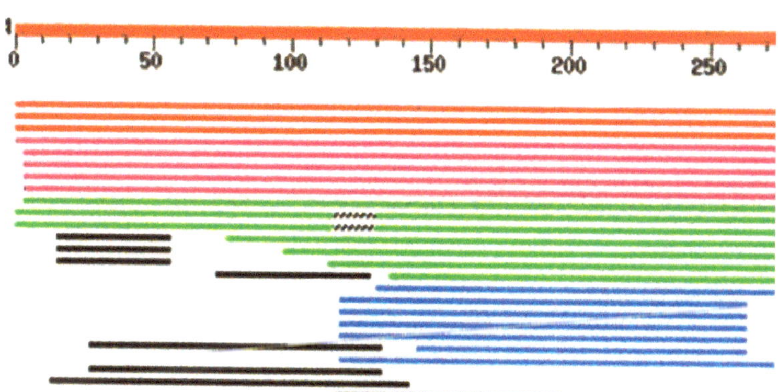

FIGURE 4.9 A visual presentation of sequence comparisons. Color key: black represents <50, blue represents 50–60, green represents 60–90, pink represents 90–250, and red represents >250.

4.3 Proposed work

4.3.1 Proposed framework for Internet of Things–based pharmaceutical data analysis

4.3.1.1 Data acquisition

It is important to collect data of body movement for analyzing the present study by using various sensors at different parts of the body. The capacitive accelerometer sensor has been used for this task, each of which consists of Bluetooth radio, microprocessor, micro-SD card, and ADXL335 accelerometer. The placement of sensors at different parts of the body was done according to some similar studies in research papers. One sensor was placed on the waist, one on the left thigh, one on the right ankle, and one on the right upper arm of the body. All the data is recorded in the micro-SD card with frequency of the sensors at 51.2 Hz.

All accelerometers were well in place with perfect working coordination among them. The activities performed by the subjects included the following:

- sitting down on the chair,
- standing up from the bed,
- sitting on a chair,
- standing, and
- walking.

The study has been conducted on four healthy subjects (two men and two women) for a duration of 8 hours (2 hours on each of the subjects) of their activity.

4.3.1.2 Feature extraction

The data collected from the accelerometers is processed and checked for any differences in the behavior of healthy normal adults and adults with some abnormality (which might result in any disease in the near future). Grouping of the samples was done, and descriptive statistical methods were used for generation of derivative features.

For each feature of wearable sensor f and every possible value $f_i \in F$, the following statistics have been used:

Max—for describing the maximum value of this feature
Min—for describing the minimum value of this feature
Sum—for giving the sum of values $\sum f_i$
Mean—for giving the mean of the values as, $\leq \mathrm{Sum}(S)/N$
Median—for finding the middle value in the data after arranging in one particular order (ascending or descending)

Standard deviation—for finding the deviation of a particular value from its mean position

$$\sigma = \sqrt{\frac{1}{N}\sum (f_i - \mu)^2}$$

Mean absolute deviation—for finding out the deviation of a value of f_i from its mean with a positive value

$$\frac{1}{N}\sum_{i=1}^{N} |f_i - \bar{f}|$$

Median absolute deviation—for finding the absolute deviation of f_i from its median

$$\frac{1}{N}\sum_{i=1}^{N} |f_i - \text{Median}(F)|$$

Coefficient of variation—for finding relative variability, σ/μ

Skewness—for finding asymmetry in the distribution of the values from the mean position

$$\frac{(1/N)\sum_{i=1}^{N} (f_i - \mu)^3}{(1/N)\sum_{i=1}^{N} (f_i - \mu)^3)^{1.5}}$$

Kurtosis—for describing the number of data points around the mean signal energy

$$\frac{(1/N)\sum_{i=1}^{N} (f_i - \mu)^4}{(1/N)\sum_{i=1}^{N} (f_i - \mu)^2)^3} - 3$$

Power—for giving the average energy

$$\frac{1}{N}\sum_{i=1}^{N} f_i^2$$

Autocorrelation—for giving the correlation between values at times t and $t+1$,

$$\frac{\sum_{i=1}^{N-1} (f_i - \bar{f})(f_{i+1} - \bar{f})}{\sum_{i=1}^{N} (f_i - \bar{f})^2}$$

After the extraction of features has been done, selection of features out of those extracted ones is important for the removal of redundant features. Hence, during the classification of these features, only the important and informative ones will be there for use in analysis.

We use Mark Halls' selection algorithm based on correlation to scale out the redundant features. This algorithm basically works on retaining those features that are closely related with the class but not correlated with each other. The class label is predicted by measuring the goodness of individual features and also by taking into account the correlation among them.

This is given by the following equation:

$$H_s = \frac{a\ \overline{r_{ci}}}{\sqrt{a + a(a-1)\overline{r_{ii}}}}$$

This is the heuristic measure of goodness.

This measure is a poor predictor of the class and hence will remove the uninformative features which are highly correlated to each other.

For the features involving ordinal or continuous class values, apart from the general nominal class values, a measure is calculated, which is based on correlation between the features and the class and also among various features. Continuous features cannot be measured as such and needs to be converted to nominal first.

If A and B are discrete random variables corresponding to the ranges R_A and R_B, the entropy is given as

$$E(B) = -\sum_{b \in R_b} p(b)\log(p(b)), \text{ entropy of } B \text{ before observing } A$$

$$(B|A) = -\sum_{a \in R_A} p(a) \sum_{b \in R_B} p(b|a)\log(p(b|a)), \text{ entropy of } B \text{ after observing } A$$

The following equation describes the dependency of B on A:

$$C(B|A) = \frac{H(B) - H(B|A)}{H(B)}$$

and hence will cut out those features from the set which are closely related to each other.

4.3.1.3 Classification

For the classification step in our dataset, we used k-NN classifier that is a supervised classification technique and hence does not require any learning process. To classify any new observation, the similarity between the training set and the new observation needs to be least in order to get added up in that cluster. The algorithm for the same is described as follows.

1. Load the data into an array arr[d]
2. Initialize the value of K as 2
3. From $i = 1,2,3,\ldots, n$

calculate Euclidean distance for each d_i using the following formula:

$$J = \sqrt{\sum_{i=1}^{k} \left(d_i - D_j\right)^2}$$

where J is the objective function, k are the number of clusters, d are the elements of the array, and D is the centroid for the cluster j

4. Measure this distance for every d_i
5. Arrange the calculated distances in ascending order
6. Extract top k rows from this sorted array of distances
7. Determine the most frequent class of these rows
8. The predicted class is then determined and returned

Thus in this way, the classification is done with minimum or no redundancies and each cluster with a different group of objects (Fig. 4.10).

Algorithm 1: IoT_BM-PDA algorithm

Step 1: Begin.
Create an array arr[z] containing all the information from all the sensors N_1, $N_2, \ldots, N_n$ deployed for collecting information about movement patterns.
Step 2: The feature extraction was done by measuring various statistical methods. This helps in measuring the differences in deviation of behavior of normal adults from abnormal adults.
Step 3: Calculate kurtosis to determine the dynamics of acceleration signal:

$$k = \frac{u_4}{std^2} - 3$$

where u_4 is the fourth moment about mean.
Step 4: Calculate crest factor to determine impulsiveness of the signal

$$c = \frac{\max(p(n))}{\sqrt{(1/N)\sum_{n=1}^{N} p(n)^2}}$$

where $n = 1, 2, \ldots, N$.
Step 5: Complexity in the movement of a body part needs to be determined before classification. Hence, for that the following formula is used:

$$\text{Energy} = \frac{\sum_{i=1}^{N} P(a)^2}{A}$$

where $P(a)$ represents the ath spectral line from the amplitude spectrum of the acceleration signal, A is the number of spectral lines.

(Continued)

(Continued)

Step 6: Next, to find out the entropy of acceleration signal, the periodicity of movement can calculated by

$$\text{Entropy} = -\sum_{i=1}^{N} p(a)\log_2(p(a))$$

where $p(a)$ is the probability of occurrence of value $P(a)$ in the amplitude spectrum.
Step 7: For feature selection, Mark Hall's selection algorithm based on correlation is used for minimizing or removing the number of redundant or uninformative features present using

$$G_s = \frac{k\,\overline{r_{ci}}}{\sqrt{k + k(k-1)\overline{r_{ii}}}}$$

It measures the heuristic measure of goodness.
Step 8: The k-NN classifier is used for classification of this data where the value of k is chosen is 2 as it was found empirically that this value of k provides the most accurate results of classification.
Step 9: The Euclidean distance is calculated to determine the distance between the sample that is tested and the remaining data vectors using the given formula:

$$\text{Euc dist} = J = \sum_{j=1}^{k}\sum_{i=1}^{n} \left|\left|d_i^{(j)} - D_j\right|\right|^2$$

Step 10: The classification for every new observation is done on the basis of the frequency with which it appears in its k nearest neighbor algorithm. The more frequently it appears, the more probability of it being added to the final classification result.
Step 11: According to this classification we can predict any kind of diseases or abnormalities that might be affecting an individual or might affect in the near future.
Step 12: End.

4.4 Implementation

This section presents the experimental framework of proceeding with analysis of the dataset obtained from the PUC-Rio Dataset that describes and classifies the movement in a person's body as sitting down, standing up, standing, walking, and sitting. This data is collected for four healthy subjects with 2 hours of study of activities of each subject, that is, a study of total of 8 hours for these four subjects. The body postures and movement is classified and on its basis the occurrence of chronic diseases can be determined in the subjects.

The attributes used in this dataset along with their description is given in Table 4.1.

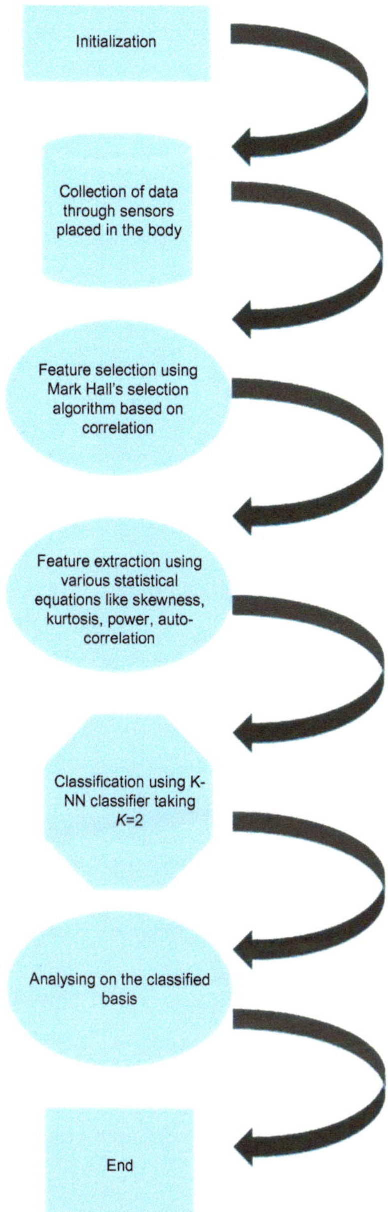

FIGURE 4.10 Flowchart for detection and classification of body movements for analysis.

TABLE 4.1 Attribute selection from UCI repository PUC-Rio Dataset.

S. no.	Attribute name	Description
1	User	String type, unique name of the patient
2	Gender	String type, male or female
3	Age	Integer type, healthy young adults
4	How_tall_in_meters	Real type, in meters
5	Weight	Integer type
6	Body_mass_index	Real type, to determine if the person is lean, normal, or obese
7	$x1$	Integer type, contains the read value of the axis x of the first accelerometer, mounted on waist
8	$y1$	Integer type, contains the read value of the axis y of the first accelerometer, mounted on waist
9	z_1	Integer type, contains the read value of the axis z of the first accelerometer, mounted on waist
10	x_2	Integer type, contains the read value of the axis x of the second accelerometer, mounted on left thigh
11	y_2	Integer type, contains the read value of the axis y of the second accelerometer, mounted on left thigh
12	z_2	Integer type, contains the read value of the axis z of the second accelerometer, mounted on left thigh
13	x_3	Integer type, contains the read value of the axis x of the third accelerometer, mounted on the right ankle
14	y_3	Integer type, contains the read value of the axis y of the third accelerometer, mounted on the right ankle
15	z_3	Integer type, contains the read value of the axis z of the third accelerometer, mounted on the right ankle
16	x_4	Integer type, contains the read value of the axis x of the fourth accelerometer, mounted on the right upper arm
17	y_4	Integer type, contains the read value of the axis y of the fourth accelerometer, mounted on the right upper arm
18	z_4	Integer type, contains the read value of the axis z of the fourth accelerometer, mounted on the right upper arm

4.5 Results and discussion

The proposed technique is compared with SVM for the dataset PUC-Rio from UCI repository. The major five attributes selected from the datasets include Body mass index, Bone density, muscle grade, height and nerve stimulation.

The RMSE and MAPE values depict the attributes against SVM and the proposed technique as shown in Figs. 4.11 and 4.12, respectively. The values

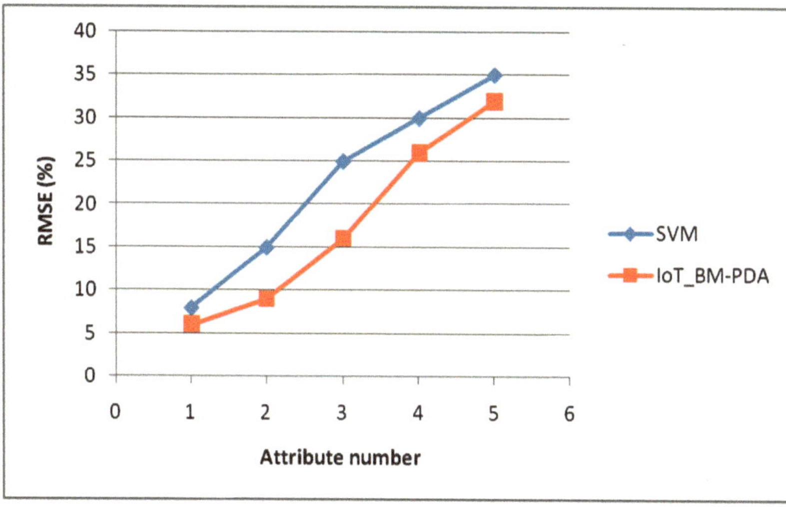

FIGURE 4.11 RMSE comparison for the proposed with the existing system.

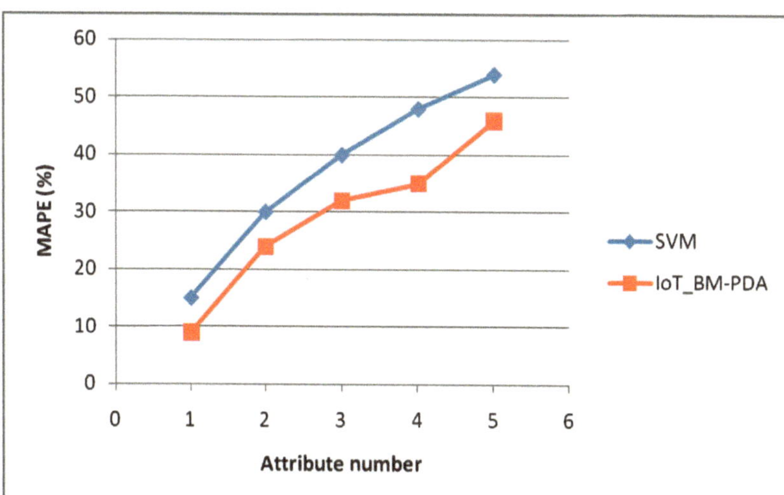

FIGURE 4.12 MAPE comparison for the proposed with the existing system.

depict that the proposed IoT_BM-PDA performs better than SVM by minimizing the root mean square error by 4%. The MAPE values also show that the proposed technique outperforms the existing SVM technique.

4.6 Conclusion

In the near future the applications of IoT in the pharmaceutical industry will grow at a rapid rate making each and every process automated, thus, allowing minimal human intervention and high-speed M2M communication for fast data processing and decision-making. In our proposed work the automated task of collecting data from the sensors placed in the body and their preprocessing is shown. It helps in detecting the early signs and symptoms to various chronic and long term diseases (so that they can be treated at their infancy stage) because of their lifestyle, physical activities, and various other factors. The proposed system is compared with the existing system for the MAPE and RMSE values and found to perform better.

References

[1] F. Samie, L. Bauer, J. Henkel, IoT technologies for embedded computing: a survey, Proceedings of the Eleventh IEEE/ACM/IFIP International Conference on Hardware/Software Codesign and System Synthesis, ACM, 2016, 8.

[2] J. Lin, W. Yu, N. Zhang, X. Yang, H. Zhang, W. Zhao, A survey on Internet of Things: architecture, enabling technologies, security and privacy, and applications, IEEE Internet Things J. 4 (5) (2017) 1125–1142.

[3] H. Atlam, R. Walters, G. Wills, Fog computing and the Internet of Things: a review, Big Data Cogn. Comput. 2 (2) (2018) 10.

[4] M. Yannuzzi, R. Milito, R. Serral-Gracià, D. Montero, M. Nemirovsky, Key ingredients in an IoT recipe: fog computing, cloud computing, and more fog computing, 2014 IEEE 19th International Workshop on Computer Aided Modeling and Design of Communication Links and Networks (CAMAD), IEEE, 2014, pp. 325–329.

[5] A. Marathe, M. Awwad, Internet of Things: Opportunities and Applications in Pharmaceutical Manufacturing and Logistics, 2018.

[6] A.C. Madhukar Saboo, M. Suranglikar, The Internet of Things: the new Rx for pharmaceuticals manufacturing & supply chains, in: Cognizant 20-20 Insights, 2017, pp. 1–12.

[7] A. Unnikrishnan, Moving towards industry 4.0: a systematic, Int. J. Pure Appl. Math. 117 (20) (2017) 929–936.

[8] N. Elgendy, A. Elragal, Big data analytics: a literature review paper, Industrial Conference on Data Mining, Springer, Cham, 2014, pp. 214–227.

[9] D.P. Acharjya, K. Ahmed, A survey on big data analytics: challenges, open research issues and tools, Int. J. Adv. Comput. Sci. Appl. 7 (2) (2016) 511–518.

[10] W. Zhang, Q. Zhao, J. Deng, Y. Hu, Y. Wang, D. Ouyang, Big data analysis of global advances in pharmaceutics and drug delivery 1980-2014, Drug Discov. Today 22 (2017) 1201–1208.

[11] S.S. Kaddi, M.M. Patil, Pharma data analytics: an emerging trend, 2017 International Conference on Smart Technologies for Smart Nation (SmartTechCon), IEEE, 2017, 1307–1311.

[12] M.E. El-Telbany, S. Rafat, E.E. Nasr, Drug design: the machine learning roles, 2014 International Conference on Engineering and Technology (ICET), IEEE, 2014, pp. 1–6.

[13] F. Van den Abeele, J. Hoebeke, I. Moerman, P. Demeester, Integration of heterogeneous devices and communication models via the cloud in the constrained Internet of Things, Int. J. Distrib. Sens. Netw. 11 (10) (2015) 683425.

[14] S.Y. Chen, C.I. Wei, Y.C. Chiu, B.C.C. Lai, A Hadoop-based Principle Component Analysis on embedded heterogeneous platform, 2017 International Symposium on VLSI Design, Automation and Test (VLSI-DAT), IEEE, 2017, pp. 1–4.

[15] H.B. Pötter, A. Sztajnberg, Adapting heterogeneous devices into an IoT context-aware infrastructure, 2016 IEEE/ACM 11th International Symposium on Software Engineering for Adaptive and Self-Managing Systems (SEAMS), IEEE, 2016, pp. 64–74.

[16] S. Hardin, Text and data mining meets the pharmaceutical industry: Markus Bundschus speaks, Bull. Assoc. Inf. Sci. Technol. 43 (3) (2017) 42–44.

[17] M. Goeminne, M. Boukhebouze, Toward a framework for embedded & collaborative data analysis with heterogeneous devices, 2017 40th International Convention on Information and Communication Technology, Electronics and Microelectronics (MIPRO), IEEE, 2017, 381–386.

[18] P. Raundale, S. Gadagi, C. Acharya, IoT based biomedical waste classification, quantification and management, 2017 International Conference on Computing Methodologies and Communication (ICCMC), IEEE, 2017, pp. 487–490.

[19] S. Dubey, A. Gambhir, S.K. Jain, A.V. Jha, A. Jain, S. Sharma, IoT application for the design of digital drug administration interface, 2017 International Conference on Information, Communication, Instrumentation and Control (ICICIC), IEEE, 2017, 1–5.

[20] D. Bansal, S. Malla, K. Gudala, P. Tiwari, Anti-counterfeit technologies: a pharmaceutical industry perspective, Sci. Pharm. 81 (1) (2012) 1–14.

[21] G.F. Barbosa, R.V. Aroca, An IoT-based solution for control and monitoring of additive manufacturing processes, J. Powder Metall. Min. 6 (1) (2017) 158.

[22] M. Pachayappan, N. Rajesh, G. Saravanan, Smart logistics for pharmaceutical industry based on Internet of Things (IoT), in: International Conference on Advances in Computational Intelligence and Communication (CIC 2016), 2016, pp. 31–36.

[23] G.J. Joyia, R.M. Liaqat, A. Farooq, S. Rehman, Internet of Medical Things (IOMT): applications, benefits and future challenges in healthcare domain, J. Commun. 12 (4) (2017) 240–247.

[24] F.H. Alqahtani, The application of the Internet of Things in healthcare, Int. J. Comput. Appl 975 (2018) 8887.

[25] S. Jayapradha, P.D.R. Vincent, An IOT based human healthcare system using Arduino UNO board, 2017 International Conference on Intelligent Computing, Instrumentation and Control Technologies (ICICICT), IEEE, 2017, pp. 880–885.

[26] P.J. Solanke, S.K. Lakshman, Smart medication and monitoring system for secure health using IoT, Int. Res. J. Eng. Technol 6 (2019) 56–60.

[27] G. Yang, L. Xie, M. Mäntysalo, X. Zhou, Z. Pang, L. Da Xu, et al., A health-IoT platform based on the integration of intelligent packaging, unobtrusive bio-sensor, and intelligent medicine box, IEEE Trans. Ind. Inf. 10 (4) (2014) 2180–2191.

[28] P.R. Lavima, G.S. Sarma, An IoT based intelligent medicine box, Int. J. Comput. Sci. Mob. Comput. 4 (10) (2015) 186–191.

[29] D. Mahalakshmi, R. Abarna, M.R. Meena, Survey on prevention of infant abduction in hospitals, Int. J. Sci. Res. Comput. Sci. Eng. Inf. Technol 3 (2018) 1352–1360.

[30] A. Gnanasekar, M. Akshaya, M. Nivedheetha, J. Nivedita, IoT based hospital sanitation system, Int. J. Curr. Eng. Sci. Res 5 (2018) 10–15.

[31] E.N. Mambou, S.M. Nlom, T.G. Swart, K. Ouahada, A.R. Ndjiongue, H.C. Ferreira, Monitoring of the medication distribution and the refrigeration temperature in a pharmacy based on Internet of Things (IoT) technology, 2016 18th Mediterranean Electrotechnical Conference (MELECON), IEEE, 2016, pp. 1–5.

[32] L. Yu, Y. Lu, X. Zhu, Smart hospital based on Internet of Things, J. Netw. 7 (10) (2012) 1654.

[33] K. Gupta, N. Rakesh, N. Faujdar, M. Kumari, P. Kinger, R. Matam, IOT based automation and solution for medical drug storage: smart drug store, 2018 Eighth International Conference on Cloud Computing, Data Science & Engineering (Confluence), 2018, pp. 497–502.

[34] S. Hiremath, G. Yang, K. Mankodiya, Wearable Internet of Things: concept, architectural components and promises for person-centered healthcare, 2014 4th International Conference on Wireless Mobile Communication and Healthcare-Transforming Healthcare Through Innovations in Mobile and Wireless Technologies (MOBIHEALTH), IEEE, 2014, pp. 304–307.

[35] A.J. Jara, M.A. Zamora, A.F. Skarmeta, Drug identification and interaction checker based on IoT to minimize adverse drug reactions and improve drug compliance, Pers. Ubiquitous Comput. 18 (1) (2014) 5–17.

[36] D.V. Dimitrov, Medical Internet of Things and big data in healthcare, Healthcare Inf. Res. 22 (3) (2016) 156–163.

[37] G. Manogaran, R. Varatharajan, D. Lopez, P.M. Kumar, R. Sundarasekar, C. Thota, A new architecture of Internet of Things and big data ecosystem for secured smart healthcare monitoring and alerting system, Future Gener. Comput. Syst. 82 (2018) 375–387.

[38] J. Ranjan, Data mining in pharma sector: benefits, Int. J. Health Care Qual. Assur. 22 (1) (2009) 82–92.

[39] A. Gandomi, M. Haider, Beyond the hype: big data concepts, methods, and analytics, Int. J. Inf. Manage. 35 (2) (2015) 137–144.

Further reading

M. Alam, R.H. Nielsen, N.R. Prasad, The evolution of M2M into IoT, 2013 First International Black Sea Conference on Communications and Networking (BlackSeaCom), IEEE, 2013, pp. 112–115.

G. Nagasubramanian, R.K. Sakthivel, R. Patan, A.H. Gandomi, M. Sankayya, B. Balusamy, Securing e-health records using keyless signature infrastructure blockchain technology in the cloud, in: Neural Computing and Applications, 2018, pp. 1–9.

M. Noura, M. Atiquzzaman, M. Gaedke, Interoperability in Internet of Things: taxonomies and open challenges, Mob. Netw. Appl. 24 (2018) 796–809.

K.K. Patel, S.M. Patel, Internet of Things-IOT: definition, characteristics, architecture, enabling technologies, application & future challenges, Int. J. Eng. Sci. Comput. 6 (5) (2016) 6122–6131.

J.D. Saffer, V.L. Burnett, G. Chen, P. Van der Spek, Visual analytics in the pharmaceutical industry, IEEE Comput. Graphics Appl. 24 (5) (2004) 10–15.

V. Sharma, R. Tiwari, A review paper on "IOT" & it's smart applications, Int. J. Sci. Eng. Technol. Res. 5 (2) (2016) 472–476.

G. Singh, D. Schulthess, N. Hughes, B. Vannieuwenhuyse, D. Kalra, Real world big data for clinical research and drug development, Drug Discov. Today 23 (3) (2018) 652–660.

Chapter 5

Reliable pharma cold chain monitoring and analytics through Internet of Things Edge

S. Balachandar[1] and R. Chinnaiyan[2]

[1]*Shell India Market Private Limited, Bangalore, India,* [2]*Department of Information Science and Engineering, CMR Institute of Technology, Bangalore, India*

5.1 Introduction

The cold chain pharmaceutical market was valued 13.4 billion US dollar in 2017. Pharmaceutical Commerce's annual market forecast projects 12.7% year on year (YOY) growth [1]. Cold chain monitoring market is to exceed 7 billion US dollar revenue by 2023 at 9.6% compound annual growth rate (CAGR) [2].

Cold chain is a temperature-controlled supply chain; an unbroken cold chain is an uninterrupted series of refrigerated production, storage, warehouse, and logistic activities [3]. Vaccination drugs are actual health measure to prevent and control the number of diseases. Vaccines is sensitive to heat and cold conditions and reacts accordingly, and the conditions need to be maintained to recommended temperature throughout the supply chain, ensuring the effectiveness of drugs. It looks more effective in terms of regulatory compliance, not only product quality but also and patient safety. Internet of Things (IOT)-based cold chain solution for pharma products should be capable of following features [4]:

1. Real-time temperature monitoring and alerts.
2. Up to date with current regulatory trends and medical norms of the emerging products.
3. Reduce costs of spoilage and cargo transfer.
4. On-time availability of products to supplier or consumer.

The logistics process begins from the manufacturer who has the approved drugs, labeled with storage, and shipping conditions. The transportation

Emergence of Pharmaceutical Industry Growth with Industrial IoT Approach.
DOI: https://doi.org/10.1016/B978-0-12-819593-2.00005-4

mode may be either through air, train, or trucks that carry the goods from manufacturer warehouse to the distributor or consumers directly.

The sealed or unsealed packages need to be kept in the cold chain equipment such as "refrigerators," "cold boxes," "vaccine carrier," "ice-packs," and "foam pads." And it needs constant monitoring of temperature during travel or transit, and it should be shifted to another storage when the container storage malfunctions or move to another truck or van when it broke down. The refrigerator or storages will not get any power for the batteries or power back-up when the truck or van is not moving for a long time.

The data comes from different devices and sensors such as "humidity," "temperature," "global positioning system (GPS)," "near field communication (NFC) chips," "cameras," and it should be processed immediately and take the necessary action in the vehicle itself. The vehicle might be connected to cloud environment that shares the data based on the bandwidth supported by the gateway or routers. Latency is the key challenge for transferring sensor data from the truck or van to cloud platform, and the transformation of data might take place at the cloud's batch layer [e.g., spark or Amazon Web Services (AWS) Glue or Databricks of Azure].

IOT Edge:

It is a fully managed service that delivers cloud intelligence of IOT to local devices or actuators or IOT gateway. It enables artificial intelligence (AI), cloud services, and applications directly on cross-platform IOT devices.

The rise of IOT Edge will help us to process data immediately and eliminates the cloud's dependency to prepare necessary action insights or alerts. There are different edge computing services available from leading cloud platform providers such as Google, Microsoft, and Amazon.

5.1.1 Statement of the problem

The cold chain logistics for pharmaceutical needs informed decisions across the distribution channel to reduce the product spoilage and supply chain cost. It also requires continuous monitoring of diverse drugs and vaccines packaged and labeled from the time of manufacturing to the concerned receiving party [5].

5.1.2 Objectives

General objectives:

The general objective of this study was to establish the influence of IOT Edge analytics in cold chain supply logistics of pharma products.

Specific objectives:

1. How does IOT Edge influence the monitoring of storage conditions of the containers during supply chain?

2. How does an IOT Edge server merge remote data from multiple IOT devices used in the distribution channel?
3. How do we build insights locally from the sensor data and share the insight to cloud?
4. How does a moving vehicle process the data without connecting to Internet and take decision?

5.2 Cold chain logistics

Logistics is the management of the flow of goods from source to destination to meet customer requirements [6]. Cold chain logistics is the system of transporting and storing medicinal goods at recommended temperature from the manufacturer to the recipients (Fig. 5.1).

The intermediary parties such as warehouse workers take care of keeping the medicinal products in a container or refrigerator and label and package it with a radio frequency identifications (RFID) tag for better tracking and identification. Different types of cold chain storage equipment [6] are being used.

1. Walk in cold rooms—storage of up to 3 months.
2. Deep freezers—(temperature of −15°C to −25°C).
3. Ice-lined refrigerator (2°C–8°C temperature), and it holds cold air better than a front-opening refrigerator.

If the vaccination or medical products need international shipment, the local customs should expect the release certificate from the national regulatory authority. Also, all the vaccinations must meet World Health Organization's recommended norms. After customs clearance, the goods will move to distributors, and they ensure the efficacy of products. The big bundle of products will be distributed to different pharmacists or medical institutions through "cold boxes," "vaccine carriers," and "day carriers" [7].

5.3 Literature review

The study was grounded on about the role of edge computing and analytics.

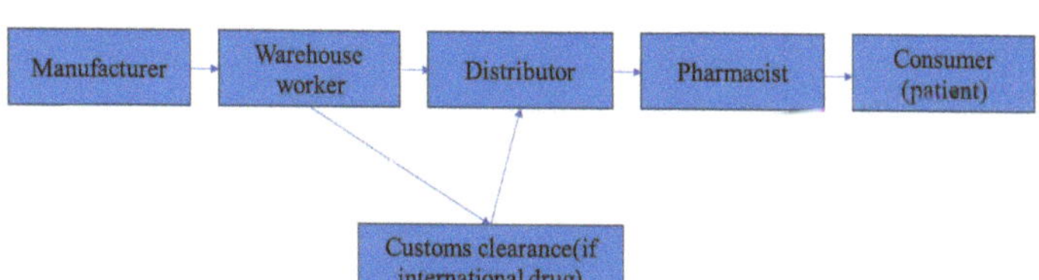

FIGURE 5.1 Cold chain logistics flow.

As mentioned in Ref. [8] that in edge computing, we want to put the computing at the proximity of data resources.

This has several benefits compared to traditional cloud-based computing paradigm. They described five potential applications of edge computing, in that "Location awareness" is a geographic-based application such as transportation and utility management; edge computing exceeds the cloud computing due to location awareness. Data could be collected and processed based on the geographic location without being transported to cloud. In edge computing, we have multiple layers with different computation capabilities. Work load allocation becomes an issue, and we need to decide which layer to handle the workload on how many tasks to assign at each part. They recommended to choose the right allocation strategy using the optimization metrics, namely, bandwidth, latency, energy, and cost.

It is evident that the emergence of edge computing in three important trends {software-defined network and associated network function virtualization, ultralow-latency [1 ms or less (e.g., 5G)] wireless networks for a new class of tactile applications, computing capabilities of wearables, smartphones, and other mobile devices that represent the Internet's extreme edge} in the computing and communication landscape, as mentioned by Prof. "Mahadev Satyanarayanan, Carnegie Mellon University" in the *Computing in Science and Engineering* magazine on January 2017. He also mentioned the usage of "cloudlet." It is a small-scale data center or cluster of computes designed to quickly provide cloud computing services to mobile devices such as wearables, portable devices, and smartphones.

As per "Smart logistics for pharmaceutical industry based on Internet of Things (IoT)" journal published in Vol. 14 CIC 2016 of *International Journal of Computer Science and Information Security* (IJCSIS) by "M. Pachayappan, Nelavala Rajesh, G. Saravanan," IOT plays a vital role for tracking and tracing the pharma products in the supply chain. They also mentioned the IOT-enabled smart container that uses "RFID," "GPS," and other sensors. Based on the nature of products, the cold chain is required physical facilities to ensure suitable temperature conditions. The physical facilities may be specialized warehouse; loading and unloading facilities and refrigeration units are required for the live temperature control. The IOT pharma system will help one to constantly monitor and track the medicine quality and safety during the whole pharma supply chain. They also referred to how two-layer network architecture of IOT platform is used, the first layer is "RFID" and "wireless sensor network (WSN)," and they explain how IOT will help one to avoid physical damage, moisture, humidity, and counterfeit drugs.

In this research paper "Supply Chain Management and Automatic Identification Management Convergence: Experiences in the Pharmaceutical Scenario," "U. Barchetti, A. Bucciero, A. L. Guido, L. Mainetti and L. Patrono" mentioned about autoidentification technologies and international standards related to goods traceability. The crux of RFID layer transponders (active and passive) and how different each of them and when to use what transponders really helps us to choose the right track and trace solution using RFID. They clearly articulated the importance of "drugs" must be constantly checked during transport and stocking. The EPCglobal consortium, mainly represented by the GS1 (Global Standards 1) organization (gs1), defines the standards for developing a universal identification system and an open architecture able to guarantee interoperability and data sharing in a complex multivendor scenario. In particular, it proposes the EPCglobal network architecture, whose main feature is the use of the Electronic Product Code (EPC), a code able to univocally identify each item. The protocol stack can be divided into three parts, namely, identity, capture, and exchange. The identity portion contains the standards for the identification of tags and the translation of tag data. The capture portion contains the standards for filtering and collecting the tag data.

The exchange portion contains the standards for storing and sharing collected and filtered EPC product data. They also explained the difference between ebXML and EPC message format, and it helps us to design the RFID tags for the pharma products efficiently.

5.3.1 Edge technologies

IOT architecture has not yet been standardized, and it is still evolving with different sensor technologies and protocols. It comprises three major layers (identification, communication, and authentication) (Fig. 5.2) [9,10].

Technologies that can identify the things in IOT are known as "edge technologies" [10].

In the identification layer, we use RFID, Zigbee, Bluetooth low energy (BLE), and Z-Wave technologies that are used in edge technologies.

- RFID is classified into two types: passive and active; passive tag cannot do any computations whereas active does. Also, passive does not have a power source, and it just transmits the signal upon receiving radio frequency energy emitted from a reader in proximity on the tag [11].

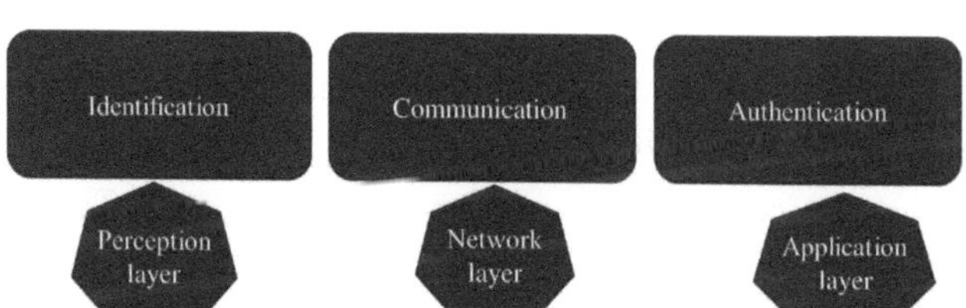

FIGURE 5.2 IOT layers. *IOT*, Internet of Things.

Dash 7 is the communication protocol that uses active RFID. NFC is a classic use case for RFID for tracking stocks and inventory.
- Zigbee is used to carry small packets of data in a relatively infrequent interval. The maximum data rate is 250 kb/s and ranges.
- BLE is used for a shorter distance (within 30 m) and transfer the data at the rate of 1 Mb/s and typically used with low energy.
- Z-Wave is used with a range of 100 m, and it transfers the data at the rate of 40 kb/s with a frequency of 900 MHz.

RFID is a better candidate among all the above protocols for edge technology in terms of power consumption as well as computational capabilities.

In the network layer, we essentially apply 2G, 3G, 4G, and Wi-Fi to exchange the data from sensors to gateway or Internet or data platform residing on-premise or cloud.

The authentication is essential to perform the authorized actions across the things that are attached and registered in the network.

5.3.2 Common sensors

Most common sensor and devices used in cold chain are as follows:

1. Temperature sensor (e.g., DHT11, sensor-mounted storage that displays the temperature value in analog or digital values)—To measure the product storage as well as outside temperature.
2. Accelerometer—To measure intensity of physical activity (e.g., packaged fragile products).
3. Location—GPS to track the vehicle details (latitude and longitude).
4. RFID—Container departure and arrival will be tracked with IOT-connected sensor (RFID).
5. Vibration sensor [12]—Using radio frequency and Bluetooth vibration sensor allow us to monitor the assets.
6. Storage security lock sensor—It will alert and send an email when the security lock is tampered or removed [13].

Nowadays QR code or barcode is being used to identify individual items; however, these item counts should tally with inventory details shared by the supplier or manufacturer.

5.4 Internet of Things edge design—conceptual framework

In Fig. 5.3, among different IOT devices, sensors are highlighted in the logistics flow of pharma goods (from manufacturer to consumer). The component mappings are explained in the following:

1. *Manufacturer*: This is the origination place where the goods get packaged and labeled. The RFID tags will be tagged to respective package, and it

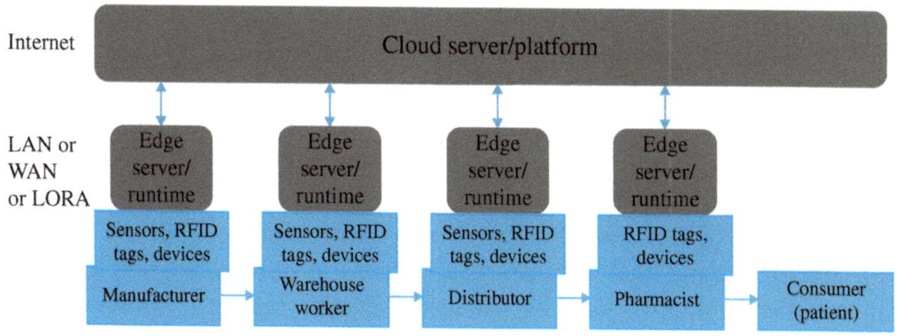

FIGURE 5.3 Conceptual framework components.

should be verified by the RFID reader and scanner before it gets shipped. The sensors such as temperature and humidity will also be plugged to track the medicines conditions. The calibration of sensor values and threshold will be set so that the goods will be measured according to defined ranges. The edge computing is essential to build real-time dashboard of all the devices conditions to decide the spoilage levels and build essential machine learning model to predict the goods availability based on demand plan. The data from different products will also get shared to cloud layer to through a designated network (e.g., WAN or LAN)

2. *Warehouse worker*: This is the second stage where it gets multiple products from different manufacturers and packers. The role of trace and track is mandatory to identify the inventory of goods available versus shipped. RFID tags will ensure that the product is available and it is not lost while loading or unloading as per the manufacturer list. They also verify the current temperature condition, and they might return the product back to manufacture if the product is damaged or crossed the defined temperature threshold. They need to closely work with the transportation team to pack the products through storage-secured lockers to prevent any chemical reactions. Here the accelerometer or vibration sensor also plays a vital role to help the warehouse worker to set the agreed vibration levels for trucks and vans. The transportation team will measure the vibration sensor and ensure that the readings are within the agreed range of values.
3. *Distributor*: They are the middlemen between the manufacturer and pharmacist. The IOT will help them to measure the stocks (using RFID, GPS tags) as well as the medicine conditions (temperature, humidity) once it gets unloaded or transferred to smaller trucks or vans based on the demand. Edge computing is essential to collect the data from different packages and trucks to recommend the shipment time to the right pharmacist or consumer. The security locker sensor will help them to read the values as per the spec; if that value is not coming as per the expectation then it will be treated as spoilage or marked for return to manufacturer.

4. *Pharmacist*: RFID tags will help them to maintain the current inventory or stock register. It will also help them to unpack multiple items based on the specification. They need to refrigerate the items based on the temperature settings and vibration settings. Necessary ice boxes will be kept ready when they deliver medicines to consumer or patients or hospitals
5. *Edge server*: Edge can physically deployed close to gateway IOT or group of sensors that can be maintained and monitored quickly. The monitoring of sensor devices will become even easier with cloud-based edge solutions such as Azure IOT Edge, AWS IOT Edge.

The following are the key benefits of processing IOT data at edge [14].

a. Reliability: Low chances of network outage or data loss as it will get processed in local network and compute.
b. Security: Easy to deploy security protocols and certificates specific to edge security zone category (e.g., if the edge is marked as high risk, additional security checks and authentication can be enabled), no dependency of inherited security model of cloud platform.
c. Flexibility: It will be flexible to scale up or scale out or scale down the resources without major impact to central data center or cloud platform.
d. Speed: Eliminates the cloud's processing and compute resources. It processes data on board or local edge server.
e. Scalability: It is less expensive on scaling out the resources (i.e., compute or storage) based on the local data processing forecast.

Deployment model: Local edge server without cloud platform connectivity

Compute server(s) required to do computations for the data collected from different sensors, tags, bar codes of respective places of the supply chain. It will eliminate the data transport from edge to cloud (Fig. 5.4). The key uses of this model are:

a. *Device management*: The devices or sensors will get registered and deregistered at plant or warehouse level, and this will help us to upgrade the security certificate or firmware directly at edge server.
b. *Monitoring*: Spontaneous control of the inventories (e.g., avoid theft) and reporting the quality control back to quality department before shipment at local warehouse or manufacturing plant.
c. *Prediction of orders*: Based on the current readings from different departmental data of the plant, it will help us to estimate the delivery to

FIGURE 5.4 Local edge components.

packaging and labeling for different products and damages. This needs a local computation using statistical model using analytical tools, such as R, Python.

d. *Cyber-attacks*: Attacks, such as a distributed denial of service attack, can be avoided as the network is localized.
e. *Network latency issues*: Based on the local network topology (mesh or star or ring), the communication between the devices and gateway or edge server will reduce the network latency–related issues compared to cloud communication and increase the responsiveness of the devices.

Following are the demerits of the preceding model:

a. The edge server is local to the plant or warehouse, and it demands for adequate fault-tolerance and disaster recovery mode.
b. It will not provide integrated data flow from an end-to-end perspective (e.g., plant to consumer or plant to stockiest) to track the conditions of the goods.
c. When the large number of devices and sensors needs to deployed, it mandates the downtime for edge server which potentially reduces production margin and increases the "unplanned" downtime of the plant or warehouse.
d. AI and machine learning: Build machine learning model at local edge and train and test the model based on the local data.

Deployment model: Local edge with cloud platform connectivity

It enables the local execution and responds to local events at the edge. It uses cloud for management of devices, analytics, and intelligence.

Leading cloud platforms, such as Microsoft Azure, Google IOT Edge, and AWS IOT Edge, support these models. The unique feature of AI and computation differs for each cloud provider; however, the intent of keeping the control and device twinning of edge devices at the cloud is same.

Three important functionalities will be benefited to keep the data at edge.

a. Real-time dashboard to track and trace the items.
b. Build analytics insights based on the edge data, which helps one to predict the goods conditions before decomposition.
c. Eliminates the data latency–related challenges between cloud and local IOT gateway. Also, it helps one to aggregate the information and send to the cloud layer based on the requirement (Fig. 5.5).

6. *Cloud platform*:

   Edge storage is limited due to the capacity of data that gets stored and processed, and data retention is also limited to a short period (e.g., days to weeks or couple of months). On the other hand, the intention of cloud storage is unlimited and elastic and ubiquitous [15] in nature, and it needs large store for complex processing of data model. It supports a diverse data format to keep it in a data lake kind of store with dynamic schema or schema less stores.

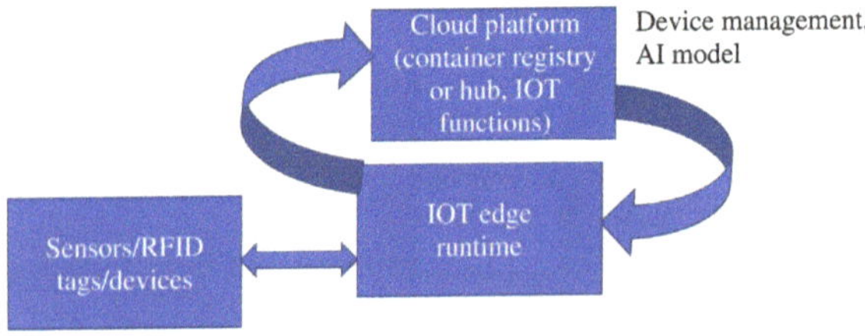

FIGURE 5.5 Local IOT Edge communication with cloud platform.

Permanent storage with scaled-out features in the cloud data platform helps one to store a large volume of data transport between devices or gateways or different applications required at supply chain. It will retain the historical data based on the legal and regulatory needs. The historical data of goods spoilage and returned during transit will help one to build the right model to recommend the sensor settings as well as calibrate the sensor values.

Sensor data from edge

RFID Tag: Usually RFID tag data is not more than 2 kB of data [16]. The historical data needs more storage such as aerospace industry that needs passive UHF (ultrahigh-frequency) tags that typically store 4–8 kB of data. The sample format of the RFID tag output is shown in the following table:

EPC	Count	Time	Date	Antenna	RSSI	PC	CRC
505400A8BC023E0000000022	3	23:02:15	02-04-18	1	1		

CRC, Cyclic redundancy code; *EPC*, electronic product code; *PC*, protocol control word; *RSSI*, received signal strength indicator.

Accelerometer

It is used to measure the acceleration forces. It monitors the movements that have the ability to record the intensity of physical activity.

Sample format

Timestamp	*X*	*Y*	*Z*
2019-04-04 20:20:12	3	5	23
2019-04-04 20:21:15	23	12	34

Acceleration values are recorded in three axes. *X*, Left or right; *Y*, forward or backward; *Z*, up or down.

Temperature and humidity sensor

Based on the sensor model (e.g., DHT11 or DHT22), the digital values will be converted to analog output.

Sample format:

Temperature: 28.2
Humidity: 45.23

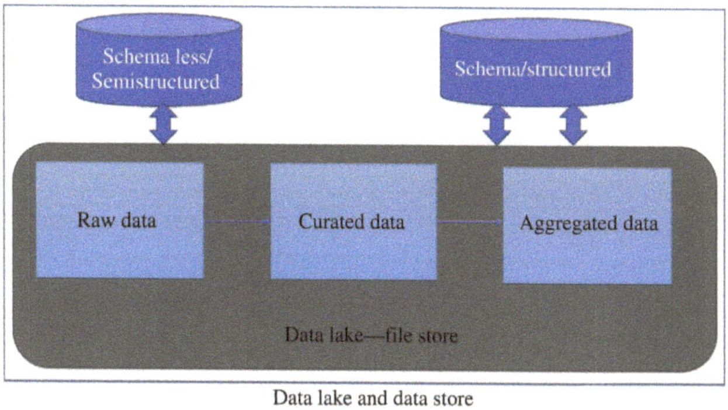

FIGURE 5.6 Data lake in cloud platform.

Data storage layer in cloud platform

Fig. 5.6 illustrates the data lake feature of cloud platform, which helps segregate the data into multiple layers.

Raw data: Data comes from the edge server or sensors directly will be kept in this layer, and it will not be transformed or curated at any time. It will help us to validate the quality of data and understand the format which needs to preserve it. Most of the data search routines or search applications will point on this layer to fetch and understand the data format, data type, etc.

The following operations will be done on this layer:

Operations	Purpose
Data ingestion	Ingest data from edge server or IOT Edge gateway or devices. The data will be come with different velocity (e.g., real time: vehicle movement and location data) Near real-time: cold chain baggage scanned data through RFID scanner or reader Batch data: Data will come from the warehouse with next day's shipment and driver information with product availability and inventory details
Data search	Search engines supported by cloud platform (e.g., Microsoft Bing, Google Search, Elastic search services, or cloud search from AWS) will help us to index, search, and query the raw data. It will help us to tag the data fields for further analytics or data analysis to be done by the data scientist
Data catalog	Query the data through a query builder or rule builder. It will help us to provide a single source of truth about the data. Also, the data stewards use this tool to standardize the data format, build the data glossary, and data dictionaries and data lineage diagrams. Data engineers and data scientist use this tool to build necessary models based on the data definition and data properties

AWS, Amazon Web Services; *IOT*, Internet of Things.

Curated data

The data residing in raw data layer is not going to be accessed directly by analytics application or reporting applications. In this layer, the following data operations will take place.

Operations	Purpose
Filtering	It will remove the unwanted columns from the raw data and apply data level filters (e.g., fetch only active flag records)
Deduplication	Elimination of duplicate or redundant information from the raw data
Data validation	Validate the data based on the metadata or data glossary
Data integration	Combine data from other source data [e.g., vehicle master data has got vehicle details and container capacity which needs a linking with driver master to build the shipment transaction along with necessary sensor data coming from the packages and vehicle specific sensor (e.g., GPS, Gyroscope, RFID tag)] Data engineers will take care of this activity and build the workflow and data mapping. There are many tools available in the cloud platform (e.g., Azure supports Azure Data Factory, Amazon supports AWS Glue, and Google Cloud supports Google Cloud Data flow and Big Query transformations)

RFID, Radio frequency identifications.

Aggregated data

In this layer, the curated data will be aggregated for high-level reporting or dashboards. The following operations will be carried out here.

Operations	Purpose
Modeling	Data residing in curated layer needs to be modeled into dimensional modeling techniques such as OLAP format. Cloud platforms, such as Azure support analytics services which help us to store the data in OLAP format (MOLAP, ROLAP, and HOLAP) and it helps us to retrieve the large volume data with less response time. The advantage of an OLAP format

OLAP, Online analytical processing.

Remote device management

It is one of the mandatory features of cloud platform, and it should maintain a list of connected devices or sensors and keep track their operation status [17], firmware upgrade or parameter configuration, and remotely command and control it (Fig. 5.7).

1. Define and design

   During this stage, we define the metadata about the devices or sensors or tags information or condition. It helps one to query the device remotely and track current metadata status.

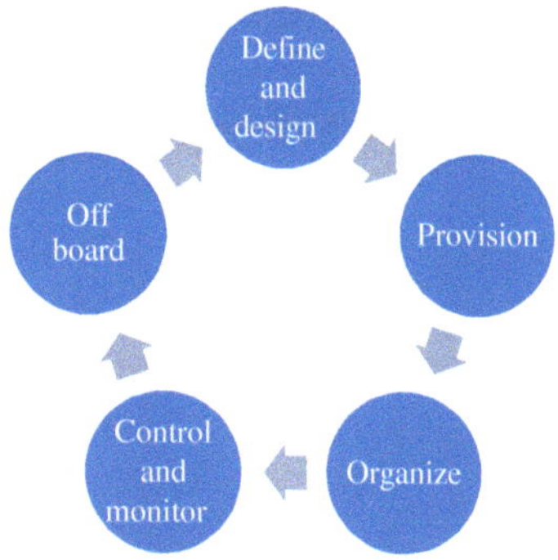

FIGURE 5.7 Device Management life cycle.

Edge devices will communicate their states using messaging protocols (e.g., MQTT) through a JSON format.

Example: In the Azure platform, device twin will take care of tracking the metadata of the physical devices and synchronize the operation status between edge device to cloud.

Similarly, in AWS platform, device shadow does the similar function.

2. Provision

It involves the creation and onboarding of devices into a cloud platform. The registration process varies in different cloud platforms (e.g., in AWS, device registration process starts after the device got templates, certificate, and necessary policies). The device specific attributes will also be captured.

```
{
"version": 1,
"thingName": "V1Pkg2Temperature,"
"defaultClientId": "V1Pkg2Temperature",
"thingTypeName": "DHT11",
"attributes": {
"model": "xyxsdf",
"volt": "25"
}
}
```

Provisioning can be done for individual or group of things

3. Organize

It is an important step to assess the health and security of the devices. We need to take care of device's firmware control and update from remote. It can be controlled through a job which broadcast and update the firmware in bulk mode.

4. Control and monitor

Monitor overall devices' health and the status of ongoing operations. Operators need more attention based on the type of alerts and notification comes from different devices; device-to-device communication will take place based on the readings or values. SQL kind of language can also be supported in a cloud platform (i.e., both in Azure and AWS) to query the device's health, status

```
Query: SELECT* FROM devices WHERE deviceid=10023;

SELECT* FROM devices WHERE properties.reported.connectivity IN
['BLE', 'wifi']

SELECT color AS t1color FROM 'd/c' WHERE temperature > 60 AND
color<> 'green'.

SELECT CASE color WHEN 'green' THEN 'go' WHEN 'yellow' THEN
'alert' WHEN 'red' THEN 'stop' ELSE 'you are not at a stop light'
END as instructions FROM 'a/b'
```

5. Offboard

   Devices will be off-boarded from the IOT network or gateway when it malfunctions or failed to respond. The offboarding will take place at the cloud platform to ensure that its properties will be set to "Inactive," and the corresponding entries and certificates must be removed from the cloud identity.

5.5 Implementations

Cold chain can be implemented in a multicloud environment; this section explains different cloud platforms of IOT Edge and its unique features.

- Google: Google IOT Edge (Edge ML, Edge IOT core)
- Amazon: AWS Greengrass (AWS Greengrass, Lambda at Edge, AWS Snowball edge, AWS CloudFront)
- Microsoft: Azure IOT Edge (Azure IOT Edge runtime, Azure container registry, Azure Functions) are different

5.6 High-level approach

This section explains about a solution approach for building a cold chain solution using IOT Edge. IOT Edge moves the cloud analytics and custom business rules or logics to devices [18]. It is made up of three major components in the edge compute:

1. Devices or sensors
2. Protocols
3. Edge modules
4. Edge hub and agent
5. Edge runtime
6. Cloud-based interface

1. Devices or sensors: As highlighted in the previous sections, the RFID tag and other important sensors bundled with the pharma product will start emitting the values once it gets packaged from manufacturing unit.
2. Protocols: Usually the communication between the sensor to the edge server to local gateway will happen based on the ranges (e.g., Wi-Fi or Bluetooth).
3. IOT Edge modules: It is a compute implemented with a software container (e.g., Docker, Kubernetes), it will help us to build applications such as monitoring, alert or notification, and analytics such as failure prediction or predictive maintenance of the equipment.

   In Azure IOT Edge [19,20], modules are plain vanilla docker containers that are mapped to a device (e.g., humidity sensor). It is built from standard docker file definitions and pushed to either public or private registry. Modules can communicate each other through a defined interface established by IOT Edge runtime. Sometime modules act as twin, which represents the physical device. They are called device twin in Azure IOT world.
4. Edge hub and agent: Edge hub is a mirror service of IOT Hub in the public cloud; it takes care of authenticating modules and communicates with public cloud through application program interface. It works like a message broker to communicate through standard protocols, such as AMQP, MQTT. Agent is responsible for maintaining desired state of devices configuration at the edge. It pulls the manifest file from public cloud (e.g., IOT Hub) and manages the interaction between cloud and Edge runtime. It also maintains the state of container (e.g., docker) configuration matching with original definition mapped with the edge device.
5. Edge runtime: It is a native binary file installed on the edge OS such as Raspbian, Core OS, Ubuntu, Cent OS, Microsoft Windows.
6. Cloud interface: It helps one to monitor the workloads running on the edge and act as control plane at the cloud (Fig. 5.8).

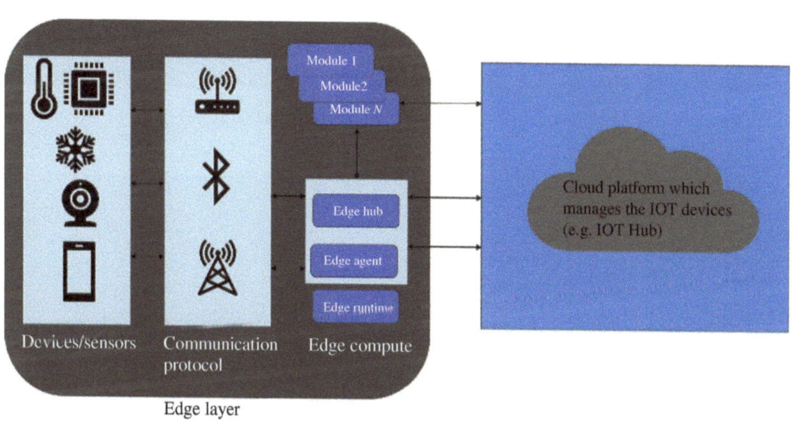

FIGURE 5.8 Edge layer and its components.

5.6.1 Methodology—experiments and results

As part of this cold chain research, we tested few of the sensors and board by connecting the Azure IOT cloud and IOT Edge and captured the results below.

Step 1: Sensors used: Tested with DHT11 sensor, RFID module, and connected with Raspberry PI3.
Step 2: Azure IOT Hub: Created IOT Edge device and added an IOT Edge Custom module as shown in Fig. 5.9.
Step 3: Configure and verify the IOT Edge at Raspberry PI (Fig. 5.10).
Step 4: Tested IOT Edge by updating another module (Fig. 5.11).

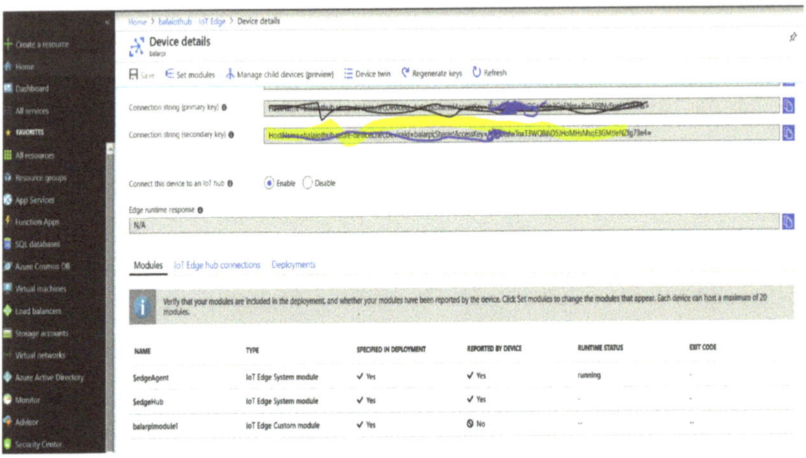

FIGURE 5.9 Azure IOT Hub. *IOT*, Internet of Things.

FIGURE 5.10 IOT Edge status as Raspberry PI. *IOT*, Internet of Things.

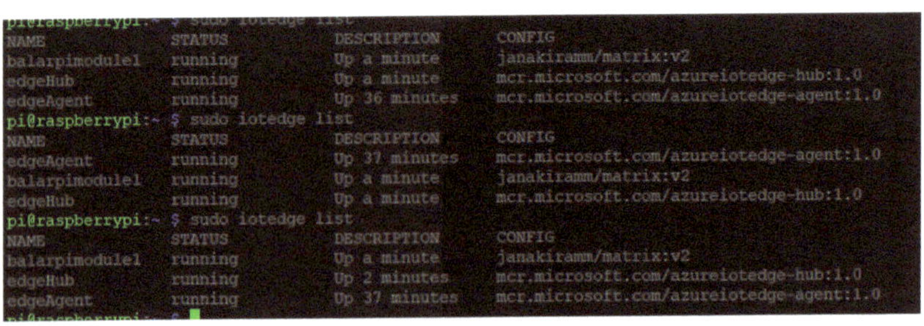

FIGURE 5.11 IOT Edge status with updated modules. *IOT*, Internet of Things.

Step 5: Verified the docker status using the docker command. This docker image can be pulled from Azure Container Services or Docker Hub (Fig. 5.12).

What a docker image contains: it will have an RFID data ingestion module that reads and validates the readings of the tags and this module runs using Node JS application; similarly, another docker image for a temperature sensor which has a machine learning module to current versus predicted the temperature of cold chain package. The machine learning module is packaged, and it will be developed using a custom code (e.g., Python or Java) and registered at a device module section of Azure IOT Edge (Fig. 5.13).

Step 6: Monitor the messages sent from an IOT Edge module in an IOT Hub. We can also configure the Event Hub to continuously receive the data from IOT Edge (Fig. 5.14).

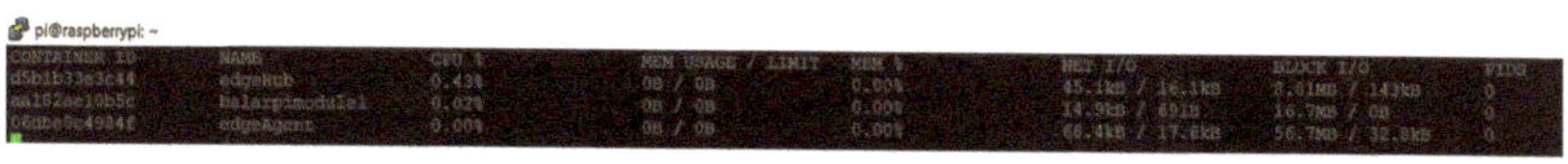

FIGURE 5.12 Docker container status running at IOT Edge. *IOT*, Internet of Things.

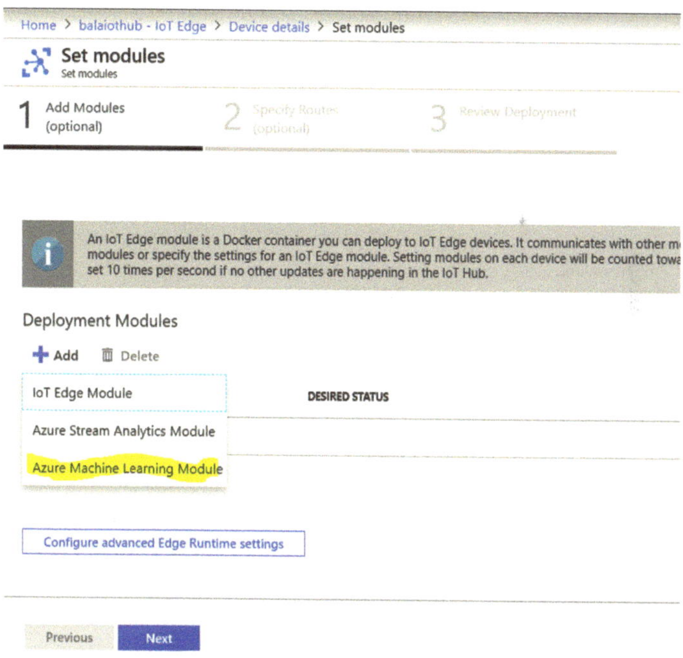

FIGURE 5.13 Azure IOT hub - Add deployment module. *IOT*, Internet of Things.

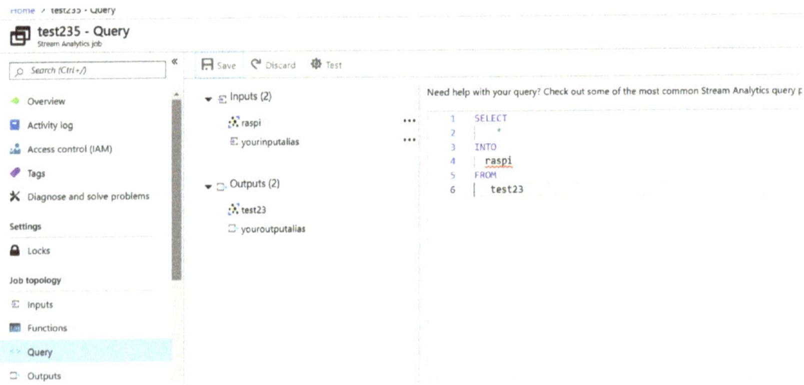

FIGURE 5.14 Device query - Azure IOT hub. *IOT*, Internet of Things.

Step 7: Once the data comes to the IOT Hub, the data will be sent to Power BI using Azure Stream Analytics, and it will be invoked through a job running in stream analytics service.

5.7 Role of containers in Internet of Things edge

The IOT Edge module is usually deployed through container at IOT Edge devices. It will communicate with another module (e.g., dockers).

Container

A container is a piece of software that packages up application (i.e., code and libraries) and dependencies, so that the application runs fast and reliably from one computing environment to other [21]. In the modern IT era, it is recommended to build a lightweight app or microservice to develop and test to facilitate seamless production deployment [22].

The container is lightweight and shares machine's operating system kernel and reduces the server and licensing cost.

IOT Edge does not have large resources or compute to handle a virtual machine, containers are best suited to run different programs or applications specific to analytics or insights at edge.

Following are the popular containers "Docker," "Rocket," "Warden," and "Garden."

Using containers also make the IOT device updates more efficient through "Device Twin" or "Device Shadow" model supported by cloud platform even if it becomes offline (Fig. 5.15) [23].

Container orchestration

It is all about managing the life cycles of container in a scalable environment. The following are the key tasks of orchestration [24]:

- provisioning and deployment;
- scale-up and removal of containers;

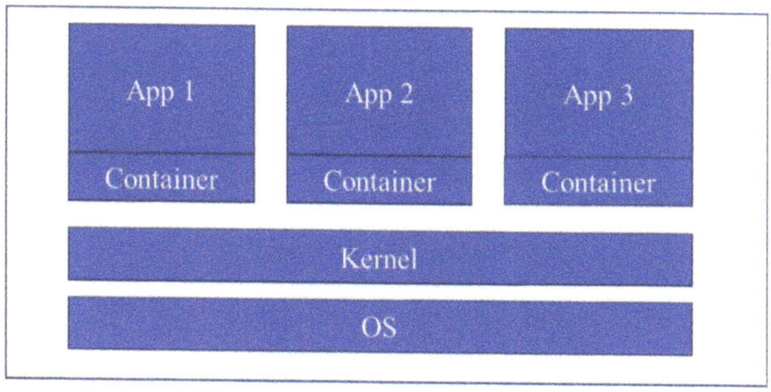

FIGURE 5.15 Container architecture.

- load balancing;
- health monitoring of containers; and
- movement of container from one machine to another machine when the machine failed or not responding.

Container orchestration tools:

The leading tools, such as Docker Swarm, Kubernetes, Mesos, or cloud-based tools (Amazon Elastic Container Service—EKS, Azure Kubernetes Services, Google Cloud Kubernetes engine), also help us to orchestrate the container lifecycle.

Prime function of container orchestration:

The container orchestration begins with an application configuration definition step that clearly describes the configuration (Fig. 5.16) [25]:

```
version: "3"
services:
vote:
build:. /vote
command: python app.py
volumes:
./vote:/app
ports:
"5000:80"
networks:
front-tier
back-tier
```

In the above example, it clearly states the build command and what application to be invoked and the corresponding application ports to be used to run the application. The storage location (docker volume) is also defined here with the network.

The below diagram depicts the configuration definition of different docker services needed to be orchestrated by the orchestration tool; we have taken the Kubernetes reference from the docker samples link (Fig. 5.17).

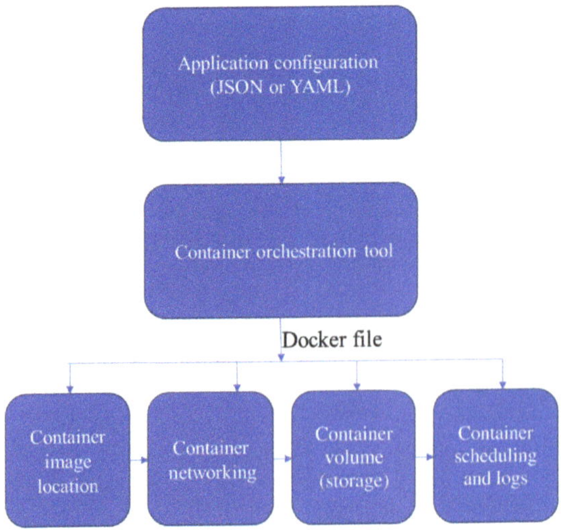

FIGURE 5.16 Container orchestration.

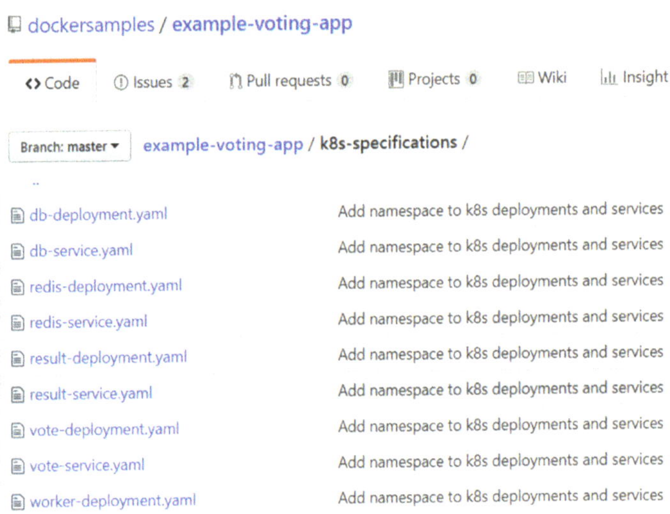

FIGURE 5.17 Git hub—Docker Samples.

The Kubernetes will manage and deploy these dockers using its framework and components.

```
kubectl create <app>
```

It will create the namespace (pod) defined in <app>, and if we give the entire k8 specification folder to create and deploy the services to Kubernetes, the following command needs to be invoked:

```
kubectl create -f k8s-specifications/
```

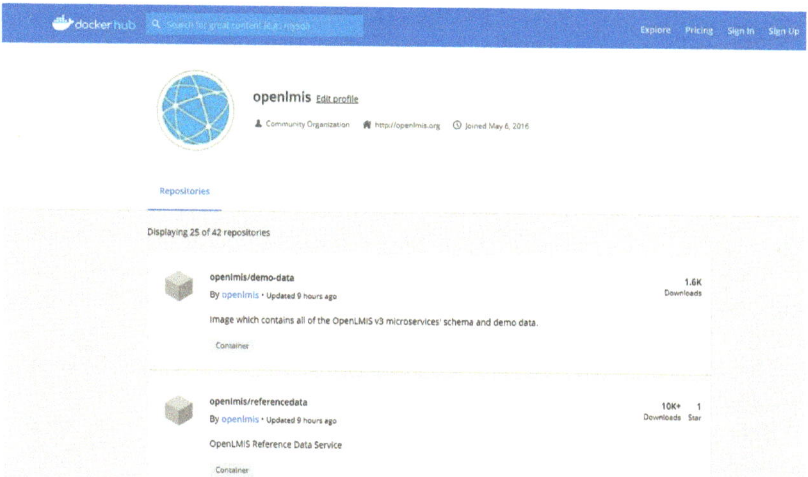

FIGURE 5.18 Docker Hub—OpenLMIS. *OpenLMIS*, Open logistics management information system.

Here Kubernetes takes care of managing all the docker services given, and it will alert and manage the volume as per our specification given in the docker file.

We can interface with Kubernetes with different *kubctl commands*

Kubectl delete—delete a pod (container) based on a name or type.

Kubectl scale—scale-up and down the deployment using this command

Cold chain example:

OpenLMIS (Open Logistics Management Information System) is software for a shared, open-source solution for managing medical commodity distribution in low- and middle-income countries [26].

Cold chain remote temperature monitoring: As of OpenLMIS 3.3, there is support for integration with the Nexleaf ColdTrace device to monitor fridge temperatures and provide status info inside OpenLMIS (*Fig. 5.18*) *[26].*

OpenLMIS uses Docker, and it can be hosted either on premise or in the cloud within an implementer's choice data center [27].

The microservice architecture of OpenLIMS allows one to run different groups of services and deploy and test it seamlessly using continuous integration and continuous deployment tools such as JIRA, Jenkins, Sonar, Docker.

5.8 Pharma—cold chain analytics

The intent of cold chain for pharma products is to improve the operational efficiency of supply chain and reduce inventory and logistics cost. Using the sensor and devices data, the IOT Edge ease the analytics. The cases classified into three key areas of cold chain: product demand forecasting, track and trace of the product, and conditional monitoring and predictive maintenance of containers, described in the following sections.

5.8.1 Product demand forecasting

The pharma manufacturing plant keep monitors the inventory of all the pharma products that are manufactured. The orders will get booked through different agents or distributors. The plant should plan the production process, raw materials, and deciding the price of the product. The role of IOT will help one to predict the real-time demand for different pharma products or product lines.

The data collected from different product tags after quality assurance will be considered are the final product or finished products.

Here we used sample data of a manufacturing plant which has the following fields (Fig. 5.19):

Field name	Description
Year part	Year with first half and second half indicator
Plant	Manufacturing plant name
Financial year	Financial year begins from Jan 1 and ends on Dec 31
Week	Week end date
Date	Transaction date
Day of week	Day of week (e.g., Monday, Tuesday)
Product group ID	Product group identifier
Product name	Name of the product
Quantity	Number of quantity in stock
Sold	Product price

To understand descriptive statistics of the sample data, see Fig. 5.20.

The following table illustrates the relationship between quantity and sold variable. We run the regression test and we could see relationship exists between these two variables.

Yea	Plan	Financial Ye	Week	Date	Dayofwe	Product	Product Name	Quantity	Sold
20172	Plant1	2017	9/30/2017	9/24/2017	Sunday	26	Albuterol	34.00	530.00
20172	Plant1	2017	9/30/2017	9/25/2017	Monday	18	Acetohexamide	1.00	17.00
20172	Plant1	2017	9/30/2017	9/25/2017	Monday	26	Albuterol	13.00	210.00
20172	Plant1	2017	9/30/2017	9/26/2017	Tuesday	26	Albuterol	16.00	319.00
20172	Plant1	2017	9/30/2017	9/27/2017	Wednesday	26	Albuterol	8.00	158.00
20172	Plant1	2017	9/30/2017	9/28/2017	Thursday	26	Albuterol	27.00	366.00
20172	Plant1	2017	9/30/2017	9/29/2017	Friday	18	Acetohexamide	2.00	35.00
20172	Plant1	2017	9/30/2017	9/29/2017	Friday	26	Albuterol	42.00	716.00
20172	Plant1	2017	9/30/2017	9/30/2017	Saturday	18	Acetohexamide	1.00	10.00
20172	Plant1	2017	9/30/2017	9/30/2017	Saturday	26	Albuterol	44.00	749.00
20162	Plant1	2016	9/24/2016	9/18/2016	Sunday	25	Metformin	1.00	31.00
20162	Plant1	2016	9/24/2016	9/18/2016	Sunday	26	Albuterol	29.00	497.00
20162	Plant1	2016	9/24/2016	9/19/2016	Monday	26	Albuterol	21.00	375.00
20162	Plant1	2016	9/24/2016	9/20/2016	Tuesday	26	Albuterol	20.00	288.00
20162	Plant1	2016	9/24/2016	9/21/2016	Wednesday	26	Albuterol	7.00	140.00

FIGURE 5.19 Sales data of pharma manufacturing plant.

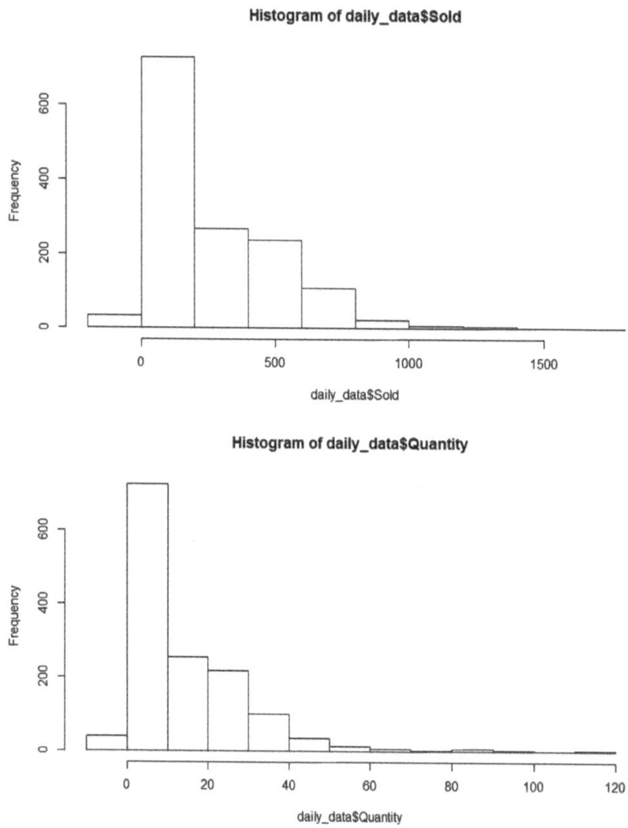

FIGURE 5.20 Histogram represents the distribution of quantity and sold variables from the sample data.

The R-squared value (0.9) clearly indicates that it fits for perfect model.

SUMMARY OUTPUT

Regression Statistics	
Multiple R	0.959982624
R Square	0.921566638
Adjusted R Square	0.921510893
Standard Error	4.292472518
Observations	1409

ANOVA

	df	*SS*	*MS*	*F*	*Significance F*
Regression	1	304603.6197	304603.6197	16531.79508	0
Residual	1407	25924.42569	18.42532032		
Total	1408	330528.0454			

	Coefficients	*Standard Error*	*t Stat*	*P-value*	*Lower 95%*	*Upper 95%*	*Lower 95.0%*	*Upper 95.0%*
Intercept	-1.004272126	0.160268884	-6.266170347	4.90861E-10	-1.318663815	-0.689880436	-1.318663815	-0.689880436
530	0.058946308	0.000458455	128.5760284	0	0.05804698	0.059845637	0.05804698	0.059845637

There are simple forecasting methods [28].

Mean method: Forecast of all future values is equal to mean of historical data Mean: meanf (x, $h = 10$).

Naive method: Forecasts equal to last observed value optimal for efficient stock markets naïve (x, $h = 10$) or rwf (x, $h = 10$); rwf stands for random walk function.

Seasonal naive method: Forecast equal to last historical value in the same season snaive (x, $h = 10$).

Drift method: Forecasts equal to last value plus average change equivalent to extrapolating the line between the first and last observations rwf (x, drift $= T$, $h = 10$).

As it clearly shows that it is a time series data with a daily transaction recorded with date. Popular Forecasting model such as ARIMA (autoregressive integrated moving average). It is easier to predict when the series is stationary; differencing is a method of transforming nonstationary time series to stationary.

A nonseasonal ARIMA model is classified as an "ARIMA (p,d,q)" model, where:

p is the number of autoregressive terms,
d is the number of nonseasonal differences needed for stationarity, and
q is the number of lagged forecast errors in the prediction equation. The forecasting equation is constructed as follows. First, let y denote the dth difference of Y, which means:

$$\text{If } d = 0\text{: } yt = Yt$$

$$\text{If } d = 1\text{: } yt = Yt - Yt - 1$$

$$\text{If } d = 2\text{: } yt = (Yt - Yt - 1) - (Yt - 1 - Yt - 2) = Yt - 2Yt - 1 + Yt - 2$$

The deseasonalize (nonseasonal ARIMA) model is shown in Fig. 5.21. It calculates the seasonal component of the series using smoothing and adjusts the original series by subtracting seasonality in two simple lines.

After evaluating model residuals and auto correlation function (ACF)/partial autocorrelation function (PACF) plots and adjusting the structure, we run the Dickey–Fuller test to on differenced rejects the null hypothesis and non-stationary.

Augmented Dickey–Fuller test
data: count_ma
Dickey–Fuller = −8.8445, Lag order = 11, P-value = .01

Alternative hypothesis: Stationary we tested and iterated the model with ARIMA (2,0,2) with nonzero mean. It will help us to predict the sold value.

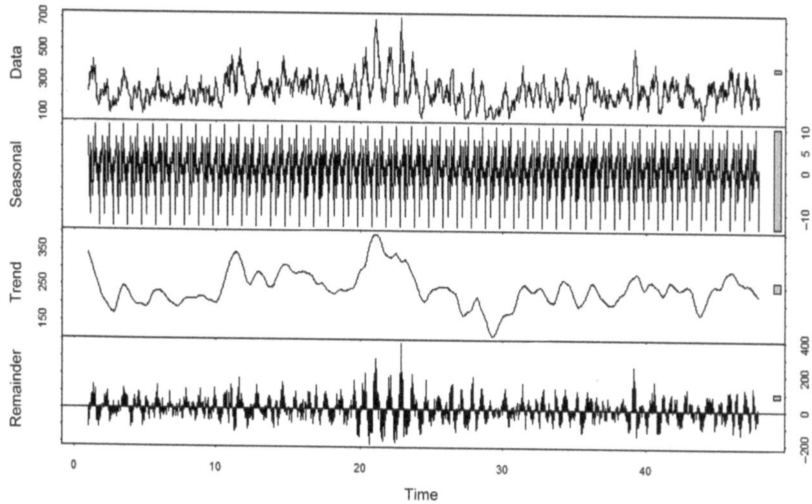

FIGURE 5.21 Deseasonalize and decomposed ARIMA model. *ARIMA*, Autoregressive integrated moving average.

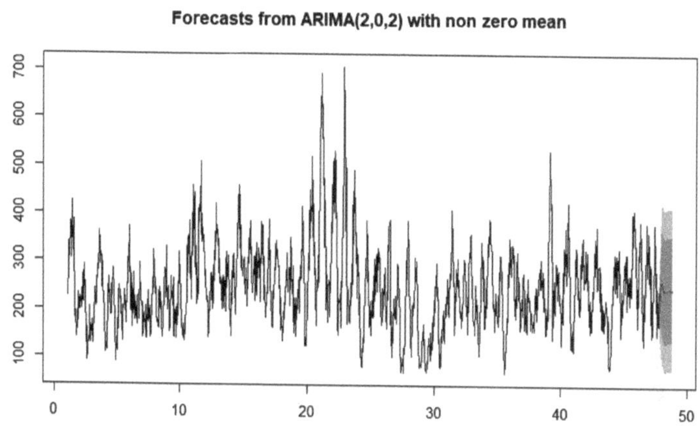

Conclusion: The demand forecasting model can be built using ARIMA with the PQD parameter value of 2,0,2, and it will help us to plan the production of those products.

5.8.2 Track and trace

Technology is playing a vital role now-a-days for transport and logistics. IOT Edge helps us to collect the transport location (geo location—latitude and longitude) and aggregates and reports to central location or customers on real time. The package status and current condition of the goods and location of the truck or vehicle are transparent to manufacturer, agents, and consumers.

The real-time geo location helps to plan and optimize the routes and efficiently help us to manage the fleets remotely.

5.8.3 Conditional monitoring and predictive maintenance of containers

The data comes from IOT sensors or devices from the product (goods) or truck or boxes or container. We usually follow the below techniques for condition monitoring:

1. vibration analysis,
2. acoustic emission, and
3. ultrasound.

Based on the above analysis and output, predictive maintenance helps us to measure the asset health and plan the maintenance in advance.

5.9 Deployment considerations and issues

5.9.1 Deployment considerations

The IOT Edge can be deployed with different technologies and different operating systems; we need to identify the device prerequisites and device management plan in advance.

1. *Production certificates*: Install and keep the necessary certification of authority files and configure it using yaml (yet another markup language) file at edge.
2. *Device management plan*: it should be essential to keep the following information ready:
   a. device firmware,
   b. operating system libraries,
   c. edge daemons, and
   d. security certificates.
3. *Memory consumption of Edge Hub*: We need to monitory the memory usage and reduce it by using streamlined capacity and low disk space.
4. *Container management*: It should allow to access the container registry and manage the versions using tagging.
5. *Bandwidth*: The edge to cloud should be communicated with a good bandwidth to reduce the data loss as well as the role of edge should do bulk transfer of data when edge becomes offline.
6. *Network connection*: Always review the network connection, protocol, and firewall rules and whitelist the application that needs to access the edge data.

7. *DevOps*: It will play a vital role on continuous deployment from cloud to edge and deploy the modules through a DevOps pipeline. It will also track and monitors the health of the container and help us to test the code (module) before deployment.

5.9.2 Known issues

1. Cyber security is a major challenge (e.g., machine phishing, hackers, or intruders) for most of the IOT developments.
2. IOT Edge security failure: The host name should not be greater than 64 characters.
3. Data might be lost when the sensor is not working, and we may not able to take any decision until the device becomes active.
4. Cold chain parameters are keeping changing based on the new arrivals, and device configuration is always an issue, and adequate metadata is to be stored and kept before we test in the edge layer.
5. Malfunctioning of sensor values: Sometimes the sensor readings will not come due to the low voltage or not enough power to the sensor.
6. Implementation will vary from cloud to cloud. The Amazon AWS IOT Edge follows different process to configure the edge where Google Cloud IOT Edge differs than Azure.

5.10 Conclusion

Cold chain is not new to us; however, the integration of cold chain with IOT Edge will reap the value for the supply chain by eliminating the spoilage rate and proactive technical monitoring to supply the goods with agreed thresholds. Many processing methods at IOT Edge will help us to take the decision quickly and keep the resiliency plan accordingly. The above research and findings are done with limited scenarios. The edge technology platform, such as Azure or Google, will claim that they have niche features.

References

[1] Pharmaceutical commerce magazine issue 8th May 2018. <https://pharmaceuticalcommerce.com/clinical-operations/the-2018-market-for-pharma-cold-chain-logistics-is-15-billion/>.

[2] Global cold chain monitoring market research report—By Component (Hardware, Software, Services), By Application (Pharmaceuticals & Healthcare, Food & Beverages, Chemicals), By Logistics (Warehousing, Transportation)—Forecast Till 2023. <https://globenewswire.com/news-release/2018/08/23/1555657/0/en/Cold-Chain-Monitoring-Market-to-Exceed-USD-7-Billion-Revenue-by-2023-at-9-6-CAGR-Industry-Forecast-by-Components-Applications-and-Logistics.html>.

[3] What is cold chain. <https://en.wikipedia.org/wiki/Cold_chain>.

[4] Development of Cold Chain Supply in India <https://www.slideshare.net/farhanbook/pgpm912-om-presentation>.

[5] The Internet of Things: the new Rx for pharmaceuticals manufacturing & supply chains <https://www.cognizant.com/whitepapers/the-internet-of-things-the-new-rx-for-pharmaceuticals-manufacturing-and-supply-chains-codex2437.pdf>.

[6] Logistics. <https://en.wikipedia.org/wiki/Logistics>.

[7] Cold chain IOT <https://www.slideshare.net/drjayeshpatidar/cold-chain-tot>.

[8] W. Shi, J. Cao, Q. Zhang, Y. Li, L. Xu, Edge computing: vision and challenges, IEEE Internet of Things J 3 (5) (2016) 637–646.

[9] L. Tan, N. Wang, Future internet: the internet of Things, in: 2010 Third International Conference on Advanced Computer Theory, 1888898 and Engineering (ICACTE) Smart Logistics for Pharmaceutical Industry Based on Internet of Things (IoT); M. Pachayappan, N. Rajesh, G. Saravanan, Int. J. Comput. Sci. Inf. Secur. (IJCSIS), Vol. 14 CIC 2016 Special Issue. https://sites.google.com/site/ijcsis/ ISSN1947–550.

[10] D. Gami, D. Nimavat, S. Sharma, Edge Technologies in IoT and Application Scenario of RFID based IoT, International Journal Of Engineering Sciences & Research Technology (ISSN: 2277-9655).

[11] RFID overview. <https://www.inlogic.com/learn-about-rfid>.

[12] Swift sensor. <https://www.swiftsensors.com/product-category/sensors/>.

[13] IOT storage security locks from IOT in a box. <http://www.coldchainsensors.com/iot-products/>.

[14] Benefits of edge computing. <https://www.vxchnge.com/blog/the-5-best-benefits-of-edge-computing>.

[15] Y. Liu, B. Dong, B. Guo, J. Yang,, W. Peng, Combination of cloud computing and internet of things (IOT) in medical monitoring systems, Int. J. Hybrid Inf. Technol 8 (12) (2015) 367–376.

[16] RFID Journal: How much information RFID tag can store. <https://www.rfidjournal.com/faq/show?66 https://www.rfidjournal.com/blogs/experts/entry?10536>.

[17] B. Umar, H. Hejazi1, L. Lengyel, K. Farkas, Evaluation of IoT Device Management Tools, Budapest University of Technology and Economics, NETvisor Ltd, ACCSE 2018: The Third International Conference on Advances in Computation, Communications and Services.

[18] U. Barchetti, A. Bucciero, M. De Blasi, L. Mainetti, L. Patrono, Traceability in the pharmaceutical supply chain, in: IEEE International Conference on RFID-Technology and Applications, 2010, pp. 17–19.

[19] Azure IOT Edge. <https://docs.microsoft.com/en-us/azure/iot-edge/about-iot-edge>.

[20] Azure IOT Edge technology primer. <https://thenewstack.io/azure-iot-edge-a-technology-primer/>.

[21] What is a container. <https://www.docker.com/resources/what-container>.

[22] R. Sairam, S.S. Bhunia, V. Thangavelu, M. Gurusamy, NETRA: Enhancing IoT Security Using NFV-Based Edge Traffic Analysis.

[23] Azure IoT Edge, machine learning and containers. <https://thenewstack.io/azure-iot-edge-machine-learning-containers/>.

[24] Container orchestration. <https://blog.newrelic.com/engineering/container-orchestration-explained/>.

[25] Docker configuration example. <https://github.com/dockersamples/example-voting-app>.

[26] OpenLMIS – cold chain. <https://openlmis.atlassian.net/wiki/spaces/OP/pages/115681666/OpenLMIS + Technical + Setup + Guide?focusedCommentId = 321126419>.

[27] Open LIMS – Technical setup. <https://openlmis.atlassian.net/wiki/spaces/OP/pages/115681666/OpenLMIS + Technical + Setup + Guide>.

[28] Time series and forecasting using R. <http://manishbarnwal.com/blog/2017/05/03/time_series_and_forecasting_using_R/>.

Further reading

B.I. Ismail, E.M. Goortani, M.B. Ab Karim, W.M. Tat, S. Setapa, J.Y. Luke, O.H. Hoe, Evaluation of Docker as Edge Computing Platform.

K. Jakobs, C. Pils, M. Wallbaum, Using the internet in transport logistics – the example of a track & trace system, in: P. Lorenz (Ed.), Networking—ICN 2001. ICN 2001. Lecture Notes in Computer Science, vol. 2093, Springer, Berlin, Heidelberg, 2001.

S.Y. Yurish, M.T.S.R. Gomes, Smart Sensors and MEMS. <https://link.springer.com/content/pdf/bfm%3A978-1-4020-2929-5%2F1.pdf https://www.sciencedirect.com/science/article/pii/S2352864817300214#s0310>, 2003.

Chapter 6

The growing role of Internet of Things in healthcare wearables

R. Indrakumari[1], T. Poongodi[1], P. Suresh[2] and B. Balamurugan[1]

[1]*School of Computing Science and Engineering, Galgotias University, Greater Noida, India,*
[2]*School of Mechanical Engineering, Galgotias University, Greater Noida, India*

6.1 Introduction

Over the past few decades, the Internet has drastically grown, which allows the world to consume incomparable services with the help of hosts through smart phones over World Wide Web. Internet of Things (IoT) embeds the World Wide Web in every day's objects and enables them to send and receive data through sensors and smart devices. IoT-enabled device is a computing device that connects the things to a network through wireless or wired fashion. The term "things" in the IoT may be a machine or wearable devices with an IP address, which automatically collect and send data over a network without any assistance. Business Insider has announced that, by 2020, the businesses around the globe may invest nearly $70 billion to develop IoT. From this it is understood that the IoT will trigger the next industrial revolution and the demand for its solutions is set to increase. IoT embraced real solutions to applications such as aviation, insurance, manufacturing, traffic congestion, industrial sector, emergency services, security, smart cities, health care, logistics, retail sector, and waste management as shown in Fig. 6.1.

6.2 Impact of Internet of Things–based wearables in healthcare

The effectiveness of IoT has opened up a world of possibilities in health care by providing smart, cost–effective, and accurate personalized healthcare service [1]. The necessity for preventive medicine and self-health monitoring is increasing rapidly due to the projected drastic increase in the number of elderly people until 2020. Wearable devices are currently at the core of just about every conversation related to the IoT. Wearable devices are small wearables that can be embedding on, in, and under accessories, body, or

Emergence of Pharmaceutical Industry Growth with Industrial IoT Approach.
DOI: https://doi.org/10.1016/B978-0-12-819593-2.00006-6

FIGURE 6.1 Internet of Things solution in various fields.

clothes of the beneficiaries. The study for the development of wearable devices through sensory and computational devices is called wearable computing. Wearable devices that operate autonomously and act as central connectors for connecting (other) devices are considered as primary wearable devices (e.g., wrist-worn fitness tracker, smartphone), and those devices that capture specific actions and report to the primary wearable devices are considered as the secondary wearable devices (e.g., heart rate monitor worn around the chest) [1]. Due to the innovation of electronic miniature components, the capability to collect and store data, to perform complex permutations in real-world environment, leads the wearable devices quickly to the most sensitive healthcare domain. The wearable devices are IoT-based things that are worn on body of the user as an accessories, or it can be embedded in the cloth. These devices are connected to Internet using Wi-Fi or Bluetooth to exchange data. The operations performed by wearable devices are sensing, analyzing, storing, transmitting, and utilizing the data depending upon the application.

Architecture of IoT-enabled wearable health care is illustrated in Fig. 6.2. The foremost layer in the architecture is the sensing layer that observers the users mental, physical, and emotional condition with the help of sensors. Data processing layer retrieves the knowledge and pattern from the sensors. Security measures are applied to protect the data confidentiality. The application layer provides judgment and suggestions based on the knowledge obtained from other three layers.

1. Sensing layer: The sensing layer is often called "device layer," which accommodates physical objects and sensor devices such as Radio Frequency Identification (RFID), barcode, infrared, wireless sensors depends on the application. These devices spot the objects and collects valuable information in the form of orientation, vibration, location, chemical changes, acceleration, humidity, and temperature. The collected information is securely transmitted to the network layer.

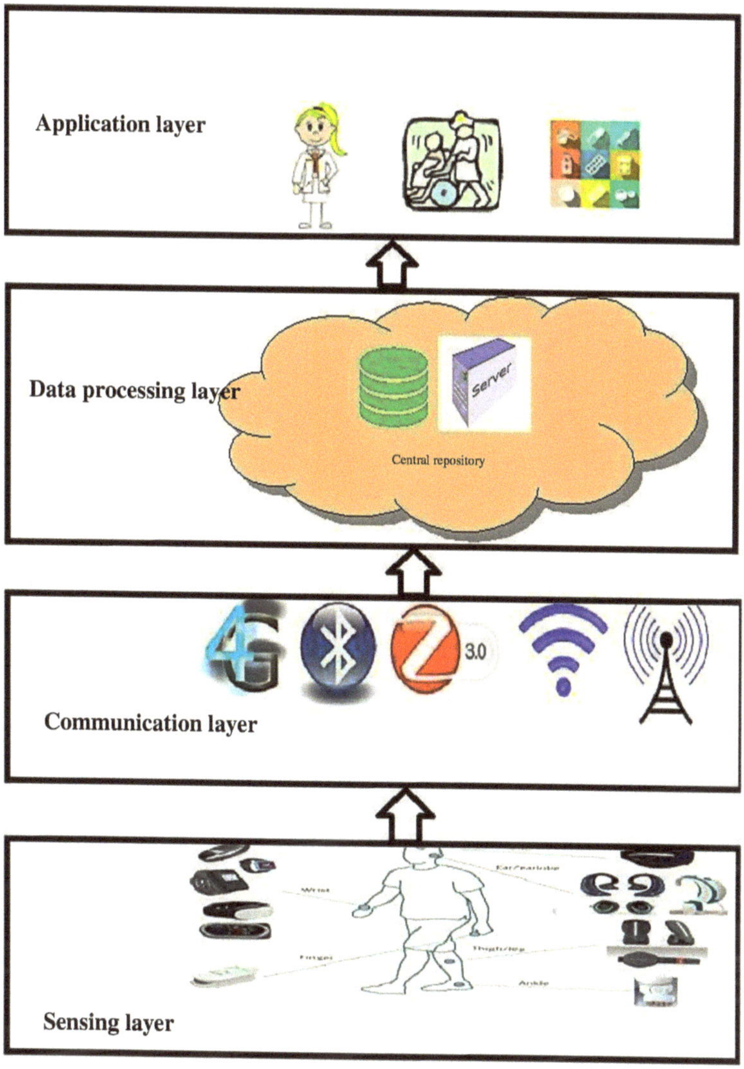

FIGURE 6.2 Architecture of IoT-enabled health care. *IoT*, Internet of Things.

2. Communication layer: The communication layer or transmission layer transmits and processes the sensor data collected from the sensor devices. The medium of transmission can be wireless or wired based upon the technologies used such as infrared, Bluetooth, Zigbee, Wi-Fi, and 4G [1].
3. Data processing layer: Data processing is otherwise called the "middleware layer" that analyzes and processes the information collected from the communication layer. The responsibilities of this layer include service management and establishing the connection with database. The

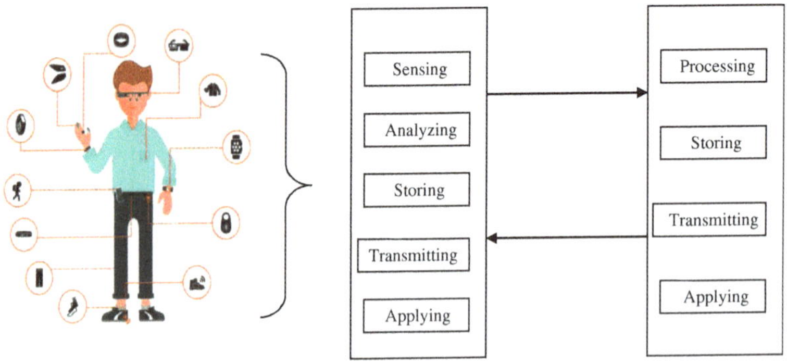

FIGURE 6.3 Situational awareness using wearable devices.

technological backgrounds for processing the huge volume of data are database, big data, and cloud computing.

4. Application layer: The users can interact with the application layer that provides application-oriented services to the users.

Fig. 6.3 shows representation of operations associated with gathering and processing data with the aid of wearable. Consider a scenario that if the wearable devices detected any poisonous gases, the sensed data is processed in the wearable, and it issues a warning. Meanwhile it may be transmitted to a remote location for testing to find accurate results to save life [2].

6.3 Taxonomy of wearables

Wearables are classified into active and passive based on the role of power supply required to operate the devices. Oximetry sensors fall under active wearables that require power to operate, whereas temperature probe is a passive wearable that does not rely on power. Based on the mode of signal transmission, the wearables can be seen as wired in which the signals are transmitted over a physical data bus or wireless that transmits the signals wirelessly to the monitoring unit. Based on the sensors, it is classified as invasive and noninvasive wearables. Invasive wearables can be further categorized as minimally invasive such as a pacemaker which needs a medical procedure to be place inside the body. Noninvasive wearables seldom require physical contact such as gas sensor to sense poisonous gases in the environment. Fig. 6.4 shows the diagrammatic representation of taxonomy of wearables.

6.4 Wearable sensors for physiological parameters measurement

Doctors in rural areas are mostly nonspecialist physicians, and hence, it is necessary for a critical patient to travel long in order to get specialized

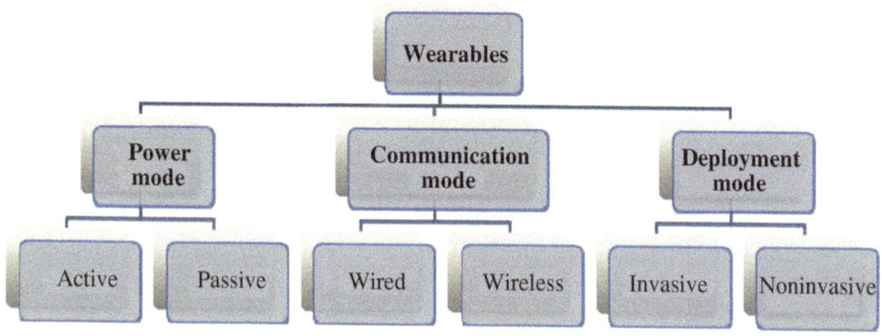

FIGURE 6.4 Taxonomy of wearables.

medical services. It is studied that most of the patients died on the way with serious illness such as lungs or heart diseases before reaching the specialist [3]. Wearable devices with remote parameter tracking can fill this gap with the help of integrated transmitter. Fig. 6.5 shows the block diagram of the parameter monitoring system for human through wearable sensors. Nowadays, many flexible user-friendly wearable sensors are available, which can perform a range of physiological and physical parameter measurements. These parameters are broadly classified into (1) physical sensors and (2) chemical sensors. These technologies can be utilized for medical prosthetics, consumer electronics, soft robotics, artificial skin, drug delivery, therapy, and health parameter monitoring. Fig. 6.5 shows the wearable sensors for biological parameters.

An interfacing unit is used to connect the wearer and the outside world. The interface is studied as input interface and output interface based on the operation. In early stage the input interfaces are keyboards or buttons that are less prone to error, but later as the complexity of wearable devices increases, writing pad and voice recognition systems are in use. In contrast to input interface, the output interfaces provides information to the wearers from the outside world. Some of the output interfaces are audio interfaces, vibrations, voice synthesis, and visual interface.

6.4.1 Physical parameters

Physical parameters are stress, motion, temperature, vibration, heart rate, acceleration, cardiovascular or neurological diseases, hypertension, and chronic obstructive pulmonary diseases (COPD). The temperature parameter obtained from human skin is providing numerous useful information regarding stroke, shock lung disease, heart attack, and infections. Human motion provides various health parameters for diseases such as osteoarthritis, heart attack, and some autoimmune diseases by considering anatomical, social, environmental, psychological, and physiological effects [4]. For example, to identify the chronic lung disease, a 6 minute walk test is a vital observation

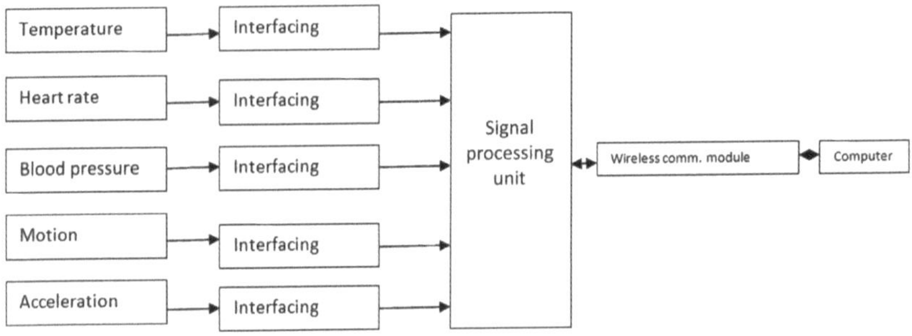

FIGURE 6.5 Wearable sensors for biological parameters.

methodology to read the lungs condition. Hence, movement or motion plays a major role in finding physical parameters. Wearable technology uses accelerometers, fabricated using piezoelectric, piezoresistive, and capacitive-type sensors to monitor detection of falls, body movement analysis, postural orientation, and motion [4]. Inertial sensors are used to find the movement of body, postural orientations, and falls. Another important sensor is the impedance sensors, which is sensitive, low power, and compact form of sensor used to capture the fluctuation of impedance of the thoracic region and heart for cardiac condition monitoring. A capacitive sensor is a thin, flexible, fabricated by conductive technique is employed to capture human activities, such as breathing rate, heart rate, gait, hand gesture recognition, and swallowing monitoring analysis [4]. These physical sensors are performing based on relative variation in their electrical parameters such as resistance, capacitance, piezoelectricity, and magnetic field. Based on the types of active sensing elements, the sensors are classified as liquid-state sensors or solid-state sensors. Liquid-state sensing uses liquid metals or ions as active elements, and solid-state sensors are fabricated using nanomaterials, such as semiconductors, polymers, carbon, carbon nanotubes, or bulk materials such as metallic nanoparticles or polymer nanofibers, metallic nanowires.

6.4.2 Biochemical parameters

Biochemical parameters include lactate, pH, electrolytes, fluoride content, glucose, the oxygen saturation of blood, the presence of ammonium, potassium, sodium, keratoconjunctivitis sicca, dinucleotide, β-nicotinamide adenine, transcutaneous oxygen of the eye, uric acid, chloride, etc. [5,6]. The chemical parameter monitoring are excreted human body fluids, such as saliva, sweat, urine, or stools. Sometimes it may be a blood sample, cerebrospinal fluid, breast milk, and bile. Cyst fluid is another form of body fluid formed due to pathological process.

6.5 Types of wearable sensors

The design and development of wearable devices mainly depend upon the sensors that collect the precise data for health monitoring system. With the advancement of technologies such as microelectronics, micromechanics have enabled the growth of many sensors to track human activities with low-power consumption. To measure the physiological parameters, the sensors are classified into two, namely, invasive and noninvasive sensors.

6.5.1 Invasive sensors

Invasive sensors require the body fluids to collect the relevant health data. Blood is a vital body fluid which can provide the essential parameters of different organs. Living cells are also needed to collect to get the status of living organs, for example, the bronchoscopy needs lung sample to identify the disease. The invasive nature of this sensor panic the patients as it is to be inserted through natural cavities or pierced into human body to take samples. Extracorporeal sensor or ex vivo sensor is an example for invasive sensor which incessantly monitors the pH and blood gases during cardiopulmonary bypass. Invasive sensors are not suitable for continuous parameter monitoring system, such as fitness-level monitoring of athletes, glucose monitoring of diabetic patients, cholesterol monitoring of heart patients, oxygen saturation monitoring for lung patients [7]. These hurdles pave the way to another type of sensors, called the invasive sensor.

6.5.2 Noninvasive wearable sensors

Noninvasive wearable sensors do not need the body fluid, and hence, it is not necessary to penetrate the body using incision or injection; hence, it is painless and more attractive. The body fluids used in this sensor may be sweat, skin interstitial fluids, saliva, and tears [8] (Tables 6.1–6.3).

6.6 Working principles of wearable sensors

The operation of the sensors depends upon different techniques, such as electrical, optical, electrochemical, and piezoelectric effect. Impedance sensors and electrochemical sensors are the important classes of wearable sensors for monitoring physiological parameter measurement. Electrochemical sensors are further classified into amperometric, potentiometric, and conductive sensors which use capacitive and resistive methodology to fabricate different sensors. Capacitors are the building blocks of the electronic world. The ability of a capacitor to store an electrical charge is called capacitance. Touch is the vital human sensory channel, and the technology which is used to respond to the physical touch is often called capacitive sensing. A capacitive

TABLE 6.1 Invasive and noninvasive sensors.

Invasive/Implantable sensor	Noninvasive/wearable sensor
Pulse oximeter	Electronic pill for drug delivery
Glucose sensor	Retina implants
Temperature	Deep brain simulator
Electromyography	Pacemaker
Electroencephalogram	Wireless capsule endoscope
Blood pressure	Implantable defibrillators
pH value	Cochlear implants

TABLE 6.2 Smart band in healthcare.

Sensor components	Measures	Applications
Altimeter	Step count	GPS tracking—track activities including calories burned, steps, streaks, distance, floor climbed, intensity, milestones, and active minutes
Ambient light sensor	Distance	
3-Axis accelerometers and gyroscope	Sleep quality and duration	Sleep monitoring—tracks sleep quality automatically and fix a silent alarm
Digital compass	Calories burned	Heart rate recording—wrist-based heart rate
GPS		Food log—tracks food intake level every day
Vibration motor		Track weight—fix a goal and track weight

sensor provides low temperature dependence, high sensitivity, low-power consumption, with the capacity of sensing various chemical and physical parameters. Different types of capacitive sensors are coaxial cylindrical, parallel-plate, fringing field, and cylindrical cross-capacitor [9]. Capacitive sensors are suitable for both invasive and noninvasive parameter measurement. The fringing field of the capacitor has the ability to sense the texture, location, and strength of the samples [10]. Electrochemical sensors are highly portable, sensitive, and low cost, useful in many hand-held analyzers

TABLE 6.3 Wearable devices in pharmaceutical applications.

Accessories	Description	Available prototypes
Smart band	Wrist-worn devices have fitness tracking capabilities and other functionalities, without a touchscreen display	Wrist-worn smoking gesture detector, wrist-worn bioimpedance sensor, ultrasonic speaker–embedded wrist piece and neck piece
Wrist watch	Wrist-worn devices with a touchscreen display	Finger-writing with smartwatch, smartwatch life saver
Smart jewelry	Smart jewelry designed with characteristics such as health-monitoring	Gesture detection ring, TypingRing
Strap	Chest straps, arm bands, belts, or knee straps embedded with sensors for health tracking	BodyBeat, pneumatic armband
Smart foot-wear	Socks, shoes, gloves, or insoles, equipped with sensors	Gait analysis foot worns, foot-worn inertial sensors, LookUp
Smart garment	Clothing items such as pants, shirts, and undergarments serve as wearables	Dopplesleep, Myovibe
Smart eyewear	Contact lenses or spectacles with sensing used as wearables	Chroma, iShadow Mobile Gaze Tracker, indoor landmark identification, Google Glass, Google Contact Lens, object modeling eye-wear, supporting wearables
Ear bud and headset	Bluetooth-enabled ear plugs or headsets. Sensor-equipped hats and neck-worn devices are also identified	
Sensor patch	Sensor patches that could be adhered to the body skin for fitness tracking	Smart Tooth Patch, DuoSkin, tattoo-based iontophoretic-biosensing system
E-skin/E-tattoo	Tattoos with stretchable and flexible electronic circuit realize wireless data transmission and sensing	

that are based on electrolytes and metabolites. The working of the sensors starts with gathering input from targeted devices, and the inputs are classified into three different categories, the target input which is the actual input measured by the sensor, interfering input refers to the sensitive input, and the modifying input which cause a change in the input–output relation of the sensor to the target and the interfering inputs. Based upon the characteristics, the wearable sensors are classified into static, for example, the body temperature and dynamic sensors. The static sensors hold the characteristics of accuracy, sensitivity, threshold, resolution, tolerance, span, linearity, short-term and long-term drift, hysteresis, response time, interchangeability, cross-sensitivity, recovery time, and yield ratio [11]. Dynamic characteristics handle the performance characteristics of the sensor inputs such as ramp, step sinusoidal, and ramp. The output response for the step input is transient, which reaches a steady state and then return to the initial value during recovery. The ramp input signal produces linear output response. The *n*th order polynomial mathematical equation relates the electrical output of a sensor with the input parameter. The electrical output may be current, voltage, and phase; the order of the equation may change according to the complexity of the sensor.

6.7 Challenges in the fabrication of wearable sensors

Security and privacy is the major challenge needed to be addressed in the deployment of wearables as the part of health care. The healthcare devices are relatively smaller in size, which store huge amount of personal information that can be hacked or stolen easily. Tracking the devices activities is one of the solutions to reduce security and privacy related issues. Personal calibration of wearable devices is another issue that falls under technical side. Every individual is different, and the cause of disease varies for each person depending upon the genetics, family history, and diet. To avoid this, machine learning-based data analysis may be used for accurate monitoring of health details of individual patients using wearable devices. The footprint of big data in health care is extremely powerful for huge amount of data, but it may mislead for individual user as data generated by the wearable device is very little, which may lead to catastrophic events and outliers.

Some of the materials used in electrochemical sensors for impedance measurement and biological conductivity are silver, chromium, copper, aluminum, nichrome, stainless steel, platinum, and gold [12]. These materials are good conductors that can react with the analyte and makes alteration in the resistance of the wire. The oxidation property of these materials also changes the resistance of the wire. One of the solutions for this issue is to use platinum electrodes that have lower impedance and do not oxidize easily [13]. Polarization of electrode is another important error for this type of sensor as charge or faradic transfer process occurs at the electrode surface. When the electrodes are immersed in a conducting liquid, it has a chance to get excited by direct current, which can neutralize the ions at the electrode surface area, leading to depositions that reduce the contact area of the electrode. The solution for this problem is to apply alternating current, which minimizes the polarization process, because the alternating electric field at the interface keeps on changing.

Tetrapolar or four-electrode measurement is the better choice in the impedance measurements. In this method the excitation current passes through the outer two electrodes, while the drop in voltage is measured between the inner two electrodes. In contrast to this, the two-electrode configuration, current passes through, and the voltage drop is measured across the same two electrodes [14]. The biofluids used to examine the physiological parameters are sometimes toxic or chemically corrosive. The acid attack changes the sensor irreversibly when it undergoes chemical reaction. The porous catalytic electrode causes modifications in pore morphology, changes the calibration of the sensor. The analytic selective film that is used to improve the selectivity and the sensitivity of a sensor can be poisoned by some nonremovable species causing drift in the output.

6.8 Small wearable antennas for healthcare system

Small antennas play a vital role in the fabrication of wearable wireless communications systems. Printed antennas are mostly employed in wearable communication systems as they are light weight, low profile, and have low production cost [1]. Fractal technology and metamaterial are used to fabricate small antennas with high efficiency [2]. The bandwidth of the antenna with metallic strips and *split-ring resonators* is around 50%. In human body the resonant frequency of the antenna with *split-ring resonators* is shifted by 3% (Figs. 6.6–6.15).

The wearable antennas are placed on the human body, which is connected to the medical system. The signals received by the antenna are transferred to the receiver, and the signal with utmost power is considered by the medical system.

(A) (B) (C)

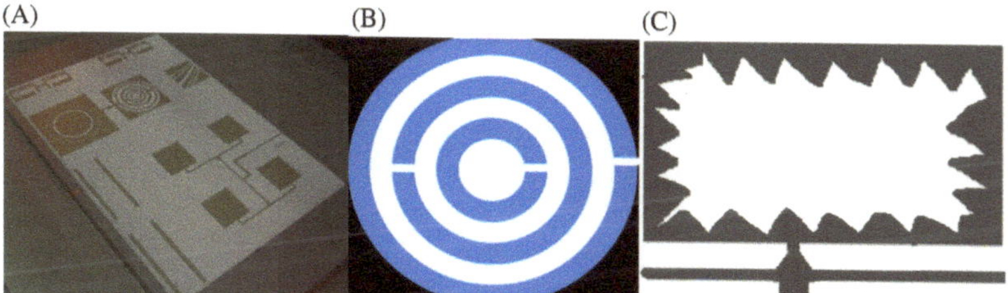

FIGURE 6.6 (A) Feed line printed antenna on paper, (B) split-ring resonator, and (C) fractal stacked patch antenna.

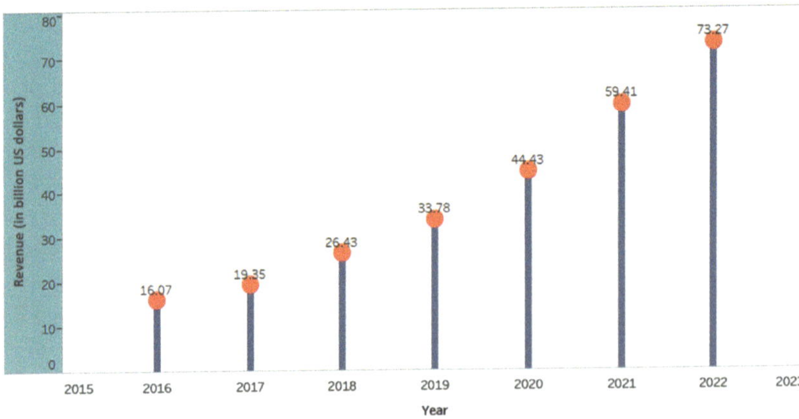

FIGURE 6.7 Wearable device revenue: world market 2016–22.

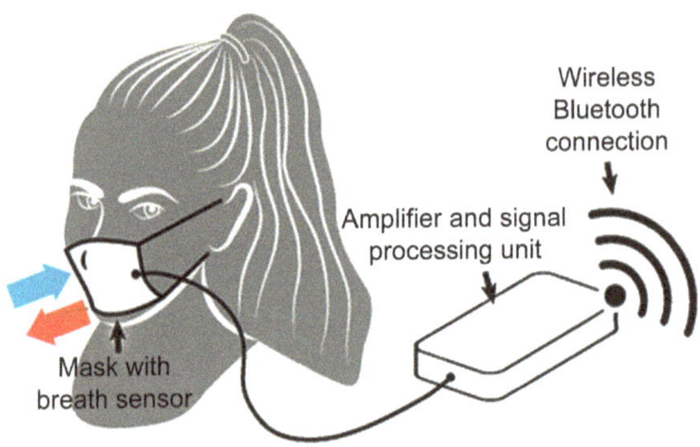

FIGURE 6.8 Electrical sensor to monitor respiration rate.

FIGURE 6.9 Body sensor temperature.

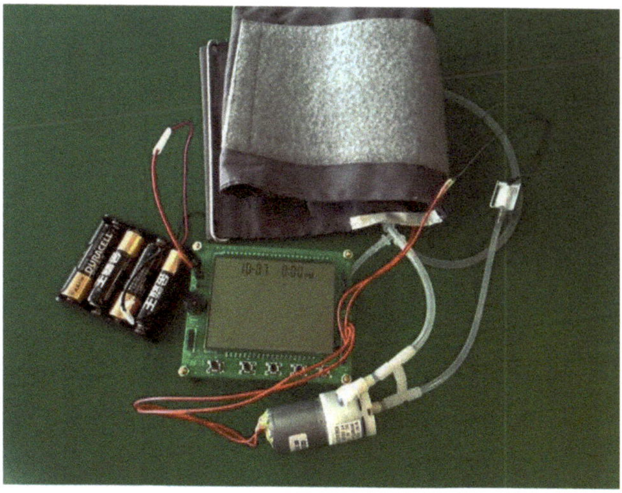

FIGURE 6.10 Blood pressure monitor.

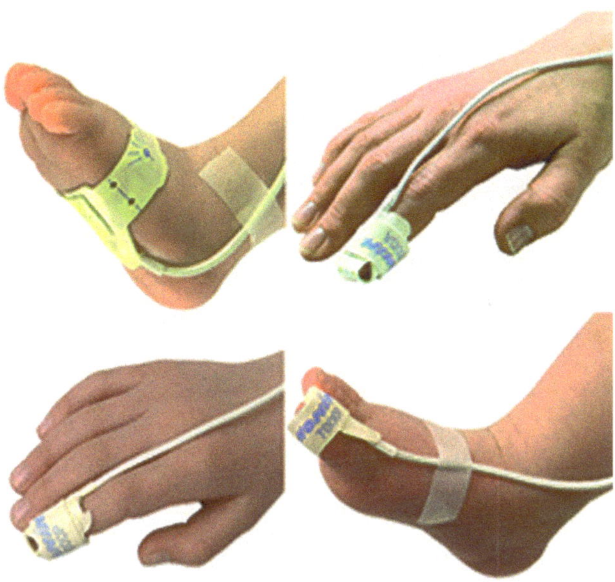

FIGURE 6.11 Pulse oximeter sensor.

6.9 Functions of wearable sensors

Nonintrusive, noninvasive sensors are the crucial components of long-term and ambulatory health monitoring systems [15]. Wearable sensors are

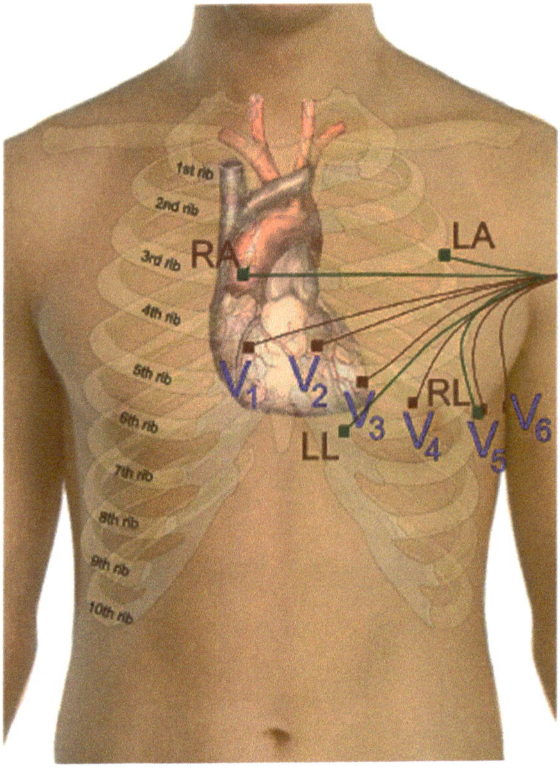

FIGURE 6.12 Wearable electrocardiogram sensor.

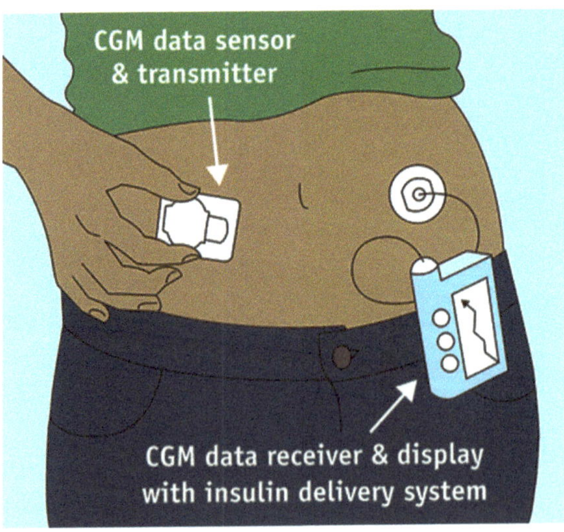

FIGURE 6.13 Blood glucose monitoring.

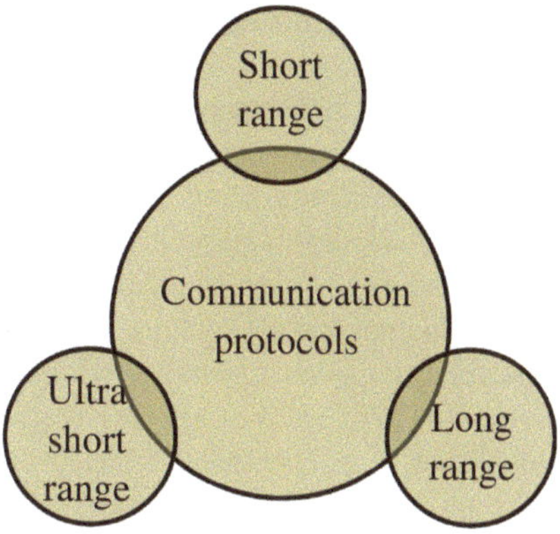

FIGURE 6.14 Wireless communication protocols.

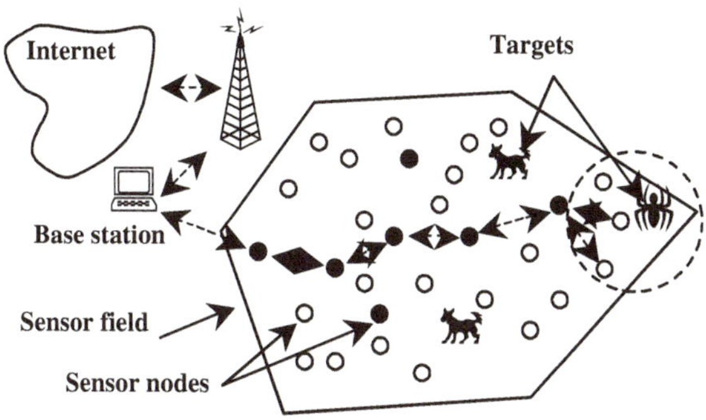

FIGURE 6.15 Communication architecture of wireless sensor network.

considered as a less obtrusive and more comfortable are suitable for monitoring patient's health without disrupting their daily activities. The sensors can be placed on different parts of the body to measure the physiological parameters. Aging in place, an application using wearable devices for aging people is being promoted by several countries which allows the individuals and senior adults with chronic conditions to stay at home, while they are remotely monitored for clinical interventions. Accelerometers are used to identify the performance of activities of daily living by senior adults in their home environment [10]. Long-term monitoring of physiological data such as blood pressure, respiratory rate, oxygen saturation, galvanic skin, body temperature, and heart rate shows the development in the analysis and treatment of various diseases. Many clinical studies have been carried out to assess and validate the smartness of the wearable sensors in monitoring physiological data over long periods of time [11]. Electrocardiograms are a noninvasive sensor application is a diagnostic tool to identify cardiac problems by measuring and recording the fluctuations of cardiac potential. Textile electrode made from silver-based conductive yarn with SpO_2 sensor and a three-axis accelerometer for fall detection is embedded in belts, and T-shirt is used to monitor heart rate, echocardiogram (ECG) and R–R interval [12–14]. Various noninvasive-based body temperatures monitoring systems are under research, and Buller et al. [15] proposed the human core body temperature determining system from the heart rate using Kalman filter. Bertolotti et al. [16] proposed a weightless, wireless wearable device for monitoring the steadiness of the body by reading the limb movements for long duration with the help of a gyroscope, magnetometer, and an accelerometer. Through a body sensor network, several units can be connected in a body for gathering more detailed measurements [17]. Yoon et al. [18] proposed a piezoelectric pressure sensor fabricated on a polyimide substrate for the estimation of heart rate by sensing the pulse wave in human artery. A piezoresistive pressure sensor constructed from a nonwoven acrylate-modified polytetrafluoroethylene sensor coated on an aluminum electrode on a polyethylene terephthalate film in a wristband is used for heart rate monitoring, having similar pattern as the ECG signal with more accuracy and less vulnerable to noise induced due to motion [19].

6.10 Wearable devices in pharmaceutical industry

Wearable devices are presently at the core regarding the discussion related to IoT. Wearable devices are the peripherals for the smart applications and rapidly growing toward a massive deployment of intelligence about everything in the environment. Wearable devices are performing different tasks related to sensing and security. For instance, wearable badges provide features such as identification and security particularly useful in the working environment. The advanced badges also have biometric capabilities (fingerprint activation), so that the badge owner can utilize it to unlock the door in the aspect of security. It can also be used for location sensing, in case of emergencies which ensures that everyone has evacuated the premises successfully. A wearable bracelet provides the reliable information about the location if it is placed in a jacket that is left on the chair.

Health and fitness wearable devices provide biometric measurements such as perspiration level, heart rate, oxygen levels in the blood flow. Nowadays, due to the technological advancement even the alcohol levels can also be tracked with the wearable device. Such devices are capable of sensing, storing, and monitoring measurements periodically and the results are analyzed efficiently. By tracking the body temperature, the device can provide the prior indication of either it is the symptom of a cold or the flu.

Smart wristband can track the perspiration level and that information can be helpful for adjusting the humidity level and the temperature.

Smartphone is acting as a centralized device for delivering such capabilities in the mentioned examples. Instead if IoT devices are communicating directly, there is no intervention of smartphone to monitor the transactions of wearable devices. Wearable devices are automatically interconnected with the devices in the surroundings. Preferred lighting adjustment can be done while watching television by sitting on a particular chair. The television can be switched on, and the lighting level can be adjusted according to the connected LED lights in the particular room. An intelligent smart home set up might support automatically to block the lighting from windows, which produces glare on the television. Perhaps the backlighting on the television screen can be adjusted to create a suitable environment to obtain the favorable experience. The interactions among devices can be done automatically once the platform is equipped well with the smartphone interface. The wearable devices such as watches, armbands etc. can be recharged easily and provide the required high-power, long-term functions. Due to the advancement in battery technology, it provides longer lifetime with small space, and it could be charged easily. The sensor-oriented wearable devices utilize the processing power periodically; however, the time consumption of wireless data transmission is minimized. Such devices should be more integrated with IoT in order to offer the wide range of features that are expected.

The following figure shows the wearable device revenue in the global market for the years 2016–22. Aruba, a network provider, conducted a research and forecasted that about 87% of the healthcare organizations will adopt wearable IoT services by 2019.

6.10.1 Wireless body area network

Wireless body area network (WBAN) is a significant component is based on IoT technology where the accurate sensors play a vital role for the successful healthcare system. Noninvasive and nonobtrusive sensors are considered for tracking the important signs of respiratory rate, pulse rate, body temperature, blood oxygen, blood pressure, etc.

Pulse sensors: The pulse rate can be used to detect various conditions such as pulmonary embolisms, cardiac arrest, vasovagal syncope. Pulse sensors are widely utilized for fitness tracking and medical purposes. Pulse can be tracked from the wrist, chest, fingertip, earlobe, etc. Fingertip and earlobe readings are more accurate; however, these are not highly wearable. Chest-worn wearables are widely used, but wrist sensors are most comfortable to use for a long-term [20]. Many fitness tracking wrist watches and chest straps are available, which provide pulse measurement functionality.

HRM-Tri by Garmin [21], Fit Bit Pure Pulse [22], H7 by Polar [23], and Tom Spark Cardio [24] are some devices that cannot be directly fixed into health monitoring system. Several such sensor types are developed and analyzed, for example, radio frequency (RF) sensors, photoplethysmographic (PPG) sensors, ultrasonic sensors, and pressure sensors. In PPG sensors, LED transmits light in the artery, and the photodiode receives the amount of blood that is not absorbed. The pulse rate can be determined by tracking the variation in the amount of light.

PPG sensors [20] are used to measure blood oxygen, pulse, and pulse rate variability with a tiny smart wristwatch. An accelerometer in PPG sensors checks the movement and the accuracy of pulse reading, which are affected due to the motion. If the motion is high, the device automatically switches into power mode, and the pulse rate is not recorded. If a person is suffering from cardiac issues at the time of exercise, the device is not suitable as the motion is too high. The accuracy of pulse rate should be improved even if the movement level is high.

In [25], two LED light intensities are used, which reduces the impact of motion in PPG sensors, and the received amount of light is compared with the photodiode. Moreover, there is a significant increase in signal quality as the motion is greatly minimized in this technique. Pressure sensors are used to track the healthcare manually by pressing the finger to read the radial pulse. The sensor is attached firmly around the wrist, and pressure is tracked continuously to capture the pulse waveform.

Ref. [26] presents the promising result using highly sensitive and flexible pressure sensor that is developed and tested used for pulse detection. Furthermore, the sensitivity is increased for better pulse detection, which automatically increases the noise level, which can be detected with the wearable sensor. The particular sensor is being tested at normal conditions, whereas more research effort is required to determine the performance during motion.

The combinations of PPG and pressure sensor [27,28], the pulse sensor modules are created with one pressure sensor and nine PPG sensors. The pulse rate is tracked from multiple sources of the wrist, and the accurate pulse reading is provided, which assists in diagnosing diseases such as diabetes.

PPG, ultrasonic, and pressure sensors are compared to investigate the diagnostic process with the help of pulse sensing [29]. The considerable accuracy was obtained by using all three types of sensors; however, different sensor categories are required to diagnose the particular disease. The pressure sensor was identified as good to detect arteriosclerosis, whilst the ultrasonic was found to be superior for detecting diabetes.

Nonconventional pulse sensor is designed using a RF array module [30] to track the different locations of the wrist, and the pulse signal is obtained at a single point, which becomes noise due to the movement. The sensible pulse readings are achieved with this technique but not clear as in traditional sensors. It is obvious that PPG sensors are mainly used for sensing the pulse, and the techniques are to be developed by focusing in the reduction of the noise impact on the signal quality.

6.10.2 Respiratory rate sensors

It assists in tracking the respiratory rate or the count of breath a person takes per minute. The process of monitoring the respiration helps in identifying various conditions such as hyperventilation, lung cancer, apnea episodes, asthma, tuberculosis, and barrier in the airway. Due to the significance of monitoring the respiration, several types of respiratory rate sensors have emerged, and one is a nasal sensor based on a thermistor [31].

The number of breath taken is calculated by tracking the rise and fall of temperature with the sensor device. Accuracy is compromised due to temperature fluctuation occurred in the other sources, for instance, if a chef is wearing the sensor and working in a kitchen. It is not widely used as it is easily identifiable and of obstructive nature. ECG signals are also used to acquire the respiration rate, and it is known as ECG-derived respiration [32]. It determines the respiration patterns, and apnea events are detected. It considerably reads the respiration rate, and it is again restricted due to wearability. It causes irritation on the skin if it is continuously used, and the ECG contacts need to be replaced regularly.

To detect the respiration rate, a microphone can be used [33], and it assists in detecting the wheezing problem—it is considered as a common symptom in asthmatics. However, it is extremely susceptible to external noises, and therefore, it may not be used as a long-term wearable device.

Fiber-optic sensor is sensitive enough to monitor the vibrations caused due to respiration [34]. The sensitive material may be susceptible to noise caused from different sources of vibration such as walking.

A pressure sensor is discussed in [35], where two capacitive plates are kept in parallel with one at abdomen. At the time of inhalation and exhalation, the plates move apart and nearer, respectively, and the respiratory rate is computed. The study reports that 95% confidence is achieved in respiratory rate computation and fairly accurate. However, the pressure sensor is susceptible to noise may be caused by some external pressures such as walking in wind.

Stretch sensor is also used for measuring the respiratory rate [36–38], and the properties of it change according to the response of tensile force, like stretched at the time of inhalation. The sensor is designed using ferroelectric polymer transducer generates the charge if a tensile force is applied. The variations in the charge are used to calculate the respiratory rate. The breathing was quite accurate for 3.3 breaths per minute if the person is sitting, and the error percentage increases if the motion was introduced. Hence, there is a limitation in these sensors that the error occurs if different movements cause tensile force to be pertained to the stretch sensor.

The significant factor in selecting the sensor type for WBAN is wearability, and the stretch sensors are highly recommended for utilizing in future due to its wearability nature. Eventually, the effort is required to concentrate in developing novel algorithms and techniques rather than developing new sensors from the scratch to enhance the robustness against the movement using these sensors.

6.10.3 Body temperature sensor

The body temperature can be used to identify fever, heat stroke, hypothermia, etc., and it is a preferable diagnostic tool in healthcare system. Recently, thermistor-type sensors are used to measure the body temperature. Commonly, positive temperature coefficient sensor and negative temperature coefficient are used. Thermistors are highly preferable to measure the appropriate range of temperature by tracking the human body with acceptable errors. The accuracy is totally dependent on how closely the sensor is kept to the human body. Thin and flexible polymers are used for developing sensors that could be easily fixed directly to human body. In recent advancement the temperature can be measured using sensors embedded in clothes with relative accuracy.

6.10.4 Blood pressure monitoring sensor

Hypertension is a major risk factor for any cardiovascular disease such as heart attack, and it is happening commonly nowadays. Blood pressure sensor is incorporated with WBAN for health care; many patients could be saved from such kind of chronic illness. Designing noninvasive blood pressure sensor still remains a challenge in healthcare IoT. Lot of research effort have been attempted to attain the accurate estimation of blood pressure by calculating pulse transit time (PTT) that is the time taken between the pulse rate at the device such as radial artery or earlobe and the pulse rate at the heart. The same could be measured between the wrist and the ear [39], and it can also be calculated between the fingertip of a hand and the palm [40]. PTT is inversely proportional to systolic blood pressure, and it could also be determined using a PPG sensor on the wrist, ear, etc. and an ECG on the chest. The results of all recent works use PTT for calculating BP yet not suitable. PTT is also dependent on various factors such as blood density and arterial stiffness [41]. The measurement read between the wrist and the ear was revealed to be accurate [42], and PTT was also said to be considerably accurate between the fingertip and the palm [43,44].

Zhang et al. [39] reviews two wearable PPG sensors, including one on the wrist and the other on the earlobe to find out the pulse arrival time between the estimated value of blood pressure and location of sensor devices. The output shows the reasonable measurements for various positions such as sitting and standing. The result is not being compared with the traditional reading based on sphygmomanometer. Such comparison would assist in analyzing the accuracy of this sensor-based system. However, there is a huge demand of such kind of system to measure the blood pressure continuously and accurately. Blood pressure is a significant parameter in health care, and it plays a vital role to improve the quality of health care in WBAN system.

6.10.5 Pulse oximetry sensors

The sensor is used to measure the oxygen level in the blood, and it assists in diagnosing the low oxygen level in the tissues of human body (hypoxia). The blood oxygen level is determined with the help of PPG signals. In particular, two LEDs such as red and infrared are directed via the human body skin. The amount of light absorbed by the hemoglobin and not absorbed is measured with photodiodes. The difference among two is used to calculate the blood oxygen level [40]. LED lights can be transmitted through a finger to a photodiode on the opposite direction, or it can be directed on the same side of the finger as light is reflected to a photodiode. These are known as absorbance mode and reflective mode sensors. Traditional pulse oximeters were preferably worn as a finger clips that are connected to a medical monitor. To make the devices more portable, many efforts have been made.

Gubbiand et al. [41] devised low-power pulse oximeter in the motive of improving wearability, and two techniques are utilized in order to reduce the power consumption. The techniques are named minimum signal-to-noise ratio (SNR) tracking to calculate the SNR and PLL (Phase Lock Loop) tracking to track PPG signal. It was concluded that only there is 2% difference in the actual and measured level of blood oxygen. Hence, it is a significant factor to improve the wearability of pulseoximeters with less error.

An in-ear reflective pulse oximeter was also designed to check the blood oxygen level if the patient is suffered from hypothermia, shock, etc., which causes blood centralization that is being undetected using finger tips. The oximeter can be fixed in the ear canal without wrapping which ensures that there is no interruption in hearing. The sensor can be used along with the finger pulseoximeters, since it obtains considerable accuracy in measuring oxygen level in the blood during clinical testing on the patients. The great concern of the wearable systems is affording remote care for the patients, and the most commonly used wearable option is the wrist-worn sensor as many can wear it as either watches or bracelets.

A reflective pulse oximeter [42] was designed to be concave in shape, which can be worn on the wrist; it blocks the light from external sources and increases robustness in terms of noise. The miniaturized size of the device made it more wearable. In addition, it can also detect skin temperature, pulse rate, since it combines three sensors into one wearable node comfortably.

ECG is a device used to check the status of heart health, and many sensors are developed to capture these signals. An armband ECG sensor [44] is also used to measure with reasonable accuracy. ECG sensors can also be integrated with chest-straps [45] and helmets [43].

The helmet is also equipped with electroencephalogram (EEG) sensor that monitors brain-related activities such as sleep disorders, seizures, and head injury. Some EEG sensors are used to detect stress management [46] and driver drowsiness [47], and it can be measured using wearable headband.

The elderly people who fall and got injuries can be monitored using fall detection. A tri-axial accelerometer [48] can be attached to a smartphone which implements machine learning algorithms for classifying different user's posture and the achieved classification accuracy is given as 99.01%.

In [49] the classification algorithm followed for detecting user's posture was less accurate while performing fall detection and alternative algorithms are required. A wearable camera [50] was used to detect falls in rapid scenery changes, and the accuracy for indoor and outdoor environment was shown as 93.78% and 89.8%. An accelerometer data was incorporated with wearable camera system and the accuracy was shown as 91% in detecting falls.

A gyroscope, magnetometer, accelerometer were used to detect falls accurately [51], and a barometer assists in detecting the variations in height more accurately [52]. Gait detective system tracks the elderly people especially in specific conditions such as Parkinson's disease (PD). Gait detection helps in tracking the patients suffered from PD or stroke [53], where foot-worn sensors measure various parameters such as walking speed and step size. The specially designed sensor for gait detection controls lower limb prosthetics [54]. Three accelerometers could also be placed on ankle, hip, and knee for the patients suffered from PD [55].

A waist-worn device consists of tri-axial accelerometer and microcontroller used to detect gait anomaly [56]. An anomaly detection algorithm detects 84% of gait anomaly periods for last 5 seconds, and it achieves reasonable accuracy. A noninvasive blood glucose monitor is currently available in the market mainly useful for diabetic patients.

6.11 Wearable devices revolutionize the entire paradigm in drug dispensing

The objective of self-injectable devices is to afford therapeutic value to physicians and patients. Self-administrative injectable drug therapies have shown an influence in the number of diseases such as rheumatoid arthritis, psoriasis, lupus, multiple sclerosis, diabetes, chronic pain, asthma, high cholesterol, mental health, COPD, hemophilia disorders, and cancers. If patients are suffered with more than one chronical diseases, they face problem in managing their treatment schedule with the self-administered drug therapies. It creates an immense pressure in lot of cases, and if emotional obstacles are not defeated, the proper outcome may not be obtained. Nowadays, the advanced drug-administrative devices are available, which overcome many hurdles such as cognitive challenges or physical dexterity, mechanical complexity, needle phobia, and pain. Self-drug administrative device aids patients to take medications without any physician's knowledge.

The challenges in drug-delivery system are today's highly expensive biologic therapies such as therapeutic proteins and monoclonal antibodies that tend to have complicated molecules with high-viscosity, and high-concentration dosage is required, which is incompatible with autoinjection devices having a dosing limit of 1 mL or traditional syringe. The handheld or advanced wearable drug-administrative devices use prefilled cartridges, and 2–10 mL of the medication could be managed well in a single-dose episode.

A wearable drug-administrative device easily adheres to the human body, and the medication can be continued over a period of time. The advanced wearable devices are able to reduce pain, minimize hassle, and ease administration compared to legacy self-injection devices and traditional syringes. The patients adhere faithfully to this type of therapeutic because of its significant factors. Lot of studies and patient interaction were undergone to improve on-going design process to make ease of using the device with minimal discomfort and steps. Less frequent dosing choice with self-administration drug therapy acts as a driving force for the ongoing research in wearable technology platform.

There is a single-dose self-administrative monthly injection choice "Repatha" for high cholesterol; it was predicted to use as wearable on-body device with prefilled cartridge namely Pushtronex in smartdose platform. Pushtronex autoinjected wearable device can be placed beneath the skin which permits 420-mg single dose of Repatha 3.5 mL in 9 minutes. The steps followed by the patient to use Pushtronex device: place prefilled drug cartridge in the device, and stick the device to the human body skin. A slight push of a button induces the needle to deliver the drug within few minutes. Once the drug administration is over, it is indicated via onboard electronics, and it is ready for device disposal. Nowadays, smartdose devices are capable of delivering 3.5–10-mL doses with high-viscosity formulation, which minimizes patient discomfort and injection timing relatively.

6.11.1 Remove hurdles and offer rewards

SmartDose wearable devices adhere patients to be pain-free and hassle-free; in addition, the reward-based approach is also introduced to refill the medication instead of reminding them with alert. Patients who are in HealthPrize platform can gather points for engaging themselves in health-related activities such as self-administration of medications, refilling prescriptions, attending health-related quizzes, and the points can be redeemed as gift cards or real-world prizes.

6.11.2 Form and functions

Device makers rely on ease-of-use feature, promoting the marketing demand of wearable devices. For developing products, the device makers combine ergonomic design with advanced techniques for data capturing and analytics, which motivates the consistent utilization of such devices. By simply pushing a button, a drug should be released, and if the dose administration is over, the patient should receive either visual or audible indication. Drastic improvement in the adherence of self-administrative therapy, because of the improvement in health outcomes, reduces long-term expense for the patients by avoiding emergency ward visits, medication, and additional hospitalization. Particularly, for drug makers, it assists in driving sales and safeguarding the market share.

The wearable drug-administrative devices act a seamless bridge between the physicians and caregivers which gains a deep insight for physicians to provide treatment under chronic conditions with advanced data collection and analytics capabilities. Physicians are highly dependent on the report changes in symptoms of patient such as blood pressure or glucose level; this tracking is often lost in patient's ongoing experience during hospital visits. In contrast, digitally equipped wearable drug dispensing devices gather, communicate, and analyze the data in a reliable manner, and it is integrated with electronic health records for promoting the efficient data access.

6.11.3 Making the data relevant

Data-enabled drug-administrative wearable devices are functioning better, and it improves health outcomes and patient adherence that ultimately promote the economic level for the stakeholders.

Data generated by the advanced wearable devices should have the below-mentioned factors:

1. Validity: The actual value of relevant parameter should be reflected.
2. Utility: Gathered data make a difference in supporting adherence goals and overcoming hurdles and demonstrates the impact of utilizing such kind of advanced wearable techniques on health outcomes.
3. Reliability: It ensures data consistency gathered from individual patient and the related group of persons.

6.11.4 Market trend

The overall desired health outcomes for patient are highly dependent on the assuring appropriate medication exploitation and quick response in detecting clinical deterioration. In recent years, expensive specialty medication is required for different diseases where reliability in drug therapy is closely associated with clinical outcomes, costly interventions, and minimizes further hospitalization. Such things are rooted directly in advanced wearable drug-administrative devices.

The patients can undergo the chemotherapy treatment with the required dosage. An adjunct therapy can be used to increase the count of white blood cells and minimize the risk of infection, avoiding the next visit to physician's clinic. A tiny, preloaded, lightweight, on-body injector is attached to the body skin, and the dose is automatically delivered within 45 minutes. The device can be discarded after completing the process, and drug delivery on time gives satisfaction to caregivers and physicians.

6.11.5 Glucose monitoring

The glucose monitoring for diabetic patients is done by taking a drop of blood and placed on a test strip. This traditional approach provides accuracy, where the test strip can be used only once in checking the glucose level. Continuous glucose monitor (CGM) assists in gaining a better insight in blood glucose patterns periodically by tracking the patient's health continuously. CGM aids patient by providing predictive alerts to identify and manage the increase and decrease in blood glucose level proactively. Wearable Guardian Sensor 3 gathers real-time glucose level in the blood, and the tiny thin sensor can be worn continuously for 7 days on the upper arm or abdomen. The glucose level is measured with the interstitial fluid beneath the skin. Bluetooth transmitter can be worn anywhere on the human body which transmits glucose readings for every 5 minutes to Guardian connect app. The app enables the user to view the glucose data and alert information via smartphone.

6.12 Safety and security issues related to wearable health care devices

Real-time health monitoring through wearable sensors is proving its dominance in reducing healthcare cost, effective management of diseases, and saving life at right time [57]. Medical sensors, when placed on human body, monitor the physiological parameters continuously, which help the patients to quickly review their medical condition. In the year 2016, about 250 millions of consumer wearables were sold globally followed by an 800% increase from 2012 sales [58]. This exponential growth is anticipated to prolong into the next decade as wearables continue to become more affordable and reliable [59]. This mammoth growth in the usage of wearable devices produces overwhelming amount of user data and consumer health data privacy concerns as well. Safety in the wearable technology might be considered more prosaically as something is a physical result of a cyber or logical event. The events that are considered as the safety impact for humans are allergic reaction due to wearable things, burning and explosion of wearable devices, sensory impacts, tumors, and infections. Literally safety is considered as the protection against random faults of an unintentional nature. Wearable devices are one of the applications of IoT, which is considered as the integration of industrial control system (ICS) and information technology. ICS includes safety instrumented systems, means it uses hardened information elements to ensure high reliability with safety. Wearable devices provide variety of applications through wireless communication protocols as shown in figure.

Wearable healthcare devices transmit extreme confidential data; it is necessary for them to protect themselves against many security threats. Wearable devices suffer from resource constraints in the form of memory, limited battery, device form factor, which restricts the implementation of more secure communication principles. In this paper the security threats to wearables are categorized as (1) availability threats, (2) integrity threats, and (3) confidentiality threats. Availability threats are the scenario in which the invaders deny services and instill enormous worthless information to overflow the storage capacity of the wearable devices. Integrity is a vital security requirement for wearable healthcare systems. Integrity involves to ensuring that the information are not changed while transmitting and being received by dedicated parties. The integrity threat falls under three categories, namely, masquerade attacks, replay attacks, and modification attacks. In confidentiality threats the invaders uses eavesdropping technique to access information.

Establishing a secured wireless connection between a wearable device and servers can be made through active protocols. Secure sockets layer or transport layer security is considered as the most secured encryption protocol for the Internet [60]. This algorithm is computationally intensive, and hence, it is not apt for wearable devices as it has feeble computation power.

6.13 Wearable devices for women safety

In the current scenario, there is a huge increase in the safety and security of women harassment issues. The thought of every citizen is that the girl should move freely without any fear about their security even in the odd hours. The wearable devices assist in automatic sense of the current situation, and the victim can be saved from the critical scenario. Security system is required to provide the security for women while facing social-related issues. The recent advancement in wearable techniques helps in detecting the location of a person that enables for immediate action accordingly based on GSM, body temperature sensor, pulse rate sensor, and alarm. There are many kinds of sensors available which precisely senses the real situation of the women in critical situations. Nowadays, smart devices for women are easy to handle and more comfortable when compared to existing solutions such as bulky belt and separate garment. The data such as body temperature, pulse rate will be communicated directly with the help of wearable devices and the movement is continuously tracked by the application that is installed in the smart device. In case of any critical situation, the particular app alerts the device to perform the below-mentioned functionalities:

- The family members are informed immediately along with the coordinates.
- Information is transmitted to nearby police station for immediate action.
- Sends information to persons in the near locality to receive the public attention.

In the densely populated cities, lot of crime against woman is occurring continuously, which threatens the women security. The solutions available are limited, and the feasible technologies are in demand for these kinds of sensitive issues. The persons who can be contacted in case of any urgent situation are acting as a community database, and the verified users can be enlisted for communication. The people nearer to the victim can be alerted using Zigbee, IEEE 802.15.4 standard, and global positioning system (GPS) tracking. Once the alert message is triggered, GSM transmits message to the individuals in the predefined community list. Meanwhile, the location details are sent to other devices in the proximity range. The broadcast receiver of the concerned victim's application checks the message that is transmitted, and the application can obtain the contacts from the community database for the persons in the range.

Clothing and other accessories incorporate advanced technologies known as wearables have seen an exponential growth from the past decade. Wearables can be worn by a user to track information such as fitness and health status. A tiny motion sensor can be fixed in the wearable devices to take snapshots and that can be transmitted to the mobile devices. These devices are not only offering information about physiological monitoring but can also be preferred for personal safety. The wireless wearable technology is designed to record and monitor women safety information. The mobile technology enables to receive the alert message on time in case of emergency.

The incidents such as theft, harassment happen on victims who are isolated out from large crowded cities and have been rising very rapidly from the past decade. The system is not supporting the immediate response in the current alert mechanism incorporated in smartphone applications. The victim's family members may be residing somewhere, whilst the nearest patrol may be little far away, informing them will not support in such circumstances. Instead, if persons around the victim are intimated about the incident, the chances for rescuing the victim would be more.

Society Harnessing Equipment is a garment that has an electric circuit that generates 3800 kV, assists the victim to get away from the critical situation. ILA (International Liberation Army) security is designed with three alarms that can disorient the attackers in such a way the victim is safeguarded from risky situation. Advanced Electronics System for Human Safety is an electronic device with GPS facility that assists in monitoring the location of the victim continuously. Smart Belt looks like a normal belt which consists of screaming alarm, pressure sensors, and Arduino board. If the threshold of the pressure sensor is exceeded, automatically the device would be activated. The siren seeks for help once the screaming alarm is triggered out.

Pulse rate sensor: The sensor gives the digital output of the heartbeat, and it is linked with the microcontroller to compute the beats per minute rate.

Temperature sensor: Body temperature plays a significant role in maintaining the health, and hence, it is compulsory to track it regularly. Several temperature sensors are available to measure the body temperature. For example, in LM35 integrated circuit sensor, it operates with +10.0 mV/°C scale factor and 0.5°C accuracy.

GPS: The longitude and latitude of a receiver is determined by computing the time variation from different satellites to attain the receiver. Approximately 12,500 miles away from the earth, 24 Medium-Earth Orbit satellites revolve 24 hours around the earth and send location every second in addition to the present time in atomic clocks. The blood flow is monitored if the human body is in touch with the wrist band for each pulse.

GSM is used to transmit data from the control unit to the base unit, and GSM 300 is operated at 900 MHz frequency. The uplink band range is from 890 to 915 MHz, whilst the down link range is from 935 to 960 MHz, and it combines the advantages of TDMA (Time Division Multiple Access) and FDMA (Frequency Division Multiple Access). At any instance, 992 channels would be available in GSM 300 [5,6].

Dual technology motion sensor: It is a sensor that tracks the moving objects, and the motion detector automatically alerts the user's movement in a specific location. Converging multiple-sensing technologies in one motion detector minimizes the false triggering, but it increases the vulnerability factor [7].

Bluetooth Low Energy connects devices with less power consumption. A Beacon software study report says that peripherals such as proximity beacons can function with a 1000 mA h coin cell battery for 1 year. Bluetooth smart protocol only sends small packets as compared to the classic one, which is suitable for high bandwidth data [8].

6.14 Open challenges and future directions

- Interoperability in IoT-based e-healthcare system is an issue to be considered in developing current solutions.
- Redundant services should be provided to the patient without any delay or data loss to improve QoS (Quality of Service) in e-healthcare services.
- Low-cost medical sensors without any toxic elements are required in IoT e-health care services. Government and regulatory bodies should suggest

guidelines to be followed for manufacturing sensors, usage, and disposal procedures.

- Low expensive and miniaturized sensor devices are significant to transmit the data efficiently in wearable platform.
- Miniaturized antennas are necessary for sensor devices in order to minimize energy consumption, interference, and maximize the transmission reliability.
- The tracking mechanism should be incorporated into WBAN to monitor the patients continuously inside and outside the hospital premises.
- The entire eco-system of health monitoring process such as data collection, analysis, and processing assists in early detection of disorders in e-healthcare.
- Individual sensors should operate stand-alone in terms of energy efficiency.

6.15 Conclusion

This chapter elaborated the recent advancement in wearable sensors for real-life applications in view of personal healthcare. Smart sensors with advanced configurations, tolerance, stretchable, and flexible can monitor human physiological signals. Due to the increase in the population of elderly people around the world, wearable devices are becoming an essential part of their daily lives. In 2016 the global health care wearable market earned a growth of over USD 5 billion, and it is expected to reach an annual growth rate of over 17% (USD 12 billion) by 2021. The objective of this chapter is to furnish a comprehensive overview of wearable sensors, its classification, applications, etc. The significance of noninvasive chemical parameter measurement using electrochemical sensors is discussed. Challenges related to the fabrication of wearable sensors, their working principle, and electrodes types are discussed.

References

[1] N. Khalil, M.R. Abid, D. Benhaddou, M. Gerndt, Wireless sensors networks for Internet of Things, Proceedings of the 2014 IEEE Ninth International Conference on Intelligent Sensors, 21–24, Sensor Networks and Information Processing (ISSNIP), Singapore, April, 2014, pp. 1–6.

[2] S. Park, S. Jayaraman, Wearable sensor network: a framework for harnessing ambient intelligence, J. Ambient Intell. Smart Environ. 1 (2) (2009) 117–128.

[3] M. Rodgers Mary, P. Vinay, S. Conroy Richard, Recent advances in wearable sensors for health monitoring, IEEE Sensors 15 (2015) 3119–3126.

[4] S.C. Mukhopadhyay, Wearable sensors for human activity monitoring: a review, IEEE Sensors 15 (2015) 1321–1330.

[5] S. Neethirajan, Recent advances in wearable sensors for animal health management, Sens. Bio-Sens. Res. 12 (5) (2017) 15–29.

[6] J. Ray Windmiller, J. Wang, Wearable electrochemical sensors and biosensors: a review, Electroanalysis 25 (6) (2013) 29.

[7] T. Islam, S.C. Mukhopadhayay, Wearable Sensors for Physiological Parameters Measurement: Physics, Characteristics, Design and Applications, IOP Publishing, 2017.

[8] A.J. Bandodkar, J. Wang, Non-invasive wearable electrochemical sensors: a review, Trends Biotechnol. 32 (2014) 363–371.

[9] W. Golding, Electrical Measurements and Measuring Instruments, fifth ed., Pitman, London, 2009.

[10] T. Islam, S.C. Mukhopadhyay, N.K. Suryadevara, Smart sensors and internet of things: a postgraduate paper, IEEE Sensors 17 (2017) 577–584.

[11] S.C. Mukhopadhyay, T. Islam, Wearable Sensors Applications, Design and Implementation, IOP Publishing Ltd, 2017.

[12] M.J. Mathie, A.C. Coster, N.H. Lovell, B.G. Celler, S.R. Lord, A. Tiedemann, A pilot study of long-term monitoring of human movements in the home using accelerometry, J. Telemed. Telecare 10 (2004) 144–151. Available from: https://doi.org/10.1258/135763304323070788.

[13] E.S. Sazonov, S.A. Schuckers, P. Lopez-Meyer, O. Makeyev, E.L. Melanson, M.R. Neuman, et al., Toward objective monitoring of ingestive behavior in free-living population, Obesity (Silver Spring) 17 (2009) 1971–1975. Available from: https://doi.org/10.1038/oby.2009.153.

[14] G. Andreoni, P. Perego, C. Standoli, Wearable Monitoring of Elderly in an Ecologic Setting: The SMARTA Project, 2017.

[15] M.J. Buller, W.J. Tharion, S.N. Cheuvront, S.J. Montain, R.W. Kenefick, J. Castellani, et al., Estimation of human core temperature from sequential heart rate observations, Physiol. Meas. 34 (2013) 781–798.

[16] G.M. Bertolotti, A.M. Cristiani, P. Colagiorgio, F. Romano, E. Bassani, N. Caramia, et al., A wearable and modular inertial unit for measuring limb movements and balance control abilities, IEEE Sens. J. 16 (2016) 790–797.

[17] A. Pantelopoulos, N. Bourbakis, A survey on wearable sensor-based systems for health monitoring and prognosis, IEEE Trans. Syst. Man Cybern. C 40 (2010) 1–12.

[18] S. Yoon, Y.-H. Cho, A skin-attachable flexible piezoelectric pulse wave energy harvester, J. Phys. Conf. Ser. 557 (2014) 012026.

[19] Y. Tajitsu, Piezoelectret sensor made from an electro-spun fluoropolymer and its use in a wristband for detecting heart-beat signals, IEEE Trans. Dielect. Electr. Insul. 22 (2015) 1355–1359.

[20] J. Zenko, M. Kos, I. Kramberger, Pulse rate variability and blood oxidation content identification using miniature wearable wrist device, Proc. Int. Conf. Syst. Signals Image Process, IWSSIP, 2016, pp. 1–4.

[21] Garmin, HSM-Tri [Online]. Available from: <https://buy.garmin.com/en-AU/AU/p/136403>, 2017.

[22] FitBit, FitBit PurePulse [Online]. Available from: <https://www.fitbit.com/au/purepulse>, 2017.

[23] Polar, H7 Heart Rate Sensor [Online]. Available from: <https://www.polar.com/au-en/products/accessories/H7_heart_rate_sensor>, 2017.

[24] TomTom, TomTom Spark Cardio [Online]. Available from: <https://http://www.tomtom.com/en_au/sports/fitness-watches/gps-watch-cardiospark/black-large/>, 2017.

[25] H. Lee, H. Ko, C. Jeong, J. Lee, Wearable photoplethysmographic sensor based on different LED light intensities, IEEE Sensors J. 17 (3) (2017) 587–588.

[26] Y. Shu, C. Li, Z. Wang, W. Mi, Y. Li, T.-L. Ren, A pressure sensing system for heart rate monitoring with polymer-based pressure sensors and an anti-interference post processing circuit, Sensors 15 (2) (2015) 3224–3235.

[27] D. Wang, D. Zhang, G. Lu, A novel multichannel wrist pulse system with different sensor arrays, IEEE Trans. Instrum. Meas. 64 (7) (2015) 2020–2034.

[28] D. Wang, D. Zhang, G. Lu, An optimal pulse system design by multichannel sensors fusion, IEEE J. Biomed. Health Inf. 20 (2) (2016) 450–459.

[29] W. Zuo, P. Wang, D. Zhang, Comparison of three different types of wrist pulse signals by their physical meanings and diagnosis performance, IEEE J. Biomed. Health Inf. 20 (1) (2016) 119–127.

[30] Y.-J. An, B.-H. Kim, G.-H. Yun, S.-W. Kim, S.-B. Hong, J.-G. Yook, Flexible non-constrained RF wrist pulse detection sensor based on array resonators, IEEE Trans. Biomed. Circuits Syst. 10 (2) (2016) 300–308.

[31] S. Milici, J. Lorenzo, A. Lázaro, R. Villarino, D. Girbau, Wireless breathing sensor based on wearable modulated frequency selective surface, IEEE Sensors J 17 (5) (2017) 1285–1292.

[32] C. Varon, A. Caicedo, D. Testelmans, B. Buyse, S. van Huffel, A novel algorithm for the automatic detection of sleep apnea from single-lead ECG, IEEE Trans. Biomed. Eng. 62 (9) (2015) 2269–2278.

[33] D. Oletic, V. Bilas, Energy-efficient respiratory sounds sensing for personal mobile asthma monitoring, IEEE Sensors J. 16 (23) (2016) 8295–8303.

[34] X. Yang, et al., Textile fiber optic microbend sensor used for heartbeat and respiration monitoring, IEEE Sensors J. 15 (2) (2015) 757–761.

[35] S.D. Min, Y. Yun, H. Shin, Simplified structural textile respiration sensor based on capacitive pressure sensing method, IEEE Sensors J. 14 (9) (2014) 3245–3251.

[36] I. Mahbub, et al., A low-power wireless piezoelectric sensor-based respiration monitoring system realized in CMOS process, IEEE Sensors J. 17 (6) (2017) 1858–1864.

[37] O. Atalay, W.R. Kennon, E. Demirok, Weft-knitted strain sensor for monitoring respiratory rate and its electro-mechanical modeling, IEEE Sensors J. 15 (1) (2015) 110–122.

[38] P. Aqueveque, C. Gutiérrez, F.S. Rodríguez, E.J. Pino, A. Morales, E.P. Wiechmann, Monitoring physiological variables of mining workers at high altitude, IEEE Trans. Ind. Appl. 53 (3) (2017) 2628–2634.

[39] Y. Zhang, M. Berthelot, B.P. Lo, Wireless wearable photoplethysmography sensors for continuous blood pressure monitoring, in: Proc. IEEE Wireless Health (WH), October 2016, pp. 1–8.

[40] J. Wannenburg, R. Malekian, Body sensor network for mobile health monitoring, a diagnosis and anticipating system, IEEE Sensors J. 15 (12) (2015) 6839–6852.

[41] S.V. Gubbiand, B. Amrutur, Adaptive pulse width control and sampling for low power pulse oximetry, IEEE Trans. Biomed. Circuits Syst. 9 (2) (2015) 272–283.

[42] G. Pang, C. Ma, A neo-reflective wrist pulse oximeter, IEEE Access 2 (2014) 1562–1567.

[43] W. Von Rosenberg, T. Chanwimalueang, V. Goverdovsky, D. Looney, D. Sharp, D.P. Mandic, Smart helmet: wearable multichannel ECG and EEG, IEEE J. Transl. Eng. Health Med. 4 (2016) 1–11. Art. no. 2700111.

[44] V.P. Rachim, W.-Y. Chung, Wearable noncontact armband for mobile ECG monitoring system, IEEE Trans. Biomed. Circuits Syst. 10 (6) (2016) 1112–1118.

[45] E. Spanò, S.D. Pascoli, G. Iannaccone, Low-power wearable ECG monitoring system for multiple-patient remote monitoring, IEEE Sensors J. 16 (13) (2016) 5452–5462.

[46] G. Li, B.-L. Lee, W.-Y. Chung, Smartwatch-based wearable EEG system for driver drowsiness detection, IEEE Sensors J. 15 (12) (2015) 7169–7180.

[47] U. Ha, et al., A wearable EEG-HEG-HRV multimodal system with simultaneous monitoring of tES for mental health management, IEEE Trans. Biomed. Circuits Syst. 9 (6) (2015) 758–766.

[48] J. Wannenburg, R. Malekian, Physical activity recognition from smartphone accelerometer data for user context awareness sensing, IEEE Trans. Syst. Man. Cybern. Syst. 47 (12) (2017) 3142–3149.

[49] P. Vallabh, R. Malekian, N. Ye, D.C. Bogatinoska, Fall detection using machine learning algorithms, in: Proc. 24th Int. Conf. Softw. Telecommun. Comput. Netw. (SoftCOM), Sep. 2016, pp. 1–9.

[50] K. Ozcan, S. Velipasalar, P.K. Varshney, Autonomous fall detection with wearable cameras by using relative entropy distance measure, IEEE Trans. Human–Mach. Syst. 47 (1) (2017) 31–39.

[51] P. Pierleoni, A. Belli, L. Palma, M. Pellegrini, L. Pernini, S. Valenti, A high reliability wearable device for elderly fall detection, IEEE Sensors J. 15 (8) (2015) 4544–4553.

[52] P. Pierleoni, et al., A wearable fall detector for elderly people based on AHRS and barometric sensor, IEEE Sensors J. 16 (17) (2016) 6733–6744.

[53] H.-C. Chang, Y.-L. Hsu, S.-C. Yang, J.-C. Lin, Z.-H. Wu, A wearable inertial measurement system with complementary filter for gait analysis of patients with stroke or Parkinson's disease, IEEE Access 4 (2016) 8442–8453.

[54] H.F. Maqbool, M.A.B. Husman, M.I. Awad, A. Abouhossein, N. Iqbal, A.A. Dehghani-Sanij, A real-time gait event detection for lower limb prosthesis control and evaluation, IEEE Trans. Neural Syst. Rehabil. Eng. 25 (9) (2017) 1500–1509.

[55] T.T. Pham, et al., Freezing of gait detection in Parkinson's disease: a subject-independent detector using anomaly scores, IEEE Trans. Biomed. Eng. 64 (11) (2017) 2719–2728.

[56] G. Cola, M. Avvenuti, A. Vecchio, G.-Z. Yang, B. Lo, An on-node processing approach for anomaly detection in gait, IEEE Sensors J. 15 (11) (2015) 6640–6649.

[57] D. Isern, D. Sanchez, A. Moreno, Agents applied in health care: a review, Int. J. Med. Inform. 79 (2010) 145–166.

[58] J. Comstock, ABI: 30M wearable sensors shipped in 2012, HIMSS Media (December 1, 2015). Retrieved 09.10.17, from: <http://www.mobihealthnews.com/19448/abi-30m-wearable-sensors-shipped-in-2012>.

[59] Gartner, Inc. Gartner says worldwide wearable device sales to grow 17 percent in 2017, Newsroom (August 24, 2017) Retrieved 09.10.17 from: <http://www.gartner.com/newsroom/id/3790965>

[60] S. Turner, T. Polk, Prohibiting Secure Sockets Layer (SSL) Version 2.0, March 2011.

Chapter 7

Internet of Things in pharma industry: possibilities and challenges

Mohan Singh[1], Smriti Sachan[1], Akansha Singh[2] and Krishna Kant Singh[1]
[1]*Department of ECE, GL Bajaj Institute of Technology and Management, Greater Noida, India,*
[2]*School of Computing Science and Engineering, Galgotias University, Greater Noida, India*

7.1 Introduction

Today's era is all about the high demand of Internet-based applications. As in the present scenario, communication through Internet becomes very convenient, so, nowadays, Internet of Things (IoT) is one of the fastest growing technologies in this 21st century. It fulfills all the requirements as per today's need. IoT is the architecture in which all the physical devices are linked through a router and transfer the information to each other. IoT is a technology, which allows devices to be controlled remotely in the presently existing network. IoT is an intelligent network, which reduces the human effort and also uses automation system that controls the device without any manual command. IoT includes hardware and software architecture, information sharing within the network and provides a variety of services [1].

IoT is a very popular technology nowadays because of its versatility in various fields such as home automation system, utilities such as smart grid, smart meters, environment-monitoring system, water-monitoring system, medical and emergency health-care system, pharmaceuticals, and electronic toll system in transportation system.

Due to the limitation of this paper, only one field is chosen to explain the utilization of IoT, that is, in pharmaceutical industry. Though pharma industry already uses different types of technologies such as artificial intelligence, three-dimensional printing, and blockchain to improve the present system, IoT becomes a revolutionary factor for the pharmaceutical industry. It has immense applications in various fields of it such as manufacturing, distribution, supply chain, and logistics, which makes the system transparent and fast.

Emergence of Pharmaceutical Industry Growth with Industrial IoT Approach.
DOI: https://doi.org/10.1016/B978-0-12-819593-2.00007-8

IoT plays an important role in the field of pharma industries. Researchers around the world are working toward the use of this promising technology in pharma. Most researchers are estimating their work in the field of revolution of next-generation pharma industry with the connectivity of each other to the Internet. This network is connected to things, household objects, vehicles, electrical devices, and industrial equipment with other utilities of Internet connectivity using dynamic data. High growth potential can be created for businesses by connecting in advance to improve nonaffiliated institutions, decision-making speed, and accuracy. Services called "Amazon Key" and in-car delivery, recently launched from Amazon, are great examples of the use of connected networks tools to deliver more efficient and hassle-free utilities to consumers. Amazon Key is the smart service provided by Amazon having smart lock, security camera, clouding cam that remotely monitors and ensures the delivery of shipment by a customer about his delivery. Many other industries are utilizing the potential of network things for the benefits of consumers [2]. Pharma is also one of those sectors that are serving customers about safe distribution of prescription drugs.

This chapter elaborates the role of IoT in the field of pharma industry related to their challenges and applications. The major contribution of IoT is to overcome the barriers of the complex problems and create their conventional solution.

7.1.1 Internet of Things

The journey of IoT was started in 1982 at Carnegie Mellon University as the first Internet connectivity appliance [2]. At that time, Transmission Control Protocol (TCP)/Internet Protocol (IP) was proposed for Internet connectivity in the field of telecommunication. Around 1990 Web 0.0 was developed as architecture of the network. The first step for web page was established toward e-commerce in 1992 using Web 1.0. IoT term was proposed in 2000 as Web 2.0 for Internet compatible devices. For full Internet connectivity, IoT was brought into the picture around 2010, which was based on technical standardization as shown in Fig. 7.1.

7.1.2 Applications of Internet of Things

Recently, everywhere everybody wants to execute every task using IoT. IoT is very useful for completion of any task smartly within the time and having minimum cost. Any device that is assigned for doing some task can be controlled easily from anywhere using the smart connectivity of IoT. Basically, IoT is the fully connected network of devices for sharing the message and transferring the data from one place to other, and managing the use of IP. IoT is also used for monitoring, tracking, and managing the system such as smart city, smart grids, etc. as shown in Figs. 7.2 and 7.3 [3].

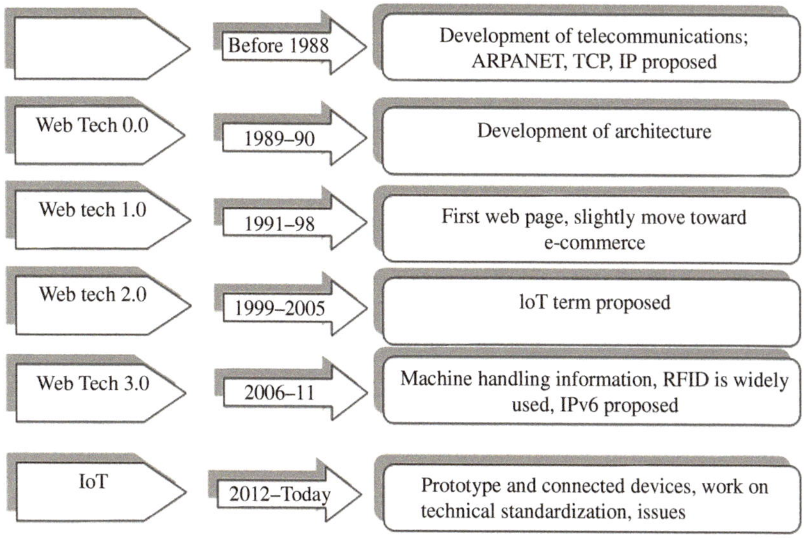

FIGURE 7.1 Internet developing stages. *IoT*, Internet of Things.

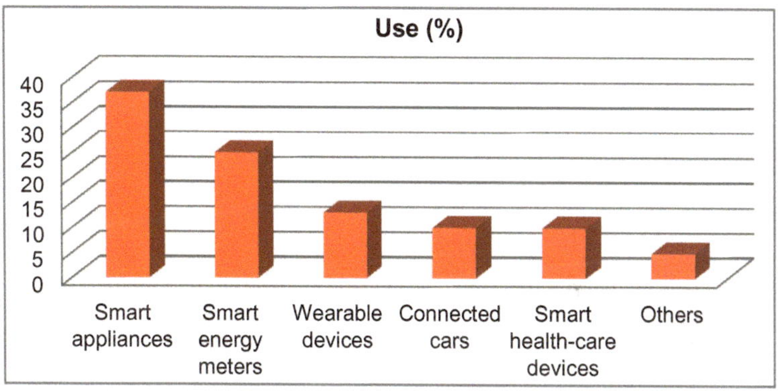

FIGURE 7.2 Percentage of use of IoT. *IoT*, Internet of Things.

Smart city is one of the main examples of IoT where all medical pharma functions of the city are connected and smoothly running with IoT and smart sensors. The grid controls of the power plants can be analyzed from far away and provide safety to the consumers. In the field of education, IoT plays very important role to pick up and transfer the huge amount of data from one place to other. Transport services such as Ola and Uber in India are easily available at the home to the user via their apps that are controlled by remote office using IoT. In the field of agriculture, IoT-based application are used for finding water level and temperature in the soil using sensors and transfer the data on the cloud where data can be shared with the help of IoT

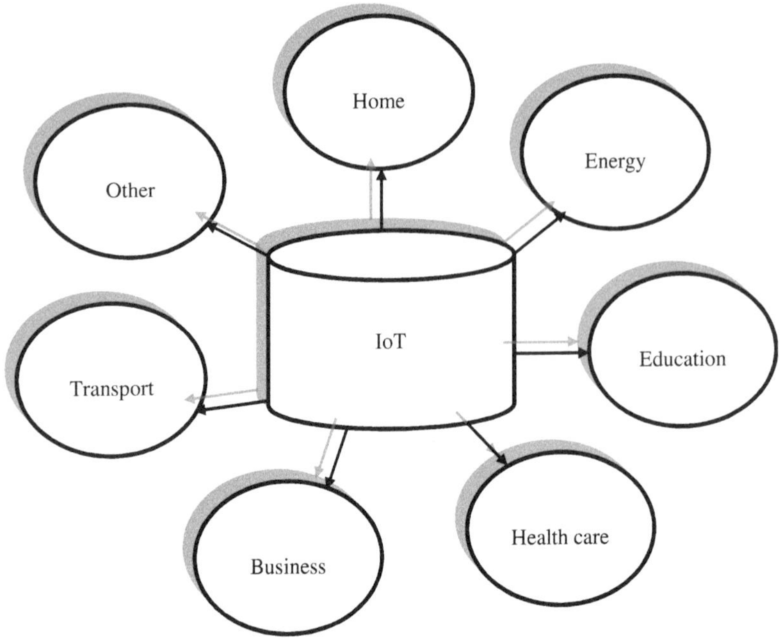

FIGURE 7.3 Application of IoT. *IoT*, Internet of Things.

service. IoT is very useful to monitor the status of health conditions and early diagnosis of patients. A doctor can excess and analyze his patients' report from far away.

7.2 Internet of Things road map in pharma

IoT is the physical connectivity of software, embedded devices, electronic sensors, which allows to exchange data from anywhere over the network. The main contribution of IoT in the field of pharmaceuticals is to monitor health problems, predict the future aspects related to the patients using evolutionary technologies, transform information, reduce cost, and save lives. The main challenges to configure IoT are type of communication, electronic sensor devices that are used to collect data from the server in any languages. The goal of a manufacturer is to synchronize the sensor digital data and go further take away from pill. The concept of pharma industry is to digitize the production of medicines and supervise all the processes using the smart connectivity of medical equipment and network topologies during the clinical observation of patients. Using advance technology, IoT can develop good relationships and reduce the ambiguity between the pharma industry and patients and can create new ways to treat disease. As the degree of IoT increases, the availability and affordability of health-monitoring technology will increase leading to automation [4].

The pharmacists will then be more focused on teaching patients how to use them. Augmented reality devices such as Microsoft HoloLens are disrupting medicine and could let pharmacists access the patient records and prescription information right before them. Technologies like these could catch problems before those become serious, providing instant communication to caregivers and real-time treatment that need not involve a trip to a lab or the doctor. Pharmacy automation and IoT technology will shift the focus of future pharmacists. Rather than counting pills from a bottle and simply filling prescriptions, they will be able to step out from behind the counter and actively engage with patients on their health [5].

The road map of the pharma sector with IoT has an important role for smooth conduction [6]. The record of medicine distribution is maintained by using IoT services. First of all, problems are identified, challenges related to problems are selected and then documented the problems for cost feature and benefits of the business that is called business case. An agreement of stakeholder is prepared for primary sensors and business need. Problems are identified using existing technology and then key success areas are defined for the stakeholder benefits. After stakeholder agreement, there is a need to design a solution for selected problems with Internet service and give more benefits to the patients [7]. Once problem is identified and solution is implemented, the pharma industry thinks about benefits by looking after all the decisions. Finally, the road map has been designed with all the challenges, which are marked with the management of protocol services for the smooth operation of all activities related to the pharma industry as shown in Fig. 7.4.

IoT pharma network topology is the arrangement of basic elements of IoT in pharma environments. The network topology is used for collecting data of body temperature, blood pressure, and ECG using the sensors and transfers this data to respective person for further treatment with the help of mobile equipment such as smartphone, laptops, and other smart electrical devices. Sensors attached with patient's body are used to capture data from patients and deliver this information on the cloud, then doctor can analyze their condition status and stored. Based on analysis, doctor can respond to his patients from any location. Supporting topology network such as WiMAX, IP, and Global Positioning System (GPS) is required for accessing services of pharma industries. Intelligent pharmaceutical is used to manage the misuse of medicine using gateway topology. For this purpose, intelligent medicines are preferred with various sensors having wireless standards. Many wireless IoT devices are connected to each other for health-monitoring purpose using the pharma IoT cloud that is for patient's diagnosis and analysis. Gateway can also be used for collecting information of a patients and storing it for display. Using pharmaceutical IoT network topology, pre- and post-services of the medical services can be used by service providers. IoT in the pharmaceutical industry is useful to get emergency medical services for a patient from a long distance without any delay with cloud computing.

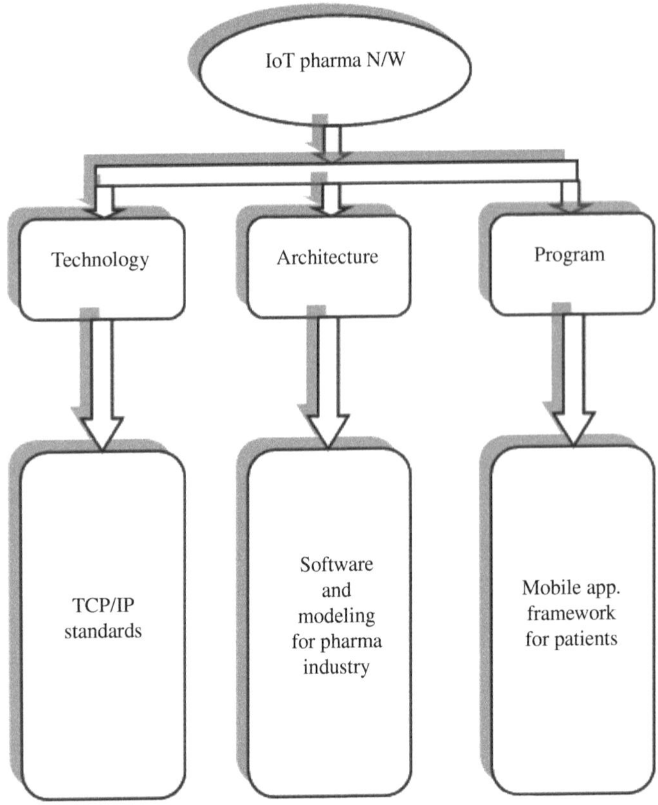

FIGURE 7.4 IoT pharma network. *IoT*, Internet of Things.

The IoT in pharma must have the pharma rule for the monitoring of the system with pharma Internet connectivity.

The IoT PharmaNet architecture has the basic physical elements such as pharma industry, TCP/IP model, WLAN, IPv6 IEEE standards, and multimedia communications. TCP/IP model has four layers as shown in Fig. 7.5.

7.3 Internet of Things in pharma industry

IoT has shown a great potential toward manufacturing and supply chain process in the pharmaceutical industry because of Radio Frequency Identification (RFID) technology, but the IoT devices are never scanned like bar codes as they transfer the real-time information to any other device through Internet. IoT is providing medical services in the field of pharma industry, so IoT plays an important role in managing all services [8]. Today, patients can get medicines through the Internet service without going to the medical store where patients can track their orders far from home, which is

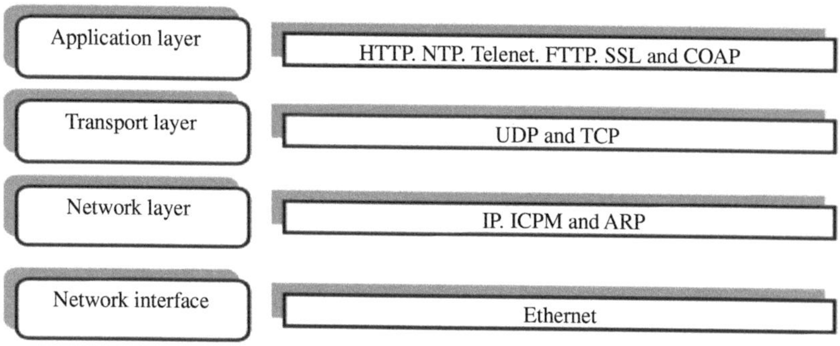

FIGURE 7.5 TCP/IP model. *IP*, Internet Protocol; *TCP*, Transmission Control Protocol.

possible only due to IoT. The product price can be recognized using bar code and sends to the patients via IoT service [9].

Nowadays, everything is to be connected with the Internet using embedded devices. In the pharma industry [5] the basic elements are patients, medicine, marketing, and pharma companies, which are interconnected to each other with the hypertechnologies using IoT connectivity. Medicines are to be distributed, and records can be transferred from one place to another place only due to IoT. Medicine-delivery monitoring and real-time tracking can be possible only due to patient's connectivity with the network protocols as shown in Fig. 7.6 [10].

In the pharmaceutical industry the production process is a very tedious task as ensuring the quality of product is an important part of the system. The process from production, distribution to till delivery to the customer has to be handled very carefully. Recently, manufacturing processes use smart sensors with RFID tags that make the production process easy and fast [11]. IoT devices ensure the specific information about the requirement and batches are managed accordingly, whereas earlier it was a complex and slow process to identify the loopholes [4].

Production of drugs also requires a sensitive eye on the maintenance and temperature-monitoring system for storage of product, so that system works efficiently, and quality of drugs to be maintained till it will be delivered to the customers. It is extensively used for temperature-sensitive products as the smart devices continuously record the temperature that is checked on regular basis and compare it with the standard thermostat, so that an alert is generated for any variation in temperature [12].

Devices using IoT technology can also track the shipment of product, which makes the pharma company easy to locate the product, real-time location, delays, etc. because the RFID tags are attached to every product. Product can be analyzed in remote areas also by using smart phones and GPS system to track the consignment, storage of products, product security, etc. by indulging IoT in it. So, nowadays IoT captures all the areas of

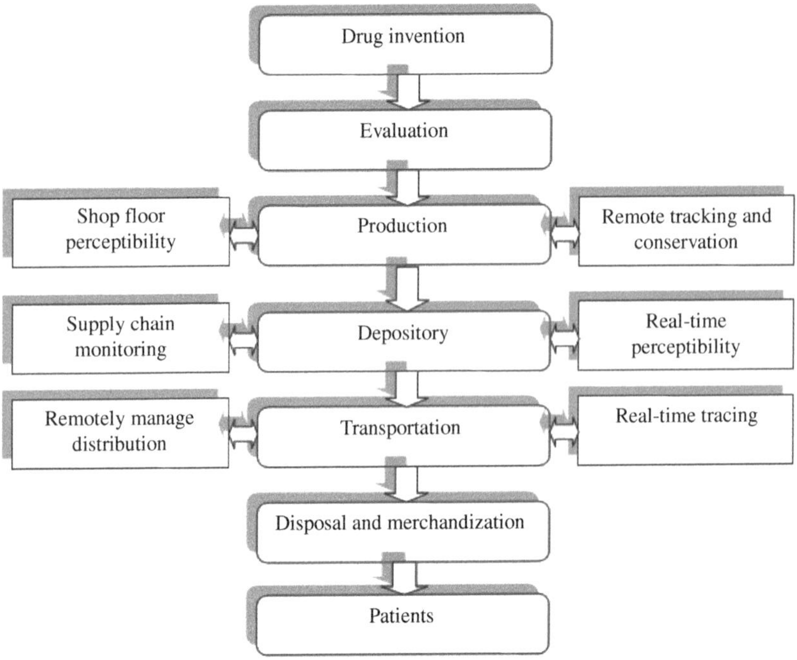

FIGURE 7.6 IoT in pharma. *IoT*, Internet of Things.

pharma industry such as manufacturing, transportation, distribution, and delivery of product.

In IoT-based pharma, every electronic device is connected with a specific sensor for capturing the information about health of patients for smart communication within the IoT system with objective connectivity. Each node in IoT system is represented as an objective for virtual connectivity. IoT can change the traditional way of manufacturing of the products in the pharmaceutical industry that is called automotive pharma plants. IoT can also improve the way of manufacturing and distribution of medicines prescribed by a doctor. Pharma industry is adapting the new technology for changing itself in the field of IoT-enabled echo system with respect to time.

7.4 Applying Internet of Things in pharma industry

There are certain stages of pharma industry in which IoT shows its wide scope, which are discussed in the following subsections:

7.4.1 Manufacturing

In pharma industry, from past several years, batch production is going on, but automatic system can be applied not only to control the machinery and material but also to increase the efficiency of manufacturing of products and other related activities. The operational data can be easily transferred to the other devices or the production engineers by IoT-enabled devices. This information helps to manage the industry in such a way that the production rate becomes higher and efficient. Manufacturing of drugs in pharmaceutical industry is divided into two series of section operations such as granulation, milling, tablet pressing, and coating. Milling is used for reducing particle size in drug powder with higher solubility. Granulation is the opposite process of milling in which small particles are to be bounded together and form larger particles called granules and are used for preventing from demixing the mixer. Drugs manufacturing process can be controlled and managed automatically by using IoT solution such as mobile and wearable products that are applicable to monitor drugs forming process. With the help of IoT-based applications the pharmaceutical industry is manufacturing clinically smart pills, which are the popular area of investment of pharma industry. IoT applications are well suited in the field of pharmaceutical company for the connectivity establishment of equipment and row material management, chain supply monitoring, and smart packaging. Drugs manufacturing in pharmaceutical industry is generally processed in batches with self-contained equipment having self-controlled automation process used for integrated life science in real conditions. All smart equipment are connected with IoT pharma network for real-time information and conditions that are not easily accessible manually such as cleaning, batches scheduling, and maintenance. These are the current issues of existing pharmaceutical industry solutions and can be resolved using smart devices and IoT-based pharma network. During the manufacturing process, continuous monitoring indicates calibration of equipment and safety of drugs, finishing of product and requires storage conditions. IoT-based technology in pharmaceutical industry allows to extend visibility into all current activities and increase drugs productivity. Sensors are the backbone of manufacturing process of such solution and used for sending information to the central system for analysis and meaningful performance of the manufacturing process.

7.4.2 Monitoring of production flow

In manufacturing of products the IoT-monitoring system analyzes the whole system of production lines to packaging of final products. It observes the operations performed and collects the real-time database that provides the scope for betterment. The close monitoring allows the system to improve, remove the lags and unnecessary work present [13]. Monitoring of the manufacturing process is one of the important things for product quality and maintenance of the pharmaceutical system. Sensors are the backbone of monitoring of any system to get the real information about the process and sent to the central system for good decisions and maintain the quality of the product easily. Every stage such as milling, coating, granulation, and packaging of manufacturing is verified when the stage is under monitoring continuously. All the environmental conditions of the pharmaceutical manufacturing process must be controlled for maintaining the quality of product that is only possible due to only monitoring of the production. IoT plays an important role to monitor the system for real quality of production from far away using smart devices and sensors. To analysis the quality of production and performance of the system, monitoring phase is necessary.

7.4.3 Controlling of environmental factors in drugs manufacturing

Environment is the major factor, which affects the manufacturing of drugs. IoT can be implemented to check on environmental conditions. IoT helps to increase the transparency in production of drugs by associating with the real-time sensors. These sensors sense the data of parameters of environment, such as temperature, humidity, radiations, and light, which can be controlled with the help of smart devices. An alert can also be generated to avoid the loss due to these environmental factors. These devices are collaborated with climate control equipment so that these devices can automatically adjust according to situation [9].

Table 7.1 summarizes different conditions of manufacturing process, parameter variations, and remedies for it, which are provided through IoT.

7.4.4 Quality control

The quality of the product can be ensured by using IoT sensors. The information collected by the sensors allows knowing the status of different stages of product cycle. The information includes the raw material used, temperature variations, wastages, and transportation. IoT smart devices can also be used to record customer feedback so that it can analyze afterward and remove the quality issues [14]. Product quality totally depends upon the real-time monitoring of manufacturing process through which the product quality is verified. The quality of a product is maintained by using IoT-based pharmaceutical applications with the observation of manufacturing process. Quality control is the crucial and vital operation of the pharma industry having the separate department in pharmaceutical industry. A number of people working in pharmaceutical industry are engaged in quality control department after research and development department in pharma industry. Quality control is to be completed in three steps such as the analysis of row material, process, and finally finished the product with the help of many instruments and IoT-based sensors. The quality of medicine in pharma industry can be measured by sampling method also.

TABLE 7.1 Enhancing manufacturing floor visibility.

Requirement of business	Particulars	Remedy given by IoT	Improvisation through IoT
• Efficiency of the manufacturing should be increased	• To check the visibility for scheduling of equipment • Decrease the runtime • Automatic alarm generation • Reduce variations and increase yield	• Smart devices gather the data regarding temperature, equipment ready for operation, and its cleaning and maintenance time • Runtime should be calculated so that it can be predicted that when the equipment suddenly shut down • Automatic sensors should be included so that it generates alarms regarding any fault, variations, or sudden changes in the process • Manufacturing floor is supervised and monitored by human efforts and with the help of technologies	• Dynamic scheduling of floor activities • Optimize the utilization of devices/ equipment and reduce the runtime • Tracking/ monitoring of fault variations • Enhanced productivity

IoT, Internet of Things.

7.4.5 Packaging optimization

To improve the packaging performance and make it cost effective, the IoT devices can be used in packaging materials and at every product. Handling multiple customers at a time will be easy to manage with tracking mechanism. It also helps in getting the insight of deterioration of quality of product by environmental factors, transportation system etc. so, IoT helps in packaging in many ways [15]. Packaging of the pharmaceutical product is the final stage of the manufacturing process that is responsible for marketing management of the pharmaceutical industry. Type of packaging depends upon the type of product that indicates the quality of product finally in the market cost effectively.

7.4.6 Warehouse operations

It has become a difficult task to check the availability of stock, temperature variations so IoT can be helpful to optimize these issues as the sensors can read the data for handling the temperature-sensitive products and check the stock availability, so that drugs can be easily supplied to the needy ones. The connectivity between different locations can also be done by using this technology so that product can be traced and reached on time. The management can also check with the unnecessarily used space of warehouses so that it can be minimized as per requirement [16].

Table 7.2 summarizes the traceability of the products in the warehouse, it shows the variety of parameter variation and their remedies provided through the IoT [17].

7.4.7 Facility management

There are certain tools and equipment which depend upon the range of temperature and vibrations. These critical machines require time-to-time maintenance so IoT sensors monitor the machinery and alert about the deviations, so accordingly management and maintenance can be done. The data gathering by sensors is also useful to check the full machine utilization. These preventive steps will help to reduce costs, enhance machine lifetime, and improve efficiency of the system (Gang et al., 2014, [18]). The constructive management of facilities in the pharmaceutical industry is the greatest reflection for restructuring the plant with respect to cost. Facilities management is the key area for contribution of the business toward the plan strategy of the pharmaceutical industry. Facilities management of the pharma industry refers asset management and portfolio.

7.4.8 Supply chain

When the drugs are manufactured, the role of supply chain comes. There are so many risks such as fluctuation in temperature, accident of vehicles, and other delays while taking the batch of drugs from the plant or warehouse to hospitals or pharmacies [19,20] as shown in Fig. 7.7. IoT-based manufacturing and supply chain management area are the most popular area for investment in the field of pharmaceutical industries. Real-time tracking and monitoring of product in the pharmaceutical industry is only possible due to the smart packing and all equipment connected with IoT echo system.

For proper management, pharma companies should know the details of route so IoT offers this extra visibility to ensure all the real-time information, so that appropriate action should be taken at the time of need and to reduce delays.

TABLE 7.2 Traceability of products in the warehouse.

Requirement of business	Particulars	Remedy given by IoT	Improvisation through IoT
• Improved efficiency of operations performed in warehouse	• Upturn the motion of warehouse resources	• Information is transmitted by the sensors attached to the products, which ensure its location in warehouse, and it is directly sent to the manager's device • Motion of material can be analyzed by the RFID tags associated with the material	• Efficient utilization of resources • Increase productivity
• Retain the required conditions for storage	• For sensitive drugs, storage conditions should be checked	• Sensors to check the environmental conditions of the warehouse are embedded within the system. These sensors generate the alarm if temperature, humidity or pressure, etc. of the warehouse increases	• Quantity of drug expiration becomes low due to the prealert
• Production and demand should be aligned	• Product movement real-time data is taken so that requirements can be aligned	• Sensors on inventory collect the data and provide a real-time database for distribution from warehouse to production planning	• Effective management of stocks

IoT, Internet of Things.

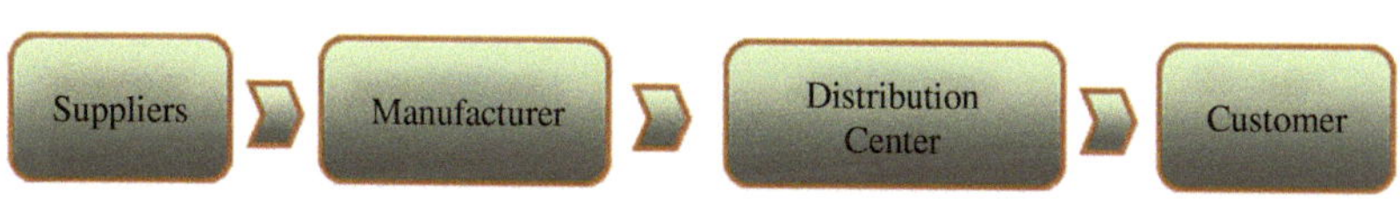

FIGURE 7.7 Supply chain.

Each product and every batch can have a RFID tag that ensures its easy identification [21]. Due to GPS-enabled devices attached to the vehicles update, the location and sensors can also combine with it to see the change of pressure, humidity, and temperature maintain the quality of drugs till reaching to the final destination. This also helps to calculate the timings of shipment, so that it can be reduced by proper management of repeated movements of the stocks and ultimately reduces the cost [22].

Table 7.3 shows the summarized details of material tracking during the supply chain that how it can be analyzed at every point and protected from duplicity and other thefts.

TABLE 7.3 Traceability of products during supply chain.

Requirement of business	Particulars	Remedy given by IoT	Improvisation through IoT
• Product accountability across the supply chain	• Authenticity check of the product at every point	• RFID tags, bar codes, electronic chip, etc. on packaging material • Information is communicated to both sides so that material can be tracked within the real-time database	• Retain the originality of the product
• Minimize the inventory cost within the partners of the supply chain	• Enhance the motion of product inventory in the hub of the supply chains	• Smart sensors used in packaging • Transmitters send the location information to the centralized system • Material is sent/received according to the database provided, and stocks out are managed in optimize level	• Efficient supply chain process ensuring stock out as per demand, which is cost effective for both the parties

IoT, Internet of Things.

7.4.9 Inventory management

Applications of IoT also include monitoring of the activities held during the supply chain. IoT systems track products on line level and provide the information to concern inventory departments. This realistic data helps to plan and organize accordingly. It makes a clear vision about the material available, progress in work, and estimated arrival timing of new material. Finally, it reduces the cost and optimizes the process of supply chain [23].

Today 80% of the pharma industries are normal, while competition is fierce [24]. To understand the concept of developing transport and capabilities, what is the factor that improves efficiency with low cost for digital systems? Digital technologies such as real-time monitoring of the situation, GPS location, and security during travel are used for digital supply. Fig. 7.8 shows how a producer can collect the performance of drivers, vehicles, and space using data collected from the vehicle. This information can be captured using the GPS and then through the General Packet Radio Service, the room is allowed to control the room and visibility in the product and transport situation, then the control panel will be able to access data, which enables one to analyze and compare [25]. It can be used to reduce marketing, protect goods destruction, and waste and reduce transport cost as given in Fig. 7.8.

Table 7.4 shows that how temperature records can be maintained in such a way that drugs wastages can be minimized.

7.5 Role of Internet of Things in challenges of pharma industry

As there is lack of transparency in the system, the pharma industries faces a lot of challenges in manufacturing and distribution of drugs in safe and secure manner. Due to this problem, the negative implications occur in the market such as drugs that are discarded and do not reach the destination in time, so revenue losses or even patient suffers a lot [26].

In today's competitive world, no one can afford the losses or nobody will allow the delays because of any reason. So, it is mandatory for pharma industry to control the activities performed inside and outside the system. Here comes the role of IoT, which makes it simpler and easy to implement.

IoT provides a lot of opportunity for pharma industries to connect with next level and have tools to fulfill the market demands on time. IoT makes it easy to connect with different people, other devices, supply chain processes into a single network with enhanced efficiency, reduced costs with safety and security [27]. Pharma industry should invest in developing good quality architecture for IoT for the benefit of handling the heavy-duty capacity of services and IoT-based security solution with predominant. Standardization is very challenging and difficult task in the field of pharma companies for

FIGURE 7.8 Management with sensors.

TABLE 7.4 Maintaining records of temperature variations.

Requirement of business	Particulars	Remedy given by IoT	Improvisation through IoT
• Maintain the quality of during the process of transportation	• Temperature of the drugs should be monitored in regular basis	• Environmental sensors are embedded within the shipment containers • Generate an alert when exceeding the range of temperature more than threshold level set	• Ensure the drug quality • Drug wastage is reduced as record of temperature is checked on regular basis

IoT, Internet of Things.

useful approach. To meet operational requirements and operational needs with key performance indicators [28], balancing the project requirements may be too heavy. At the same time, they can challenge the management to change a sensitive issue that people, processes, and responsibilities should be kept in mind. For successful transition in relation to stakeholders, the pharma industry needs to create a communication network to establish every explanation related to employees and stakeholders. Companies should review their investments, and along with the display, more value should be added to the business and their customers in relation to innovation, promotion, efficiency, and process [29]. Pharma industries can be prepared by preparing IoT using the best practices for competing with today's demand and hyperconnectivity of the global network.

Some challenges of pharma industries are explained in the following subsections.

7.5.1 Plant safety and security

Health and safety are the major aspects of every industry. IoT collaborated with big data analysis is very helpful to analyze the overall health and safety of a worker by checking the standard parameters of health and safety such as percentage of illness or injuries, absenteeism in long term and short-time period, accidents, damage in property in regular operations. So, effective and efficient monitoring ensures better safety. In pharmaceutical industry, there are two aspects that protect system using encrypt network. When the manufacturing process of drugs is going on using IoT-based network, it is necessary to maintain the system in order to plant safety and its security.

Huge amount of data is to be transferred from one place to another. Confidential data for any pharmaceutical industry need to be kept safe and secure from any unauthorized fraud. Drug formulation is the confidential information of any pharmaceutical plant, which is necessary to keep secure. Today's most of all pharmaceutical industries are connected with IoT-based networks, and they are selling their medicines online to the customer all over the world. Consumers made their payment through gateway using net banking or debit/credit card to the pharmaceutical company. At the time of payment, reliability of the IoT-based network comes into the picture after safe and secured transaction of the payment.

7.5.2 To overcome the short supply of drugs

It is very important concern of pharma industry to supply drugs on time so that it can be utilized in efficient manner. IoT technology is very helpful to optimize the inventories based upon the standard norms made for business, so in a planned manner decisions can be taken about shipment of desired material to the required place and about manufacturing of the drugs. The short supply of drugs in the market can be overcome by using the IoT-based network through which information can be shared easily by a pharmaceutical company. After getting the information of short supply of the drugs in a particular area, company generates new supply toward the retailer or medical store and problems get resolved easily with the help of IoT and connected electronic sensors. Customer can generate their quotation online to the company directly, and company delivers medicines to the customer. In this way, short supply of drugs can be overcome using IoT-based IP network.

7.5.3 Security of supply chain

Security for supply chain is an important parameter, and it has become stronger by using IoT as it provides communication exchange from both sides. The tracking of product during transit and by checking the current status in reference with the location is very useful in terms of security. These can be ensured by RFID tags, smart devices, 2D bars, etc. It helps in tracing the product from inventory to stakeholders.

It is also very important to record the temperature variation of the store products as in transit time, if the drugs got saturated, then it is of no use, so by the help of IoT sensors it can be recorded and checked as it is mandatory to have stable temperature of drugs till it reaches its final destination. IoT sensors are also inbuilt into the packets to trace the environmental conditions. This would decrease the wastages due to increase in temperature in the boxes of drugs.

7.5.4 Theft of drugs during transportation

In between the transportation, there is a big loss to pharma industry due to stealing of drugs. This type of activity reduces the revenue of pharma industry and also decreases the stakeholders as required amount of drugs are not supplied to them as per time. IoT technology (using GPS) helps one to locate the vehicles, and the RFID tags help one to know the current status of product, which ensures the security of whole consignment. So, logistics team would able to analyze the real-time situation of shipment from anywhere which reduces the stealing of drugs. The quotation is generated by a customer to the company, and then company deliver package to the customer using the RFID tags to prevent package from the stolen. The drug stolen during the transportation is the big loss of any pharmaceutical industry. It can be stopped when the shipment package is to be connected with IoT-based network through which vehicle location during the transportation can be traced. The current location of the consignment can be detected and monitored using IoT-based smart devices.

7.6 Conclusion and future scope

It's time for digitization everywhere, and, therefore, the pharma industry is adopting it as soon as possible. IoT actually has an important role to digitize an industry, but it is necessary for the pharma industry to include IoT as part of the focus. IoT will help us in finding and implementing chain components with adoption of potential candidates. For this purpose the pharma industry needs to take strict measures in favor of the reevaluation of systems, processes and to develop new business models with the help of the next upcoming architecture. IoT pharma could change business model with industry quality. On the other hand, IoT promotes new innovation and creates tremendous opportunities for the new era of change in the pharma industry. All over the world researchers have initiated to explore new technologies and solutions to promote pharmaceutical industry with existing problems by mobility of IoT applications. In this chapter, IoT-based pharma technologies network architecture and platforms are introduced, which assist to access the transmission and reception of pharmaceutical data with the help of IoT-based connectivity network. Comprehensive research projects have been proposed for solving the problems related to pharma and patients how the IoT is useful for medical problems such as chronic disease monitoring, care of pediatric, and management of health and fitness. For better intuition into pharma industry trends and empowerment of IoT-based technologies, this chapter provides a broader view on how the pharma industry is motivated for current and continuous facilitates in sensors, electronic devices, and Internet connectivity for pharmaceutical services and explores the use of IoT-based services for additional progress into pharmaceutical area with satisfactory solution. To understand the security issues in IoT-based pharmaceutical, this chapter will help one to mitigate research problems and challenges related to pharma security risks. The conversation of several issues such as pharma standardization, network topology, QoS, and pharmaceutical information preservation is anticipated for future continuous research on IoT-based pharmaceutical industry services. This chapter presents e-Pharma based on IoT policies and ordinance to provide more benefits to the patients using IoT-based pharma technologies and usefulness for engineers, researchers working in the area of IoT-based pharmaceutical technologies. For pharmaceutical industry, IoT enhances virtual visibility of the real-time pharma business such as manufacturing, distribution, and consumption. It is the right time for dynamic regulatory pharmaceutical industry to advance utilization of IoT and their implementation of solutions.

References

[1] B. Cortés, A. Boza, D. Pérez, L. Cuenca, Internet of things applications on supply chain management, International Journal of Computer, Electrical, Automation, Control and Information Engineering 9 (12) (2015) 2493–2498.

[2] K. Rose, S. Eldridge, L. Chapin, The Internet of Things: An Overview, The Internet Society (ISOC), 2015, pp. 1–50.

[3] D. Guinard, V. Trifa, E. Wilde, A resource oriented architecture for the Web of Things, in: Proc. Internet Things (IoT), 2010, pp. 1–8.

[4] J. Macaulay, L. Buckalew, G. Chung, Internet of Things in logistics: a collaborative report by DHL and Cisco on implications and use cases for the logistics industry, in: DHL Trend Research and Cisco Consulting Services, 2015.

[5] L. Da Xu, W. He, S. Li, Internet of Things in industries: a survey, IEEE Trans. Ind. Inf. 10 (4) (2014) 2233–2243.

[6] I. Toma, E. Simperl, G. Hench, A joint roadmap for semantic technologies and the Internet of Things, in: Proceedings of the 3rd STI Roadmapping Workshop, Crete, Greece, 2009.

[7] H. Sundmaeker, P. Guillemin, P. Friess, S. Woelfflé, Vision and challenges for realizing the Internet of Things, Cluster of European Research Projects on the Internet of Things, vol. 3, European Commission, 2010, pp. 34–36.

[8] I. Vazquez, Social Devices: Semantic Technology for the Internet of Things, Week@ESI, Zamudio, 2009.

[9] F. Hu, D. Xie, S. Shen, On the application of the Internet of Things in the field of medical and health care, in: 2013 IEEE International Conference on Green Computing and Communications and IEEE Internet of Things and IEEE Cyber, Physical and Social Computing, IEEE, 2013, pp. 2053–2058.

[10] Z. Pang, Technologies and architectures of the Internet-of-Things (IoT) for health and well-being (M.S. thesis), Dept. Electron. Comput. Syst., KTH-Roy. Inst. Technol., Stockholm, 2013.

[11] B.C. Hardgrave, M. Waller, R. Miller, RFID's impact on out of stocks: a sales velocity analysis, in: Research Report from the University of Arkansas, 2006.

[12] K. Romer, B. Ostermaier, F. Mattern, M. Fahrmair, W. Kellerer, Real-time search for real-world entities: a survey, Proc. IEEE 98 (11) (2010) 1887–1902.

[13] H. Choi, J. Song, K. Yi, Brightics-IoT: towards effective industrial IoT platforms for connected smart factories, in: 2018 IEEE International Conference on Industrial Internet (ICII), IEEE, 2018, pp. 146–152.

[14] T. Banerjee, A. Sheth, IoT quality control for data and application needs, IEEE Intell. Syst. 32 (2) (2017) 68–73.

[15] R. Beica, Enabling information age through advanced packaging technologies and electronic materials, in: 2018 Pan Pacific Microelectronics Symposium (Pan Pacific), IEEE, 2018, pp. 1–5.

[16] M. Wazid, A.K. Das, M.K. Khan, A.A.D. Al-Ghaiheb, N. Kumar, A.V. Vasilakos, Secure authentication scheme for medicine anti-counterfeiting system in IoT environment, IEEE Internet Things J. 4 (5) (2017) 1634–1646.

[17] G. Kortuem, F. Kawsar, D. Fitton, V. Sundramoorthy, Smart objects as building blocks for the Internet of Things, IEEE Internet Comput. 14 (1) (2010) 44–51.

[18] G. Yang, L. Xie, M. Mäntysalo, X. Zhou, Z. Pang, L. Da Xu, et al., A health-IoT platform based on the integration of intelligent packaging, unobtrusive bio-sensor, and intelligent medicine box, IEEE Trans. Ind. Inf. 10 (4) (2014) 2180–2191.

[19] V. Naidu, K. Mudliar, A. Naik, P. Bhavathankar, A fully observable supply chain management system using block chain and IoT, in: 2018 3rd International Conference for Convergence in Technology (I2CT), IEEE, 2018, pp. 1–4.

[20] M. Chamekh, S. El Asmi, M. Hamdi, T.H. Kim, IoT based tracking system for supply chain management, in: 2018 6th International Conference on Wireless Networks and Mobile Communications (WINCOM), IEEE, 2018, pp. 1–5.

[21] A.C. Madhukar Saboo, M. Suranglikar, The Internet of Things: the new Rx for pharmaceuticals manufacturing & supply chains, in: Cognizant 20-20 Insights, 2017, pp. 1–12.

[22] T. Bocek, B.B. Rodrigues, T. Strasser, B. Stiller, Blockchains everywhere-a use-case of blockchains in the pharma supply-chain, in: 2017 IFIP/IEEE Symposium on Integrated Network and Service Management (IM), IEEE, 2017, pp. 772–777.

[23] J. Höller, V. Tsiatsis, C. Mulligan, S. Karnouskos, S. Avesand, D. Boyle, From Machine-to-Machine to the Internet of Things: Introduction to a New Age of Intelligence, Elsevier, Amsterdam, 2014.

[24] A. Katasonov, O. Kaykova, O. Khriyenko, S. Nikitin, V. Terziyan, Smart semantic middleware for the Internet of Things, in: Proceedings of the 5th International Conference on Informatics in Control, Automation and Robotics, Funchal, Madeira, Portugal, 2008.

[25] J. Pasley, How BPEL and SOA are changing web services development, IEEE Internet Comput. 9 (3) (2005) 65–67.

[26] G. Santucci, Internet of the future and Internet of Things: what is at stake and how are we getting prepared for them? in: eMatch'99-Future Internet Workshop, Oslo, Norway, September 2009.

[27] W. Yao, M. Wu, J. Wang, Internet of Things in centralized management of medical equipment, in: 2018 11th International Congress on Image and Signal Processing, BioMedical Engineering and Informatics (CISP-BMEI), IEEE, 2018, pp. 1–5.

[28] C.E. Turcu, C.O. Turcu, Internet of Things as key enabler for sustainable healthcare delivery, Procedia 73 (2013) 251–256.

[29] L. Tan, N. Wang, Future Internet: the Internet of Things, in: Proc. 3rd Int. Conf. Adv. Comput. Theory Eng. (ICACTE), vol. 5, 2010, pp. V5-376–V5-380.

Further reading

C.S. Liu, S.X. Yan, T.H. Lai, The establishment of an integrated automation production line with multiple monitoring and control loops, in: 2018 IEEE International Conference on Applied System Invention (ICASI), IEEE, 2018, pp. 968–971.

L.D. Xu, W. He, S. Li, Internet of Things in industries: a survey, IEEE Trans. Ind. Inf. 10 (4) (2014) 2233–2243.

Chapter 8

Internet of Things technologies for elderly health-care applications

Jinesh Padikkapparambil[1], Cornelius Ncube[2], Krishna Kant Singh[3] and Akansha Singh[4]

[1]Higher College of Technology, Dubai, United Arab Emirates, [2]British University in Dubai, Dubai, United Arab Emirates, [3]Department of ECE, GL Bajaj Institute of Technology and Management, Greater Noida, India, [4]School of Computing Science and Engineering, Galgotias University, Greater Noida, India

8.1 Introduction

Nowadays, life expectancy has increased notably as compared to earlier decades. However, life expectancy of women and men differs in most cases in favor of women. Due to this cause, a main part of the elderly people live alone. The Middle East countries have witnessed the increased population of elderly people. The concern of elderly population becomes more severe. Health conditions of the aged people, including the ability of maintaining gait balance, neurological conditions, and cardiac function, are waning. Health care and safety observation of the aging individuals is turning out to be the latest research problem that needs research solutions. A major concern of progressively increasing elderly populace in several nations is the efficient supply of Medicare that is frequently problematic due to the declining state in their neurological state. Elderly people suffered from dementia, Alzheimer' or other health problems are required a health-monitoring system. This health-monitoring system can be made by Internet of Medical Things (IoMT). IoMT-based assistive system for elderly people plays a vital role in their life saving and alarms the patient to danger of life. The assistive IoMT system consists of various biomedical sensors and artificial intelligence (AI)-based algorithms connected with expert through Internet. The application of IoMT-based health care–monitoring frameworks helps aged persons get information about their health condition and find services provided by the health-care center without going outside the home. The health-care Internet

Emergence of Pharmaceutical Industry Growth with Industrial IoT Approach.
DOI: https://doi.org/10.1016/B978-0-12-819593-2.00008-X

of Things (IoT) based on medical digital devices makes the home health monitoring for the elderly possible. By setting up an IoT-based home care–monitoring system, the elderly can be aware of their medical conditions and avail the medical support and services by staying at their home. It would also make the government and the society able to cushion the blow of the aging population. In the home care–monitoring system the smart home gateway collects signals from the body sensor network and transmits them to the health-care server. The development of home gateway–based home care–monitoring systems has been through three stages. In the first stage the telephone modem acts as the home gateway, and data was transmitted through the telephone line. There are methods of monitoring systems by using modems for patients with chronic respiratory failure. This kind of health-care systems can transmit a limited amount of data with a limited transmitting speed, which restricts the expansibility of the system. Meanwhile, when the data needs to be transmitted, it needs to be manipulated by the patient, which seems not user-friendly. The popularization of the personal computer (PC) drives the health care–monitoring system to its second stage. In the second stage, PCs were used as the home gateway, and data was transmitted through broadband. There is no doubt that PCs have enough operation ability to process data, while they consume large electric power. One of the difficulties of an inexorably maturing populace in numerous nations is the compelling conveyance of social insurance administrations, which is regularly scrambled due to their deteriorating neural control situation. Specific consideration of the elderly individuals leads to incredible matter of concern for the family members, particularly, in the events that are not accompanied by any caretaker and are all by their own in their homes, where the likelihood for unknown conditions is significant. The option in contrast to staying within their comfort zone, particularly, in their own house is basically avoiding them to stay in care centers with high expenses; this increments more whenever particular consideration is given in an in-house arrangement. Endowing expansions for free living by the more seasoned individuals, such as the Ambient Assistive Living (AAL) frameworks, has evolved as a choice to upgrade help made available in a practical way. With the advent of technological advances in media transmission, processing, sensor availability, and the cell phone universality, start-to-finish and self-governing smart surrounding helped living have now turned into a conceivable truth. These frameworks ensure consistency along with ongoing observation about the surroundings and tenant conduct by event-based astute framework, appraisal, also enabling appropriate help when required. As a developing territory of research, it is fundamental to examine the methodologies received in creating AAL frameworks in the writing to recognize ebb and flow practices and bearings for impending research and improvement. As a rule, most systems concentrated on action observing for evaluating impending dangers, while the open doors for incorporating ecological variables for investigation and basic leadership, specifically for the long-haul care, were frequently neglected. Socio-social perspectives, for example, difference among gatherings, adequacy, and ease of the use of AAL frameworks were ignored. Likewise, there is a particular absence of solid supporting clinical proof in most as of now executed helped living innovations. In addition, the possibilities for wearable gadgets and sensors, just as appropriated stockpiling and access through the web, are still to be completely investigated for elderly health care.

8.2 Elderly population distribution

The aging individuals' populace on the planet is expanding because of advances in innovation, general well-being, nourishment, and drug [1]. Rising future and deteriorating birth rates will keep on impacting this huge move in socioeconomics globally, although at an alternate pace. People in the age-group of 60 or beyond that have increased to 11.5% of the total populace. Within the coming decades, this rate is relied upon to be multiplied, where 33 countries possess an excess of 10 million individuals in the age-group of greater than 60. Nations would encounter a noteworthy increment in their more established individuals' populace, just as a precarious decrease in their working-power populace. For instance, the level of the populace of more than or equal to 65 years of age expanded will increase drastically over the years. Open exchanges, especially for human services, assume a significant redistributive job to support the dimensions of utilization among more seasoned people in some high-salary nations. Then again, in low-pay and lower center salary nations, more seasoned people money the vast majority of their social insurance utilization all through off-take uses. The low dimensions of general well-being use in these nations add to an absence of well-being security and second-rate care for more established people. Older people's welfare is identified with the offer of utilization financed by open exchanges.

Internationally, the number of individuals above the age of 80 is developing considerably quicker than that of more established people by and large. In 2000 there were 71 million individuals matured 80 or more around the world. From that point forward the amount of most established old increased to 77% to 125 million in 2015, an additional increase of about 61% is estimated in the upcoming 15 years, coming to around 202 million of every 2030. Forecasts reveal that in 2050 the most established old will reach 434 million in total; this is a significant increase since 2015. About 66% of the world's more established people live in the creating districts, and their numbers are becoming quicker there than in the created areas. The more created locales were home to 38% of the world's more established people in 2000, and in any case, that rate tumbled to 33% in 2015 and is predicted to even go down. With the end goal, in 2030, 27% of the total populace matured 60 years or over will dwell in the more created locales. The rate of growth of the more established populace of the more created locales is forecasted to be adequate in the coming years. Interestingly, in the creating districts, the development of the populace matured 60 years or over is quickening. The amount of more seasoned people in the less created districts developed from 376 million in 2000 to 602 million out of 2015—a growth of 60%—and it is forecasted to develop by 71% for each penny somewhere in the range of 2015 and 2030, when forecasted 1 billion individuals matured 60 years or more will dwell in the less created districts.

In the continuous past the more settled masses of the least made countries were growing more steadily than in the diverse less made countries. In some place in the scope of 2000 and 2015, the amount of individuals developed 60 years or more at all-made countries extended by 54% stood out from 61% in the diverse less made countries. Regardless, advancement in the amount of increasingly settled individuals is enlivening even more quickly at all-made countries, with the true objective that, somewhere in the range of 2015 and 2030, the foreseen 70% extension in the people developed 60 years or over is practically unclear to that foreseen in the distinctive less made countries (71%). Notwithstanding such quick improvement in any case, the least made countries all things considered are foreseen to speak to simply 6.3% of the overall masses developed 60 years or more in 2030 and 8.9% in 2050, up from 5.8% in 2015 (Table 8.1 and Fig. 8.1).

In the future years, expanding future, a deteriorating birth rate, and the maturing of the time of increased birth rates age will significantly build the number and extent of the US populace beyond 65 years old. This maturing of the populace introduces various difficulties and unreciprocated inquiries, comprising where individuals can stay and methods through which they can acquire the help and care they will require as they age while holding however much autonomy as could reasonably be expected. Most seniors show that they would want to age setup, either remaining in their present home or looking over a scope of reasonable, age-fitting lodging choices inside their locale. A 2010 AARP review found that 88% of respondents aged more than 65 years needed to stay in their homes for whatever length of time that conceivable, and 92% said that they needed to stay in their communities. These alternatives can be made practical by adjusting homes and networks so as to cater the requirements of maturing inhabitants, make accessible moderate lodging choices appropriate for maturing occupants, and interface seniors to the administrations they need in the spots that they live.

A mix of open strategies, open and private key activities, and commercial center advancements try to meet the well-being and lodging needs of the rising senior populace by encouraging maturing setup and by utilizing lodging as a stage for getting to medicinal and nonmedical administrations.

8.3 Societal adaptions

Similarly as with home change, the network condition can be adjusted to encourage maturing setup both through retrofitting and new plan. Most

TABLE 8.1 Population demographics.

	Persons aged 60 years or over (millions)				Percentage change		Distribution of older persons (percentage)			
	2000	**2015**	**2030**	**2050**	**2000–15**	**2015–30**	**2000**	**2015**	**2030**	**2050**
World	607.1	900.9	1402.4	2092.0	48.4	55.7	100.0	100.0	100.0	100.0
Development groups										
More developed regions	231.3	298.8	375.2	421.4	29.2	25.6	38.1	33.2	26.8	20.1
Less developed regions	375.7	602.1	1027.2	1670.5	60.3	70.6	61.9	66.8	73.2	79.9
Other less developed countries	341.9	550.1	938.7	1484.9	60.9	70.6	56.3	61.1	66.9	71.0
Least developed countries	33.9	52.1	88.5	185.6	53.8	70.0	5.6	5.8	6.3	8.9
Regions										
Africa	42.4	64.4	105.4	220.3	51.9	63.5	7.0	7.2	7.5	10.5
Asia	319.5	508.0	844.5	1293.7	59.0	66.3	52.6	56.4	60.2	61.8
Europe	147.3	176.5	217.2	242.0	19.8	23.1	24.3	19.6	15.5	11.6
Latin America and the Caribbean	42.7	70.9	121.0	200.0	66.1	70.6	7.0	7.9	8.6	9.6
Oceania	4.1	6.5	9.6	13.2	56.2	47.4	0.7	0.7	0.7	0.6
Northern America	51.0	74.6	104.8	122.7	46.4	40.5	8.4	8.3	7.5	5.9
Income groups										
High-income countries	230.8	309.7	408.9	483.1	34.2	32.0	38.0	34.4	29.2	23.1
Upper middle–income countries	195.2	320.2	544.9	800.6	64.0	70.2	32.1	35.5	38.9	38.3
Lower middle–income countries	159.7	237.5	393.9	692.5	48.8	65.9	26.3	26.4	28.1	33.1
Low-income countries	21.2	33.2	54.0	114.8	56.2	63.1	3.5	3.7	3.9	5.5

(Continued)

TABLE 8.1 (Continued)

	Persons aged 80 years or over (millions)				Percentage change		Distribution of oldest-old persons (percentage)			
	2000	**2015**	**2030**	**2050**	**2000–15**	**2015–30**	**2000**	**2015**	**2030**	**2050**
World	71.0	125.3	201.8	434.4	76.5	61.1	100.0	100.0	100.0	100.0
Development groups										
More developed regions	36.5	59.1	85.2	127.8	61.8	44.1	51.5	47.2	42.2	29.4
Less developed regions	34.4	66.2	116.6	306.7	92.1	76.3	48.5	52.8	57.8	70.6
Other less developed countries	32.0	61.4	108.2	285.9	91.6	76.3	45.1	49.0	53.6	65.8
Least developed countries	2.4	4.8	8.4	20.7	99.2	75.4	3.4	3.8	4.2	4.8
Regions										
Africa	3.0	5.7	9.3	22.2	85.7	64.3	4.3	4.5	4.6	5.1
Asia	30.9	60.0	103.7	255.7	94.0	73.0	43.6	47.9	51.4	58.8
Europe	21.2	34.6	46.1	71.0	63.0	33.2	29.9	27.6	22.8	16.4
Latin America and the Caribbean	5.1	10.3	18.7	44.8	101.4	81.4	7.2	8.2	9.3	10.3
Oceania	0.7	1.1	2.0	3.6	69.8	76.8	1.0	0.9	1.0	0.8
Northern America	10.0	13.6	22.0	37.2	36.1	61.7	14.1	10.9	10.9	8.6
Income groups										
High-income countries	37.0	60.9	90.9	145.4	64.5	49.3	52.2	48.6	45.0	33.5
Upper middle–income countries	19.0	37.2	66.6	182.5	96.2	79.0	26.7	29.7	33.0	42.0
Lower middle–income countries	13.5	24.4	39.3	94.8	80.9	61.1	19.0	19.5	19.5	21.8
Low-income countries	1.5	2.7	4.9	11.3	83.6	80.9	2.1	2.2	2.4	2.6

Courtesy:United Nations, Department of Economic and Social Affairs, Population Division (2015).

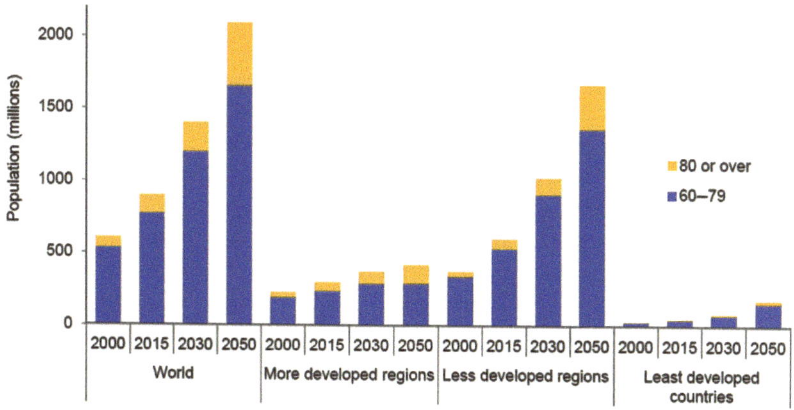

FIGURE 8.1 Population distribution. *United Nations (2015).*

families with occupants matured 65 and more seasoned are situated in a suburb. Suburbs, in any case, with their generally dispersed living arrangements that are frequently removed from supermarkets, specialists' workplaces, and different administrations and comforts, are ill-suited for seniors, particularly, the individuals who cannot drive. Ellen Dunham-Jones and June Williamson, creators of Retrofitting Suburbia, recommend that rural spaces can be repurposed to address the issues of maturing inhabitants—for instance— an empty strip shopping center could turn into a "restorative shopping center" as a one-stop goal for medicinal services. Similar adjustments are likewise fitting for rustic and urban regions. Network organizers imagine the structure of "deep-rooted neighborhoods" that are steady with savvy development standards and that can oblige inhabitants of any age by consolidating availability, person on foot access and travel, neighborhood retail and administrations, and open spaces for social collaboration. Getting ready for deep-rooted neighborhoods incorporates adaptable zoning statutes that can grow potential roads for maturing in spots, for example, embellishment abiding units (independent living units adjoining or inside a solitary family staying), cohousing, and multifamily lodging and would enable private and business regions to be arranged nearer together. Open travel offers a potential answer for seniors' versatility hindrances, yet customary travel frameworks are ordinarily designed for the necessities of suburbanites. As per an AARP investigation of the 2009 National Household Travel Survey, individuals aged more than 65 years made just 2.2% of their treks by open travel contrasted and in an excess of 87% via vehicles and 8.8% by walking. Paratransit administrations—way to entryway—request responsive administrations required by the Americans with Disabilities Act—could be an option in contrast to open travel; however, an expected 58% of more seasoned individuals don't meet all requirements for ADA paratransit administrations since they don't have a genuine disability. These administrations are likewise pricey; in 2011 the normal expense of a single direction paratransit trip was $34.59.

Whenever private and business utilizes are isolated and networks need availability, walkability, and sufficient open transportation, seniors become subject to their capacity to drive or get rides from others. Without trustworthy and reasonable transportation choices, seniors can experience issues getting to important merchandise and ventures and can turn out to be socially isolated. A recent report found that more seasoned Americans who do not drive make 15% less outings to the specialist, 59% less treks to shopping and cafes, and 65% less excursions for social or religious exercises than the individuals who do drive. Programs that give transportation through volunteer drivers or taxi endowments—or that help seniors keep driving securely for whatever length of time that conceivable—can enable more seasoned Americans to conquer versatility obstructions even in networks that are not especially walkable or well served by open transit. Techniques of improving existing homes, of consolidating generally helpful highlights in new homes, of structure attentive new networks, and of retooling existing neighborhoods must be comprehensively coordinated into our locale building methodologies at the nearby dimension over the United States, composed by Former Secretary of HUD Henry Cisneros. All these intercessions, and likely more, might be important to meet the different needs and expanding requests of a maturing populace.

8.4 Connected homes

Effective maturing setup relies upon access to required backings and administrations, both restorative and nonmedical. Various current models interface seniors with administrations and conveniences in their homes and networks; however, these may not be adequate to satisfy developing need. The essential methods for associating seniors to the help and care they need are through casual parental figures—companions, family, and neighbors—with only an expected 5% of more established individuals bolstered just by paid caregivers. The AARP Public Policy Institute appraises that the financial estimation of unpaid providing care achieved a stunning $450 billion in 2009. Research demonstrates that casual providing care enables seniors to defer or maintain a strategic distance from systematization even as their requirement for consideration grows. Although these discoveries bolster the attestation that casual providing care can enable seniors to age setup, other research proposes that unreasonable guardian stress frequently prompts the confirmation of a consideration beneficiary to a nursing home. Policymakers might be keen on supporting casual parental figures to diminish their pressure. Studies demonstrate that such help should concentrate on showing adapting abilities to manage "issue conduct" of the consideration collector, and they likewise demonstrate that extra backings may need to separate amid parental figures who are grown-up youngsters and the individuals who are spouses. The worry of providing care that presently falls on many children of postwar America may likewise have consequences for their own well-being as they age. One examination finds a relationship between providing care and weakness practices among parental figures that put their long-haul well-being at risk. Some investigations additionally propose that the time of increased birth rates companion will be less inclined to have a life partner or grown-up kids to give casual consideration and in this way will be bound to require nursing home care. Albeit all seniors are about to get support from casual guardians, some of them pick lodging alternatives that incorporate paid providing care.

"Hypothesis and proof help a job for sheltered and available lodging and administrations as an approach to keep up most extreme well-being, working, and freedom in the more seasoned populace and possibly delay or abstain from nursing home position, which is least favored by more established individuals and all around exorbitant for open projects," composed by Spillman, Biess, and MacDonald. A bunch of observational investigations discover proof of improved well-being results, upgraded profitable commitment, and open cost funds owing to different mediations that support maturing in place. Continuous assessment of existing activities and new projects is important and is as of now in progress. Maturing setup has turned into a "central idea in the insightful field of gerontology"; various scholastic establishments and research organizations are committing in regard to maturing issues, and—in another indication of the issue's striking nature—three of the five beneficiaries of MacArthur Foundation How Housing Matters concedes in 2012 are directing examinations identified with the lodging of more seasoned adults. For its part, HUD is at present assessing Vermont's SASH program, the previously mentioned HUD and HHS exertion to arrange HUD and CMS information, and the seniors and services of demonstration venture, which assesses the viability of models for associating seniors in sponsored lodging with steady administrations. Given the statistic elements of populace maturing, research and assessment should proceed close by advancement and practice in the arrangement of age-proper lodging and strong administrations. Progressively thorough research will be expected to recognize the most gainful and financially savvy projects to encourage maturing setup. On the off chance that effective, such activities will empower seniors to stay in their homes and networks, use lodging as a stage for well-being and different administrations, improve well-being and by and large personal satisfaction, and decrease the open expense of long-haul care.

8.5 What is Internet of Things?

The IoT is the trend in the next-generation technologies, which interconnects the smart devices in today's Internet structure with more associated benefits. The major advantage of the system includes the connectivity of these smart devices with the services and system that exceed the machine-to-machine structure. IoT plays a major role in building smart pervasive blueprints. A variety of application domains have the usage of IoT, mainly health care. The IoT revolution is reframing modern health care with assuring technological, social, and economic prospects. The count of elderly people who stay alone is increasing exponentially. Advancements in terms of monitoring them remotely are the need of the time, and IoT growth has leveraged the same. Elderly persons face a lot of challenges in conducting their day-to-day life, the challenges range from memory loss, susceptibility to diseases and impairments making them live in a survival mode. IoT assistance can be deployed for elderly monitoring with the aid of smartphones, vital signal sign monitors, smart wearables, audio–video sensors, smart devices, and smart TV. IoT techniques combined with Big Data and Cloud can be used effectively to give numerous solutions to the problem of remote elder monitoring. There are several manual aids that already exist, which are either slow or inefficient to be used with elders. The IoT-based remote monitoring can be used to monitor elders without any safety concerns and in an effective and uninterrupted way by deploying all the implementations in a typical living scenario. Right now, there are frameworks that encourage self-care and broaden the freedom of maturing populace. Such frameworks are frequently known as helped living frameworks. The fundamental points of the frameworks are to empower old individuals to autonomously live longer in their very own homes, to upgrade living characteristics and to decrease costs for society and general well-being frameworks. Helped living frameworks can help bolster old people with their day-by-day exercises so as to enable them to keep up sound and well-being while at the same time living freely. In this part, IoT advancements for older human services will be nitty gritty. You may have just gotten a feeling of how IoT gadgets can make life simpler for seniors from the past models. One of the greatest ways they can help is by making regular undertakings simpler. For example, purchasing more bathroom tissue can be as basic as requesting that your keen speaker buy the thing and after that hanging tight for it to touch base via the post office. It can likewise lessen the need to get up in the night, on account of gadgets that can turn on a fan or the cooling through your telephone. Savvy autos can help by sparing your as often as possible visited addresses and exploring for you.

8.6 Ambient assistive living systems

A major concern in health care is identified with the arrangement of economical consideration to the developing number of more established individuals within their premises or in assisted premises. The guarantee of these frameworks is the consistent observation of the surroundings and the patient with the help of an event-based system, with a point of contact for observing, evaluation, and activation of requisite aid whenever needed. These empowering advancements, alongside safeguard measures for solid and dynamic maturing, are commonly considered the path forward from the viewpoints of well-being and social consideration. The method of reasoning is that solid and dynamic maturing can bolster autonomy that empowers the more established individuals to lead a healthy life and consistent well-being.

Huge advances in media transmission, processing, and sensors scaling down, just as the universality of versatile and associated gadgets, are affecting the improvement of AAL frameworks. In spite of their ongoing advancement and exhibit of beneficial outcomes on more seasoned individuals' everyday living, a few restrictions of the examination and routine with regard to encompassing helped frameworks are distinguished [2]. It is essential to think about when managing AAL innovation the advancement of tasteful medical proof, for genuine enhancement in personal satisfaction that is accomplished by presenting this innovation [3]. Different difficulties can be identified with the meaning of the dimension of end clients' acknowledgment of the innovation, convenience, usage, protection, and moral concerns [4]. In particular, the requirements and requests of more seasoned individuals as innovation clients are not explicitly tended to where numerous activities are structured dependent on the specialists' presumptions. Well-being and care laborers in charge of the more established individuals are not in every case all around educated related to the effect of the actualized AAL frameworks in the viewpoints that influence the effort rehearses [5].

8.7 Requirements of activity recognition

Typically, AAL frameworks innovation that can bolster more established people groups' life ought to be able to:

- observe the exercises of individuals in the surroundings to ensure their well-being;
- understand their physical conditions to keep up their well-being and health;
- warn parental figures and relatives in case more seasoned individual is in challenges or showing critical misery;
- facilitate in-house restoration of the more established individuals utilizing programmed sounds or visual analysis; and
- mechanize some assignments that the more established individuals are unfit to work individually.

The previous discussion reveals that merely human exercises will demonstrate a few examples, rhythms, and patterns; yet these are not as standard as machines. Some sporadic exercises could be a piece of routine exercises. One of the significant difficulties to recognize rates by a mechanized observing framework, for example, a fall or meandering is to recognize genuine occurrences from nonfrequency exercises. This is the reason sensors increase can help for grouping of information through increasing data to sensors estimations, which might be significant in characterization. Current AAL frameworks for the most part intend to give client explicit help inside the home condition, for example, mechanized task of warming, ventilation, cooling framework (HVAC), lights, and cautions for medication. A few frameworks perform explicit assignments that require association with outside specialists or frameworks; for example, paying bills and requesting staple goods [6]. Now and again, AAL frameworks offer help for dull work, for example, home robots that help moving articles or displaying nourishment. For more seasoned individuals with psychological hindrance, the help ought to be responsive that is, the subject's day-by-day exercises ought to be observed to initially recognize his/her day-by-day exercises and afterward give the ebb and flow task important help. Some researchers examined the standards of action acknowledgment and showed that it may be very well extended to accomplish expanded societal advantages, particularly, in human-driven applications, for example, more seasoned individuals care. Their investigation concentrated on perceiving straightforward human exercises, while perceiving complex human exercises is as yet a test and a functioning region of research. The idea of the issue, which implies analyzing individual activities, needs a comprehension of the individual's activities. Of different procedures the first they tried for movement acknowledgment depended on an underlying customized model where, a calculated action stage should exist as the initial step to construct an inescapable recognizable proof framework. In another procedure, they tried was centered on using likelihood calculations to create a model for action acknowledgment; well-known strategies utilized in the design are the conditional random field along with HMM. Le et al. [7] outlined a technique that empowers action acknowledgment of a more established extrovert staying all by his own. They contemplated the instance of individuals staying alone in their homes, outfitted with nonintrusive nearness sensors, to distinguish and survey the lack of self-governance by considering the level of exercises carried out. The method initially recognized the subject's versatility states grouping in various areas around the space. At that point, through these conditions, they separated expressive standards to choose exercises that most affected the subject's independence. An action acknowledgment framework utilizing fluffy rationale in home situations with the assistance of a lot of physiological sensors, for example, heart recurrence, act, fall recognition sensing devices. They approved their methodology on a genuine domain and utilized this movement recognizable proof way to deal with fabricates a model for nervousness, with expanding or diminishing certainty as per the condition of every sensor utilized. They effectively installed the qualities of the information obtained from various sensors utilizing fluffy rationale, which permitted acknowledgment of everyday living exercises for nonexclusive medicinal services applications.

In another framework the accelerometers were used to recognize and characterize human movement into the following classifications: pushing ahead, upstairs, first floor, and bouncing developments. Their ID framework relied upon three distinct highlights: standard deviation, top plentifulness, and relationship amid various tomahawks, which utilized as contributions to a fluffy distinguishing proof framework. Fluffy principles and information/yield enrollment capacities were characterized from the exploratory estimations. Their outcomes bolstered the case that a fluffy induction framework outflanks different sorts of classifiers.

Papamatthaiakis et al. [8] utilized information extraction methods to fabricate a shrewd framework that can perceive human exercises. They contemplated regular indoor exercises of an observed subject. Their test results demonstrated that for certain exercises, the acknowledgment exactness beats different techniques depending on information mining classifiers. They guarantee that this strategy is sufficiently precise for dynamic situations.

A strategy for indoor action recognizable proof that connects the subject's movement and position information is also present in literature. The method connected a latency sensor that recognizes the introduction in three measurements to the subject's correct thigh for movement information accumulation and utilized an optical position framework to get the subject's area information. The optical situating framework can be supplanted by some other area recognition framework. This mix kept up high recognizable proof precision, while being less obtrusive. They used two neural systems to distinguish fundamental exercises. Right off the bat, Viterbi calculation for finding the no doubt grouping of shrouded states was utilized to perceive the exercises from movement information just, framing a coarse characterization arrange. Second, Bayes' hypothesis was connected to refresh the perceived exercises from movement information in the principal organized. They manufactured a fake condo to direct their trials. The obtained outcomes demonstrated that this strategy is compelling and creates satisfactory outcomes for movement acknowledgment.

8.8 Internet of Things–based technologies

In the previously discussed scenario of elderly health care, the IoT, an innovation that interfaces an assortment of ordinary gadgets and frameworks (e.g., sensors, actuators, machines, PCs, and cell phones), can give exceedingly circulated clever frameworks so as to associate a few gadgets and trade data with individuals and gathering the related information, in this manner speaking to a compelling answer for configuration keen home with incorporated e-well-being and helped living innovation. The use of IoT technologies and frameworks could assume a pivotal job in toppling the medicinal services framework for the old.

8.8.1 Activity recognition

An activity recognition framework ordinarily comprises two subframeworks:

1. a sensor framework that can distinguish what occurs in the ambience and
2. a savvy model that can perceive exercises from sensor data.

The point of encompassing knowledge is to enhance the surroundings with present day sensor gadgets interconnected by a correspondence system to frame an electronic worker, which detects changes in the environment, at that point reasons the reasons for these changes, and chooses the suitable activities important to profit clients of nature.

Chen et al. [9] directed a complete review looking at the improvement in sensor-based movement ID frameworks. They displayed a survey of the significant attributes of video-based and sensor-based action recognizable proof frameworks to feature the qualities and shortcomings of these procedures and to look at among information- and vision-driven action acknowledgment systems. Direct detecting includes the following factors that are identified with the subject himself, though roundabout detecting centers around recognizing the ecological state. Both immediate and aberrant frameworks are utilized in research and practice for catching human conduct. Direct detecting incorporates sound catch, camcorder, and movement sensors just as wearable body sensors. Crude information/signals from these sensors are exchanged to the database. Detected information are ordinarily clarified and frequently joined with one another to distinguish human conduct in the later phase of investigation, where health-related AAL frameworks can be partitioned into six fundamental classifications:

- Physiological assessment: these incorporate heartbeat rate, breath, temperature, circulatory strain, sugar level, gut and bladder yields, and so forth.
- Utilitarian assessment: these incorporate general movement level estimations, movement, walk ID and feast consumption, and so on.
- Well-being monitoring: these are identified with the investigation of information that distinguishes natural perils, for example, gas spillage. Security help incorporates capacities, for example, programmed task of restroom/hall lights, lessening excursions, and falls.
- Security monitoring: these are identified with estimations that recognize human dangers, for example, gatecrasher caution frameworks and reactions to distinguished dangers.
- Social association: these are identified with frameworks comprising video-based correspondence to help intervened association with family and virtual support in exercises and so on.
- Intellectual checking frameworks: these are identified with subjective help advances, including those of programmed updates and other psychological guides, for example, mechanized prescription and key locators.

They additionally incorporate verbal errand guidance advancements for machine task and sensor help advances that help clients with shortfalls, for example, sight, hearing, and contact [3].

Disseminated processing empowers more extensive arrangement of innovation in regular day-to-day existence. Shrewd sensors, gadgets, and actuators have turned out to be increasingly reasonable, amazing, and simple to introduce. Fast improvements in inserted frameworks and specifically the system on chip low-power processing engineering, for example, ARM empowered the installing of insight in regular gadgets and gear. Therefore individuals would now be able to be watched and aided their very own home as opposed to assembling them to clinics, bringing about practical and secure consideration supervision [10]. In addition, highlights-rich PDAs can have bidirectional correspondence with cloud framework to offload process substantial assignments, offering open doors for rich functionalities. They can be utilized to draw in older folks' considerations to specific activities, prerequisites, or direction, while experiencing their day-by-day life exercises, just as impart certain data to supporters and relatives in basic circumstances. Therefore these innovations can decrease human services costs altogether just as the physical weight on medicinal services supporters and relatives [11].

The difficulties of the AAL frameworks impact on clients were explored by Allameh et al., where they distinguished that a client's acknowledgment of individual space changes relies upon client's needs and way of life inclinations. Their work ordered the improvements in AAL frameworks into three sections: encompassing wise space (AmI-S), physical space, and virtual space (VS), which can be incorporated together to help autonomous life. In addition, their model takes into account changes in ways of life because of changes in client's movement design. At present, there is an enthusiasm for progressively point-by-point examinations on the linkage amid AAL frameworks and client's ways of life [12].

In certain applications, relatives and crisis administrations are likewise connected to the framework for moment cautioning in explicit circumstances. Sensors that produce twofold flag are ordinarily simpler to introduce and require less alignment than those with constant sign yield.

8.8.2 Wearable systems

Helal et al. [13] delineated a programmed circumstance age strategy to make a compelling sensor framework to screen exercises. Their framework establishes a 3D graphical UI to accomplish virtual spatial projection from mimicked sensors arrange in a computer-generated experience condition. This framework gives clients reenactment information to add to movement acknowledgment legitimately connected to a specific domain. Their work demonstrated how a 3D test system named Persim can be utilized for exercises recognizable proof purposes in an augmented simulation area to meld the datasets required for ongoing action acknowledgment application. Their framework is organized dependent on a PC interface utilized for producing information with respect to exercises completed by a virtual character in a VS utilizing Persim 3D's instinctive graphical UI [13].

On the investigation of a framework called Centinela was represented. This framework joins the subject's body quickening estimations with his indispensable signs to create very exact action recognizable proof framework. The framework focused on five principle exercises, strolling, sitting, running, plunging, and climbing stairs. Their proposed plan comprises a versatile recognizing gadget and a cell phone. Subsequent to testing three diverse time window sizes and eight distinct classifiers, results demonstrated that the Centinela stage can accomplish around 95% exactness, which beats different strategies when tried under similar conditions. Besides, the outcomes demonstrated that imperative signs estimations are significant in separating amid various kinds of exercises. These discoveries reinforce the case in which fundamental signs blended with movement data structure, a viable strategy to perceive human exercises by and large superior to relying upon movement information as it were. The situation of the sensors was a significant point in the investigation, where researchers distinguished that finding the movement sensor at the chest of the more seasoned social butterfly disposes of contentions that may come whenever appended to the wrist. Notwithstanding action acknowledgment, the framework displayed a continuous essential signs checking interface, adding simple well-being conditions observing to the movement acknowledgment target. In a recent technique the developers built up a remote and nonintrusive sensor framework that can catch the vital action data from the succession of sensor framework estimations. In this investigation, they proposed and assessed a sliding time window way to deal with distinguish exercises in a streaming design. To separate amid various exercises, they consolidated the alleged time rot relationship weighting of sensor estimations inside a period window. They finished up from their examination that consolidating joint data of weighted current sensor estimations and past logical data creates the best performing spilling movement recognizable proof framework. Tended to the issue of building up an action ID framework for helped living innovation application from the perspective of client acknowledgment, individual protection, and framework cost. The fundamental point of this exploration considered was to plan a movement recognizable proof framework for acknowledgment of nine distinctive day-by-day life exercises of a more established individual subject considering these angles. The investigation proposed a movement acknowledgment framework for a more established social butterfly utilizing minimal effort wrist-worn sensor gadgets. Their trial discoveries demonstrated that their framework can accomplish arrangement exactness that surpasses 90%.

They performed further factual tests to help this case, where they demonstrated that by consolidating estimation information from an accelerometer with the temperature sensor perusing, movement grouping exactness can be essentially improved.

8.8.3 Ready-to-use products

Numerous ready-to-use products are available in the market that can be readily used for remote health-care monitoring. GreatCall Responder is a little, GPS-empowered gadget that can without much of a stretch be joined to a keychain, tote, or rucksack [14]. It gives a simple and advantageous approach to protect an old at home and in a hurry. The client can speak with a prepared administration specialist by squeezing a catch on the responder. The specialist at that point surveys the circumstance and takes further fundamental activities. The framework additionally enables the client to contact the EMS legitimately. MobileHelp utilizes a GPS-empowered wearable framework and offers comparative benefits as GreatCall offers. GrandCare gives in-home social insurance and parental figure administrations for their customers. The framework speaks with the remote sensors introduced in the habitation over the web. Parental figures can sign into the GrandCare site to check the well-being status of the occupants. The parental figures are told if any uncommon exercises are distinguished. They additionally offer a wide scope of administrations counting correspondence and amusement administrations to their customers. BeClose remote observing framework, which is as of now possessed by Alarm.com, is intended to keep old in close contact with their family and parental figures. The framework utilizes discrete remote sensors put at various areas in the home to follow the day-by-day exercises of the old. The parental figures or relatives of the old can likewise screen his/her exercises utilizing a private and secure website page. The framework can tell the parental figures by telephone calls, messages, or instant messages if there should be an occurrence of any crisis. CareSmart Seniors Consulting Inc. (Kelowna, BC, Canada) offers remote checking administrations for the older by utilizing a remote observing framework from Care Link Advantage. The framework uses cameras to follow the exercises of the older living at home or to decide the dimension of criticalness. They counsel with the older and his/her family to distinguish the regions of concern and program the framework for producing warnings as needs be. On account of an issue, warnings are sent to the relatives and the guardians by means of messages, instant messages, and voice messages. Independa offers cloud-based old consideration benefits through a product stage that utilizes a savvy TV to associate the old with the guardians or the relatives [166]. The inhabitant utilizes a conventional remote controller to switch between TV shows and one of the Independa administrations. It offers correspondence administrations, for example, video talk, photograph sharing, message, and ready call between the old and the family individuals. It additionally reminds key occasions, for example, significant exercises of day-by-day living (ADL), arrangements with specialists, social commitment, and timetable of medicine.

As of now, many driving correspondences and media organizations, for example, Rogers Interchanges, Bell Canada, AT&T and British Telecom are offering savvy home answers for their clients. In spite of the fact that these arrangements offer brilliant administrations for observing the well-being and security and controlling the earth and the machines of the home, regardless they need far reaching social insurance checking administrations. Be that as it may, these arrangements are basically intended for enormous scale clinical conditions.

Samsung, one of the pioneer innovation organizations, has been attempting to make a brought together stage for elderly care. The stage is intended to be interoperable among Samsung and other gadgets. With the guide of this brought together stage, they are hoping to give customized, basic, also simple-to-utilize medicinal services arrangements, in this way offering better consideration, autonomy and improved way of life for the seniors. Alongside guaranteeing standard correspondence between the old relatives and the human services staff, they will likewise offer consistent availability between SMART TVs and apparatuses, therapeutic alarm administrations, estimation and observing of home condition, physiological signs, and exercises of the seniors.

8.9 Existing systems

"AAL homes" is a common terminology regularly used to depict the existing condition in which data and correspondence advancements are acquainted all together with help occupants' day-by-day living exercises, for example, moving furnishings, coordinated prescription, eating, dressing, and communicating. The beginning period of executing AAL homes ventures has an attention on disturbing framework in crisis circumstances, for example, the rate of fall and the framework was created principally by clients. It is apparent that most activities fundamentally go for appropriate ecological and individual situations observing, prior to accomplishing any needed help work. It got its name from the fundamental thought of home computerization, which utilizes an appropriated tactile framework to gather data identified with the condition of the earth where people are situated inside, at that point because of this data chooses certain activities and initiates explicit actuators to work certain home gadgets, plays out specific capacities, and trades information with outside spaces. An AAL home might be otherwise called a savvy space, a mindful house, or one utilizing synergistic encompassing insight. AAL homes that have these capacities can convey to more established individuals different sorts of home help, controlled drug, fall anticipation, and security highlights. Such frameworks produce a safe inclination for the more established individuals inside their homes. In addition, AAL will assist relatives with observing their more established individuals from anyplace with a web association [15]. Different research facility preliminaries, ventures, and modern exhibits concerning AAL homes are accessible around the globe; a ton of them share numerous highlights. After investigating the goals, these ventures are planning to accomplish, they utilize different mechanical advancement, data choice, approval technique, and results affirmation. In this regard, as of now accessible advances for AAL homes could be exhibited in three classifications:

- Social connectedness systems. Those focusing on encouraging social exercises, long-range interpersonal communication, and recognizable proof of social efficiencies.
- Safety enhancement AAL homes. Those focusing on fall identification, individual crisis, and drug the executives frameworks.
- Health monitoring AAL homes. Those focusing on overseeing endless issues. It additionally incorporates dynamic tele-well-being empowered remote association with the individuals under monitoring to collect their medical data.

AlarmNet is a venture created to give medicinal services observing to free living, as remote sensor-based AAL framework [16]. The framework utilizes varied gadgets that comprise wearable body sensors, conveyed remote sensors, UIs', just as database and basic leadership rationale. A portion of the correspondence and disturbing gadgets are versatile, and the remaining are static. Versatile body-worn sensors give physiological detecting to circulatory strain, beat rate, and accelerometer information. Data are gathered, sifted, collected, and utilized in regard to the prerequisites of the home occupants. The framework could be adjusted to the patient's ailments and can be customized to give certain warnings to indicated clients. Emplaced sensors gadgets are disseminated in the ecological to gather information, for example, surrounding temperature, dust rate, light power, and occupant's neighborhood position. For instance, weight sensors can be set on the ground to screen their stride's example to recognize expanded danger of fall, while a lot of bed sensors can screen breathing rate, pulse, and dimensions of development amid dozing. AlarmNet adaptability permits framework extension as more sensors gadgets are conveyed or new emerging situations need checking. Notwithstanding the inventive movement examination technique, AlarmNet framework engineering has its issues. AlarmNet is a shut engineering, without center to help outsider sensors gadgets or programming, which confines the answer for the components utilized in the plan [17].

Helped cognition environment venture was created to target concentrating the utilization of AI methods to improve and offer help for more established individuals' everyday life who are living with intellectual issue [18]. The created framework gives the capacity to detect the encompassing condition, patient's area, translate patient's personal conduct standards and offer help to the patient through verbal and physical intercessions. The framework additionally can give certain alarms to guardians if the circumstance requires this.

Mindful home venture aimed at conceptualizing a home setting of more established individuals. This is accomplished by acquainting omnipresent figuring with give significant data to their relatives. The proposed framework makes use of a ground response framework model for monitoring the activities of the elderly. The movement data was analyzed with Hidden Markov Model [19].

The CareWatch framework concentrated on observing dozing examples of intellectual weakened individuals and initiating a few warning frameworks for consideration suppliers. Dozing design change is one of the regularly happening side effects in psychological decay and dementia people, which influences their psychological and related conditions and conveys weights to the concerned consideration suppliers [20]. CareWatch venture expects to educate parental figures to give the fundamental required consideration data for more seasoned individuals who have subjective debilitation staying in their house. It is intended to avert normal houses ways out, particularly, amid the night, and to discharge a portion of the weight from the consideration suppliers amid evening times. It has demonstrated to be versatile in various home conditions. The framework is intended to expand the personal satisfaction for both the consideration beneficiary and the parental figure in the meantime, particularly, amid evening times. BioMOBIUS venture is shaped from open structure programming connected with equipment that permits brisk plan and usage for medicinal applications. BioMOBIUS was created for examination on signal acknowledgment, development investigation. BioMOBIUS stage consists of detecting framework for screening physiological features, an information handling stage that utilizes various strategies, and a savvy operator that changes over estimations into valuable expressive data for the clinicians. The objective is to screen circulatory strain, stride dependability, chance readiness, and social action. The framework is versatile for an assortment of equipment through its conventional utilization of blended-wired and remote interface systems. CASAS was developed for distinguishing the personal conduct standard of more seasoned individuals living with dementia, and dissecting their particular conduct by utilizing AI strategies. The exercises of more established individuals who are psychologically sound and those determined to have dementia issues are checked by movement and different kinds of sensors, while the gathered information are investigated to characterize "subjective well-being" and "dementia" conduct in the test condition. The outcomes demonstrated that the learning calculation utilized can recognize the contrasts amid the exhibitions of exercises; in any case, it cannot recognize the reason for these distinctions, regardless of whether it is a consequence of perplexity because of dementia or essentially only a mix up.

Another method utilized an alternate observing strategy to distinguish the anomalous conduct of dementia people. They utilized arrangement of checking signals from various areas so as to depict the progression of the inhabitant's action inside the home, close by the time length of these sign. These are recorded for dissecting the example of day-by-day exercises by utilizing a bunching strategy. Their strategy outlines a superior refinement in capacity than the earlier techniques. Casattenta framework comprised a lot of area with stationary sensors circulated close by the checked condition, a lot of wearable gadgets, and a correspondence stage [21]. The framework's fundamental target was to follow occupants' well-being and day-by-day live development planning to help more established individuals staying aloof. The combination of every framework components depended on the celebrated ZigBee information correspondence remote strategy, which enabled the framework to follow and perceive basic circumstances for more established individuals, for example, threat of falls and fixed status circumstances. CodeBlue remote sensors for therapeutic consideration venture were developed [22] and analyzed the use of remote tactile systems innovation utilization with the scope of medicinal services for individuals with cardiac problems and calamity reaction. The undertaking utilized a remote tactile system (WSN) that comprises battery-controlled sensors gadgets improved through suffice calculation and correspondence blocks. The wireless sensor network utilized requirement for imperative signs programmed accumulation, handling, and incorporation into the patient consideration record framework for constant restorative use. Numerous monetarily accessible remote therapeutic sensors were used [17]. Safe programming foundation is created for remote medicinal gadgets to verify data trade with patient's therapeutic recorders, PDAs, PCs, and other checking gadgets that could be utilized to screen patient's well-being. Gator-Tech shrewd house venture was created by the versatile and unavoidable figuring research center. The Gator-Tech brilliant home is a savvy situation intended to help the more seasoned individuals in their day-by-day live exercises (Helal et al., 2005). The Gato-Tech shrewd home undertaking depends on certain steady highlights dispersed in the home area, for example, brilliant machines, attachment, and savvy walking area for activity monitoring. The general framework has a nonexclusive plan for savvy space condition comprising meanings of administration related to the various sensing devices conveyed in the observed condition to shape the requisite more established individuals support. Georgia-Tech mindful brilliant home task was created to address the mindful assistive living [23]. In addition, indoor position following was executed utilizing basic RFID sensors and vision arrangements. The Georgia-Tech task's exploration bunch made action recognizable proof tending to general inhabitants' exercises, for example, sitting in front of the TV, perusing, setting up a supper, or blood glucose checking with the end goal of consistently movement observing. The general framework planned to take care of the psychology of the general population who are unable to take self-care appropriately and need assistance. GERHOME utilized programming stages for programmed acknowledgment of human practices utilizing ongoing video reconnaissance joined with different kinds of sensor information. The venture displayed a correspondence foundation that permits simple reconciliation of various sorts of sensors inside a current framework structure, in light of clever operator design. GERHOME comprises connected and remote sensors embedded in furniture and machines around the house to gather information on the utilization of these offices to derive exercises of the more established individuals inside the home condition. The gathered information are broken down so as to select uncommon practices or recognize altering patterns in conduct. It planned to mechanize a remotely located therapeutic guidance for postponing more established individuals' access to nursing homes. To address this objective, three activities should have been considered. Right off the bat, the framework expected to play out an evaluation of the more established individuals' slightness utilizing multisensors examination for action acknowledgment to fabricate a knowledge base from 3D geometric data of the individual under observation. Besides, the framework requires distinction disturbing circumstances. Another point to be considered is that any nonconformity from the normal behavior should be monitored and acted upon to identify any unusual conditions.

I-LivingTM framework aimed at giving meaning to structure an elderly care foundation that permits circulated remote sensor gadgets of various correspondence conventions to cooperate in a protected way [24]. The UI was intended to give different kinds of administrations to empower more seasoned individuals with various capacities to upgrade their freedom and additionally helped living prerequisites. The point was to utilize as of now financially accessible module innovations in detecting undertakings, for example, RFID, restriction, and nearness ID modules, while utilizing remote correspondence organizing advancements, in open-framework design. MavHome is a venture to accomplish robotization of the living place for helping everyday living exercises in which the example of person's exercises was demonstrated by AI [25]. So as to accomplish the objective, the framework perceives and predicts the everyday activities of the inhabitants. MavHome utilizes distinctive shrewd action ID calculations that use natural sensors and actuators to accomplish this objective. Their distinguishing proof calculation design comprised four layers, a data layer that gathers and spares data from sensors, an information correspondence layer that controls information trade amid layers, a basic leadership layer for executing activities, and a physical layer comprising actuators conveyed about situations to accomplish some assignments.

SOPRANO venture aimed at developing an "encompassing helped living" condition to help more seasoned people groups' to help them live freely. A subjective procedure is created dependent on experience and application looking into destinations to distinguish concerns and requirements of more established individuals' living movement in the network. More seasoned individuals are urged to join the exploration as members in center gatherings, singular meetings, and appraisal process all through the examination time. MITHouse concentrated mainly on the structure components along with related innovations for a smart place of living to all the more likely serve the future for more established individuals. A research center office outfitted with sensors in different areas was developed close MIT for test preliminaries. A product stage was actualized that was utilized to create inventive kinds of UIs. The venture examined the requirements for natural conditions checking, proactive social insurance, biometric observing, indoor air quality, and new development arrangements requirements for well-being and movement observing [9]. Different kinds of sensors gadgets were inserted in the earth, for example, infrared transmitters, camcorders, receivers, and biosensing devices altogether combined for gathering different information related to the clients and the vicinity of their stay. Sensing devices are used to screen exercises in the lab first with the goal that analysts had the option to contemplate how individuals respond to new gadgets situated in the earth around them. Information perception and UIs to the framework were built up to enable numerous versatile gadgets to impart and communicate with the framework. ORCATECH is a venture that was committed to examine the advancement of innovations that help autonomous living for a wide scope of prerequisites in more established individuals' well-being observing and home consideration support. The framework included savvy bed sensors that had the option to follow more seasoned individuals resting design and help them manage themselves by switching the lights based on the sensing of bed status. The framework likewise offers remote controlled tele-nearness with the end goal of providing elderly people who stay individually, to give well-being care just as real world collaborations with distant social relations [26]. The framework centers extensively around the utilization of purchaser gadgets to upgrade the personal satisfaction of more seasoned individuals and furnish them with fundamental assistance to accomplish dynamic and autonomous life. The task group tried a few regular living applications, for example, the area of items, utilizing drug gadgets, checking individual imperative signs, identifying design anomalies, sending explicit proper notices, and using automated stages to improve more seasoned people groups' aptitude and achieve capacity. A few off-the-rack innovations were used in these tests, for example, RFID labels, cross examiners, advanced mobile phone, and other similar devices. Brilliant Medical Home is a framework designed to progress intelligent advances utilized for home social insurance [27]. The task is gone for creating advances to increment forward location and expectation of the patient's well-being and ailment. The framework utilized an intelligent medicinal warning framework to cooperate with the concerned in the brilliant restorative home just like the prevailing consideration suppliers to give the dimension of help needed by the individual under observation. Utilizing discourse acknowledgment and man-made consciousness methods together with patient's accessible medicinal information, the intelligent framework encourages inhabitants to identify conceivable sickness utilizing organized intuitive inquiries and replies continuously. The framework likewise gives inhabitants data in regard to conceivable drug that can be utilized, their reactions, and other medical problems and along these lines helps individuals and care suppliers better to comprehend a doctor's guidelines.

Another application used a combination of multiple sensors for elderly care [28]. The essential testing office was worked in 2002 as a brilliant condo space, which was stacked with different sorts of sensing devices put to identify and regulate the ecological factors. A wide variety of sensors were used, for example, indoor regulators, accelerometers, receivers, attractive switches, RFID labels, movement sensors, and keen get bars. The TAFETA venture used weight delicate floor cushions, bed tangles, and seating pads to screen development ceaselessly in the condo space. Similarly mattresses are utilized for screening the breath to recognize resting nature of tenants. The framework additionally given cautioning signals if there should be an occurrence of conceivable well-being risk issues dependent on the past medical transactions and records of the individual under observation. WellAware venture gave a coordinated structure that utilized tangible framework and UI to empower proficient parental figures just as relatives to remotely screen and convey backing to more seasoned individuals [24]. The framework utilized numerous vicinity and movement finders that were conveyed in the keen house and utilized the ZigBee remote convention to speak with the principle PC. Significant parts of WellAware are identified with sensors that trail the development of the individual, remote information organizing, and controlling programming with UI monitoring more established individuals' exercises to typicality. The framework likewise gives site access to parental figures for remotely monitoring the state of the more established individuals and select before intercession necessities for genuine well-being conditions.

TeleCARE is a venture showing a nonexclusive engineering for encompassing assistive living condition [29]. The venture gives reflection to both equipment and programming utilized without determined data about managing outsider equipment drivers. Besides, there is no unmistakable data of the equipment prerequisites for the structure. This is a significant issue while going for an answer intended to help more seasoned individual residents and to accomplish constant correspondence among them and their relatives. CAALYX venture is basically worried about how the more seasoned individuals will utilize the framework. CAALYX venture sensors are situated in an interesting equipment structure to adjust to more seasoned individuals necessities. The CAALYX venture depends on cell phones, which creates an issue related with the telephone battery. It is hard to ensure that the more seasoned individuals will dependably make sure to charge their telephones, which can without much of a stretch trade off the framework. Then again, CAALYX demonstrates an increasingly open arrangement, which is anything but difficult to design using the TV set.

8.10 Conclusion

Thus from the discussion presented in the previous sections it can be concluded that IoT is a upcoming technology for elderly health care. Numerous researchers have worked on this technology and have proposed many innovative frameworks in this area. However, significant research scope is still required in the field.

References

[1] J. Beard, S. Biggs, D.E. Bloom, L.P. Fried, P.R. Hogan, A. Kalache, et al., 2012. Global Population Ageing: Peril or Promise? (No. 8912), Program on the Global Demography of Aging.

[2] A.J. Bharucha, V. Anand, J. Forlizzi, M.A. Dew, C.F. Reynolds, S. Stevens, et al., Intelligent assistive technology applications to dementia care: current capabilities, limitations, and future challenges, Am. J. Geriatr. Psychiatry 17 (2) (2009) 88–104.

[3] G. Demiris, B.K. Hensel, Technologies for an aging society: a systematic review of "smart home" applications, Yearb Med. Inform. 3 (2008) 33–40.

[4] C.K. Or, B.T. Karsh, A systematic review of patient acceptance of consumer health information technology, J. Am. Med. Inf. Assoc. 16 (4) (2009) 550–560.

[5] C.M. Blaschke, P.P. Freddolino, E.E. Mullen, Ageing and technology: a review of the research literature, Br. J. Soc. Work 39 (4) (2009) 641–656.

[6] G. Van den Broek, F. Cavallo, C. Wehrmann, AALIANCE Ambient Assisted Living Roadmap, vol. 6, IOS Press, 2010.

[7] X.H.B. Le, M. Di Mascolo, A. Gouin, N. Noury, Health smart home for elders—a tool for automatic recognition of activities of daily living, in: Engineering in Medicine and Biology Society, 2008, EMBS 2008, in: 30th Annual International Conference of the IEEE, IEEE, 2008, pp. 3316–3319.

[8] G. Papamatthaiakis, G.C. Polyzos, G. Xylomenos, Monitoring and modeling simple everyday activities of the older people at home, Consumer Communications and Networking Conference (CCNC), 2010 7th IEEE, IEEE, 2010, . 1–5.

[9] L. Chen, J. Hoey, C.D. Nugent, D.J. Cook, Z. Yu, Sensor-based activity recognition, IEEE Trans. Syst. Man Cybern., C: Appl. Rev. 42 (6) (2012) 790–808.

[10] S. Dengler, A. Awad, F. Dressler, Sensor/actuator networks in smart homes for supporting older people and handicapped people, in: IEEE, 21st International Conference on Advanced Information Networking and Applications Workshops, 2007, AINAW'07, vol. 2, 2007, pp. 863–868.

[11] M. Fahim, I. Fatima, S. Lee, Y.K. Lee, Daily life activity tracking application for smart homes using android smartphone, in: 2012 14th International Conference on Advanced Communication Technology (ICACT), IEEE, 2012, pp. 241–245.

[12] E. Allameh, M.H. Jozam, B. de Vries, H. Timmermans, J. Beetz, Smart home as a smart real estate: a state of the art review, in: 18th Annual European Real Estate Society Conference, Eindhoven, The Netherlands, 2011.

[13] A. Helal, K. Cho, W.J. Lee, Y. Sung, J.W. Lee, E. Kim, 3D modeling and simulation of human activities in smart spaces, in: Ubiquitous Intelligence & Computing and 9th International Conference on Autonomic & Trusted Computing (UIC/ATC), 2012 9th International Conference on, IEEE, 2012, pp. 112–119.

[14] Health Apps, Cell Phones for Seniors, Medical Alert|GreatCall. Available from: <https://www.greatcall.com/> (accessed 10.06.2019).

[15] P. Cheek, L. Nikpour, H.D. Nowlin, Aging well with smart technology, Nurs. Adm. Q. 29 (4) (2005) 329–338.

[16] Alarmnet, www.cs.virginia.edu/wsn/medical/, 2011 (accessed 21.05.14).

[17] A.D. Wood, J. Stankovic, G. Virone, L. Selavo, Z. He, Q. Cao, et al., Context-aware wireless sensor networks for assisted living and residential monitoring, Network IEEE 22 (4) (2008) 26–33.

[18] H. Kautz, D. Fox, O. Etzioni, G. Borriello, L. Arnstein, An overview of the assisted cognition project, in: AAAI-2002 Workshop on Automation as Caregiver: The Role of Intelligent Technology in Elder Care, 2002.

[19] C.D. Kidd, R. Orr, G.D. Abowd, C.G. Atkeson, I.A. Essa, B. MacIntyre, et al., The aware home: a living laboratory for ubiquitous computing research, Cooperative Buildings. Integrating Information, Organizations, and Architecture, Springer Berlin Heidelberg, 1999, pp. 191–198.

[20] M. Rowe, S. Lane, C. Phipps, CareWatch: a home monitoring system for use in homes of persons with cognitive impairment, Top. Geriatr. Rehabil. 23 (1) (2007) 3.

[21] E. Farella, M. Falavigna, B. Riccò, Aware and smart environments: the Casattenta project, Microelectron. J. 41 (11) (2010) 697–702.

[22] T. Codeblue, Wireless sensor networks for medical care. Available from: <http://fiji.eecs.harvard.edu/CodeBlue>, 2011.

[23] J.A. Kientz, S.N. Patel, B. Jones, E.D. Price, E.D. Mynatt, G.D. Abowd, The Georgia tech aware home, CHI'08 Extended Abstracts on Human Factors in Computing Systems, ACM, 2008, . 3675–3680.

[24] M. Bal, W. Shen, Q. Hao, H. Xue, Collaborative smart home technologies for senior independent living: a review, in: Proceedings of the 2011 15th International Conference on Computer Supported Cooperative Work in Design (CSCWD), 2011, pp. 481–488.

[25] D.J. Cook, M. Youngblood, E.O. Heierman III, K. Gopalratnam, S. Rao, A. Litvin, et al., MavHome: an agent-based smart home, in: null. IEEE, 2003, . 521.

[26] J. Nehmer, M. Becker, A. Karshmer, R. Lamm, Living assistance systems: an ambient intelligence approach, Proceedings of the 28th International Conference on Software Engineering, ACM, 2006, . 43–50.

[27] V. Ricquebourg, D. Menga, D. Durand, B. Marhic, L. Delahoche, C. Loge, The smart home concept: our immediate future, in: 2006 First IEEE International Conference on E-Learning in Industrial Electronics, December 2006, pp. 23–28.

[28] TAFETA, Tafeta smart systems for health. Available from: <www.tafeta.ca/home2.html>, 2011.

[29] P. Whitten, B. Collins, F. Mair, Nurse and patient reactions to a developmental home telecare system, J. Telemed. Telecare 4 (3) (1998) 152–160.

Further reading

N. Agoulmine, M.J. Deen, J.S. Lee, M. Meyyappan, U-health smart home, IEEE Nanotechnol. Mag. 5 (3) (2011) 6–11.

BiomobusI, BioMobus research platform. Available from: <http://biomobius.trilcentre.org/index.php>, 2011.

U. Bischoff, V. Sundramoorthy, G. Kortuem, Programming the smart home, in: Third IET International Conference on Intelligent Environments (IE 07), 24–25 September 2007.

L. Chen, C. Nugent, Ontology-based activity recognition in intelligent pervasive environments, Int. J. Web Inf. Syst. 5 (4) (2009) 410–430.

D. Oliver, C. Foot, R. Humphries, Making Our Health and Care Systems Fit for an Ageing Population, King's Fund, London, UK, 2014.

P. Rashidi, D.J. Cook, L.B. Holder, M. Schmitter-Edgecombe, Discovering activities to recognize and track in a smart environment, IEEE Trans. Knowl. Data Eng. 23 (4) (2011) 527–539.

Y. Wu, T.S. Huang, in: A. Camurri, G. Volpe (Eds.), Vision-Based Gesture Recognition: A Review, Gesture-Based Communication in Human-Computer Interaction, Springer, 1999.

Chapter 9

An insight of Internet of Things applications in pharmaceutical domain

Sushruta Mishra[1], Anuttam Dash[1] and Brojo Kishore Mishra[2]
[1]*KIIT Deemed to be University, Bhubaneswar, India,* [2]*GIET University, Gunupur, India*

9.1 An overview of Internet of Things

IoT, also known as the Internet of Things, is the internetworking of physical nodes that contain electronic devices embedded into the proposed architecture in directive to sense and communicate interactions with each other or in correlation to the external environment. IoT can be termed as a new revolution in the digital world. Objects have gained the capability to self-recognize themselves and have obtained intelligent architecture by enabling or making context-centric decisions. These objects are also able to transfer information about own self with each other. The information that is pooled by some other objects can be accessed by them, or there could be proponents of several other complex types of services. Such alteration has proved to be connected with the onset of abilities of cloud computing and the evolution of the Internet in the direction of IPv6 addressing standard with a nearly boundless amount of addressing capabilities. In the upcoming years, IoT-based technologies will offer advanced levels of services which, in turn, will change the way people lead their day-to-day lives. Power, medicine, agriculture, gene therapies, smart homes, and smart cities are a few of the distinct examples where IoT is extensively used and which will get more advanced in the upcoming days. Around 9 billion number of "Things" (physical objects) are now connected to the Internet. In the coming days, this number is estimated to rise to a whopping 20 billion mark.

IoT-based models permit end users to accomplish analysis through in-depth automation along with proper integration with a computer node. They extend the outreach and accuracy of these domains. IoT makes use of a combination of emerging as well as existing technologies used in networking, robotics, and sensor technologies.

Emergence of Pharmaceutical Industry Growth with Industrial IoT Approach.
DOI: https://doi.org/10.1016/B978-0-12-819593-2.00009-1

IoT makes effective use of developments in the reduction of the price of hardware units, software, and latest concepts in the direction of technology. The new and advanced elements of the IoT architecture have brought significant variations in the delivery of furnished products and services, and the socioeconomical and political importance of those alterations.

There are mainly four following components used in IoT:

- *Low power–based embedded systems*
  High-performance and less battery consumption are the inverse factors that play a noteworthy role in the designing of electronic systems.
- *Cloud computing*
  A massive amount of data is collected from the IoT devices that are further required to be stored in a reliable centralized server. This is where we need cloud computing. The data are learned and processed, giving additional opportunities to us to determine where things such as electrical errors/faults can occur within the system.
- *Availability of big data*
  IoT depends deeply on sensors, particularly real time. As these end-nodes spread in every field, their usage is going to generate a huge flux of big data.
- *Networking connection*

Internet connectivity especially where each and every physical object is signified by an IP address is a must for the communication. However, according to IP nomenclature, only a few numbers of IP addresses are available. Due to a rapid increase in the number of connected devices, the existing IP nomenclature can no more be feasible. Therefore researchers are in search of another naming system in order to meet the growing number of connected devices. The IoT infrastructure is illustrated in Fig. 9.1.

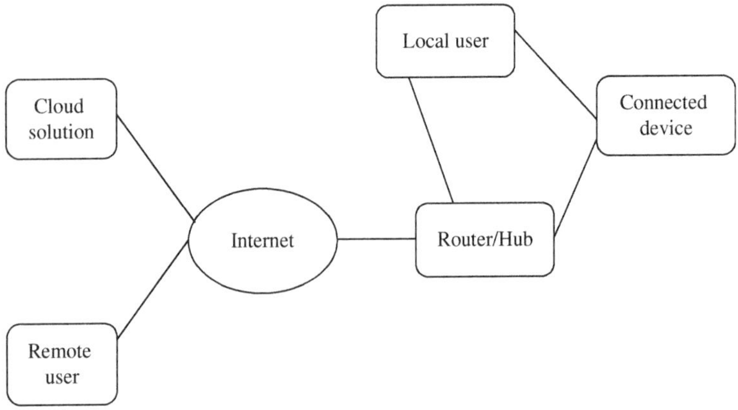

FIGURE 9.1 The general overview of an IoT infrastructure. *IoT*, Internet of Things.

9.2 Characteristics features of Internet of Things

The widely known characteristics of IoT comprise interconnectivity, artificial intelligence (AI), sensory devices, miniature node use, and active engagement. These features have been mentioned briefly in the following:

- *AI*

  IoT has the capability to virtually make everything smart, thereby improving various aspects of one's life with the influence of computational intelligence algorithms, networks, and collection of data. It implies that it is somewhat as simple as improving your cabinets and refrigerators to determine when your favorite cereal or milk is about to finish and then ensure to have the food item ordered from your preferred dealer.
- *Connectivity*

  New supporting technologies in the field of networking, specifically networking in the IoT domain, mean that networks are not anymore exclusively knotted to big players in the list of network providers. Networks are now available for a much less and inexpensive scale, still being relatively more practical. IoT has the ability to create such kinds of small-scale networks within its system devices.
- *Sensors*

  Without sensors, IoT can lose its dissimilarity. Sensors are important devices, which transmute IoT from a normal passive form of the interconnectivity of devices to active form of system, which is capable of integrating with the real work environment.
- *Active engagement*

  A major part of communication with the connected world of technology in today's world happens in the mode of the passive form of engagement. IoT presents a novel methodology for active kind of product, content, or service management.
- *Small devices*

  As predicted earlier, devices are now more cost effective, powerful, and smaller with due course of time. IoT makes use of such tiny nodes to provide their scalability, precision, and versatility.

  The fundamental characteristics of IoT architecture have been mentioned next:
- *Interconnectivity*

  According to IoT, anything may be interconnected around the globe for data sharing, communication, and information frameworks.
- *Things-related services*

  Thing-related services revolve around the restraints of things, such as consistency between physical devices and their related virtual things on the basis of semantics, and privacy protection, which can be effectively provided by IoT. In order to provide these forms of services around the restraints of things, both the physical and information world technologies will change.

- *Heterogeneity*

  IoT platform consists of heterogeneous devices on the basis of various distinct networks and hardware interfaces. They are capable of interacting among many service platforms or other nodes through various kinds of networks.
- *Dynamic changes*

  These devices are capable of changing their states dynamically, for example, waking up and sleeping, whether disconnected or connected as well as the context of the node that includes location and speed. Moreover, the total count of devices could also increase in a dynamic fashion.
- *Enormous scale*

  The total number of devices that communicate within each other and that need to be managed will be almost at an order of magnitude that is way higher than the cumulative quantity of all nodes that are currently available online. The management of the produced data and understanding for practical applications can be even more critical. This narrows down to data semantics and data handling in an efficient manner.
- *Safety*

  We should never forget about safety as we get more and more benefits from IoT. We must design and improvise the safety in IoT as both the recipients and creators of IoT systems. This includes both our physical well-being safety and personal data. Safeguarding the networks, end-nodes and the data moving over the networks mean generating a scalable security paradigm.
- *Connectivity*

It allows accessibility and compatibility of networks. Adding up onto a network is termed as accessibility while compatibility is the ability of data production and data consumption.

9.3 Advantages of Internet of Things

Basic benefits of IoT range to each and every domain of business and lifestyle. The list of benefits that IoT offers has been mentioned next:

- *Improved customer engagement*

  The present analytics suffer from significant accuracy flaws and blind spots which thereby make the engagement passive. IoT transforms this completely to achieve more effective and richer engagement with a wide range of audiences.
- *Technology optimization*

  The data and technologies that improve the experience of the customer also help in improving the usability of the device which gradually aids in more potential technology improvements. IoT has unlocked a world of field data and critical functions.

- *Reduced waste*

  The areas of improvement can be made clear by using IoT. Present analytics give us apparent insight, but IoT helps us to understand real-world information foremost to more operative resources management.
- *Enhanced data collection*

  Modern methods of collection of data suffer from their designs and limitations for passive use. IoT helps in breaking the data away from those kinds of spaces and placing them where humans actually need to go to examine the real world. It gives a clear-cut picture of every facet.
- *Tracking*

  The computers keep track of both the visibility and quality of the devices and things at home. The perception of product expiration date before someone consumes it improves the quality and safety in life. In addition, there will be continuous storage of products allowing you to consume them in time of emergency.
- *Time*

  The time saved in covering the number of trips and monitoring done can be otherwise tremendous.
- *Money*

The best advantage is the financial aspect. This technology can replace the human force required for the maintenance and monitoring of the supplies.

9.4 Architectural framework of Internet of Things

There are varieties of technologies that are grouped to form IoT architecture. This design provides service to many types of technologies that are relevant to each other. IoT construction consists of different layers of expertise supporting this tremendous concept of the IoT. It provides many services to show how different innovations classify with one another and to suggest the versatility, measured quality, and arrangement style with organizations in various situations. Fig. 9.2 demonstrates the point-by-point engineering architecture of IoT. The usefulness of each layer is portrayed as follows:

- *Smart device/Sensor layer*

  There are some powerful sensors in a sensor layer, which are capable of input, output, and doing minor processing in the lowest layer of IoT organization. The digital antenna device empowers the connectivity of the corporeal and advanced universes, enabling continuous data to be collected and handled. Many types of digital nodes perform multiple types of tasks with the capacity to do calculations such as assess the temperature degree, constituents of air, fastness, moistness, weight, stream,

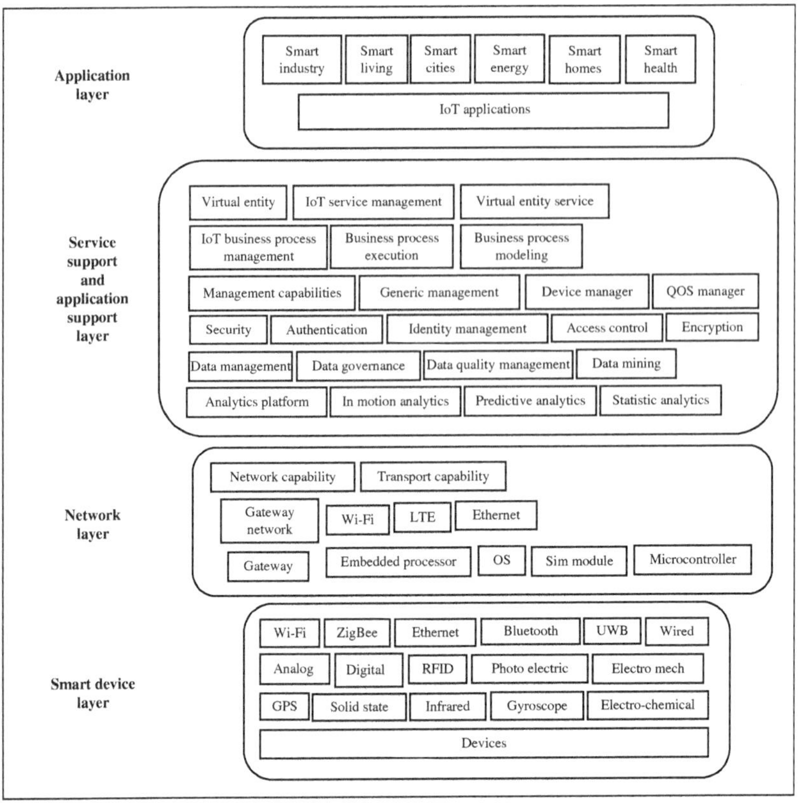

FIGURE 9.2 Basic IoT architecture. *IoT*, Internet of Things.

development, and power, and so forth. The digital nodes can weigh the physical belongings and convert it into a flag that can be comprehended by a gadget. Digital hubs are collected by their remarkable basis, for instance, ecological digital nodes, corpse digital hubs, home engine digital hubs and vehicle telemetric central points, and so forth. Most of the digital hubs expect connectivity to the digital hub entryways. This can be as a small network, such as local area network (LAN), Ethernet, and wireless connections, or PAN (personal area networks), such as ZigBee and Bluetooth mini-networks. For digital hubs that don't expect communication to digital hub aggregators, their connectivity to backend servers/applications can be given to use in wide area networks. Digital hubs that utilize low power and lower connectivity normally structure networks generally known as inaccessible digital hub networks. Wireless sensor networks (WSN) are picking up many challenges as they can suit incontrovertibly more digital hubs while holding reasonable battery life and covering magnanimous regions.

- *Gateways and networks layer*

  The gigantic volume of data will be made by these minor digital hubs, which requires a solid and tip-top wired or remote framework establishment as a vehicle medium. Current frameworks aim to help machine-to-machine architectures and their applications. With intrigue expected to provide an increasingly broad extent of IoT organizations and utilities, for instance, quick esteem–based organizations and setting careful applications, various frameworks with different advances and accesses shows rely upon each other in a heterogeneous setup. These frameworks can be as personal, open, or creamer models and are attempted to help the correspondence essentials for torpidity, information exchange limit, or security.
- *Management service layer*

  The administration organization renders the getting ready of information possible through examination, security controls, process showing, and the board of devices. One of the fundamental features of the administration organization layer is the business and methodology rule engines. IoT brings affiliation and association of articles and systems together by giving information as events or important information, for instance, the temperature of items, current territory, and traffic information. A bit of these events requires filtering or guiding to posttaking care of systems, for instance, getting of discontinuous substantial information, while others anticipate that response should be the brief conditions, for instance, reacting to emergencies on patient's prosperity conditions. The standard engines support the arrangement of decision bases and trigger clever and robotized strategies to enable an inexorably responsive IoT structure.

  In the sector of examination, different inspection gadgets are used to isolate relevant information from a gigantic proportion of unrefined information and to be dealt with at a much speedier rate. Examination, for instance, in-memory examination, empowers gigantic volumes of information to be held in subjective access memory (RAM) rather than set away in physical circles. In-memory examination reduces information question time and develops the speed of fundamental initiative. Spilling examination is another sort of examination where the examination of information, considered information in-development, is required to be finished continuously with the objective that decisions can be made in just seconds.

  Information management is the ability to direct the information stream. With the information of the board in the administration organization layer, information can be received, joined, and controlled. Higher layer applications can be shielded from the need to process pointless information and lessening the threat of security revelation of the information source. Information filtering techniques, for instance, information anonymization, information blend, and information synchronization, are used to disguise the nuances of the information while giving simply fundamental information that is usable for the significant applications. With the use of information consideration, information can be removed to give a regular business viewpoint on information to build increasingly conspicuous mastery and reuse transversely over territories. Security must be executed over the whole segment of the IoT designing legitimately from the sharp thing layer appropriate to the application layer. Security of the system foresees structure hacking and deals by unapproved personnel, thus decreasing the probability of risks.

- *Application layer*

The IoT submission consists of challenging situations and important factors in areas, for example, transportation, construction, city, everyday life process, retail, cultivation, plant, supply chain, emergency, health care, user association, culture, and the travel industry, atmosphere, and power.

9.5 Application areas of Internet of Things

Among a number of potential outcomes of using IoT, which formulate its feasibility to develop different functions subject to it, only two or three functions are more popular. In the following points, out of many, a few of the fundamental points of reference uses of IoT has been discussed:

- *The industry of aerospace and industry of aviation*

  The potential of IoT is that it can increase the efficiency of insurance and security of things and organizations by safely perceiving phony things and segments. Developed business, for instance, is susceptible to the issue of unreliable and unsafe parts. It is difficult to deal with such issue by exhibiting such mechatronic families for explicit orders of carrier sections, which file their root and security fundamental issues in the midst of their life cycle (e.g., changes), and securing these families inside a decentralized database similarly as on radio-frequency identification (RFID) marks can be checked before foundation. Thus prosperity and operational relentless nature of carriers can be developed slowly.
- *Automotive industry*

  Smart vehicles, locomotives, and transports more similarly as bicycles are getting the opportunity to be outfitted with bleeding edge advanced center points, actuators with extended dealing with services. Functions in the vehicle commerce fuse the usage of sharp effects to display and tale various parameters from weight in tires to the immediacy of various locomotives. RFID advancement was recently used to restructure vehicle age, improve collaborations, increase quality control, and improve customer organizations.
- *The industry of telecommunications*

  IoT will make the probability of combining different media transmission development and make innovative organizations. A symbolic form is the exercise of GSM, and communication such as near field communication, low control Bluetooth, wireless LAN, frameworks, geographic positioning system, and advanced center point arrange together through SIM-card development. In these sorts of usages the peruse (for instance label) is a bit of the personal digital assistant (PDA), and particular applications contribute to the SIM card. The near field communication (NFC) engages trades between things in a fundamental and protected course presently by having them almost everyone. The PDA can along these lines be utilized as an NFC-peruse and pass on the read data to an essential service provider computer.

- *Medical and health-care industry*

  The technology based on IoT supports various technologies in the therapeutic administrations subject, associating the probability of including the mobile phone with RFID-computerized center point limits as a phase for seeing helpful parameters and drug movement. The great position grabbed is in balancing activity and straightforward seeing of afflictions uniquely delegated finding and giving brief restorative thought in examples of mishappening. Devices that can be planted and addressed remote contraptions can be used to preserve prosperity details that can set aside a person's life in critical conditions, in particular for people with the diabetic disease, threatening development, knock, constant obstreperous pneumonic affliction, scholarly deterrents, seizure issue, and Alzheimer's disease.

- *Independent living*

  IoT functions and organizations will critically influence free living by offering assistance for a developing people by recognizing the activities of consistently living using wearable and enveloping advanced center points, checking social affiliations using wearable and encompassing computerized centers, watching never-ending affliction using wearable central signs computerized center points, and in body advanced centers. With the ascent of model acknowledgment and AI computations, the things in a patient's surrounding would in all probability keep an eye out and care for the patient.

- *Pharmaceutical industry*

  In IoT perspective, associating splendid names to medicines, finishing them the creation arrange, and checking their status with advanced centers have various potential points of interest. For example, things requiring express limit conditions, for instance, use of wrong medicine, can be unendingly watched and leftover if the situation was manhandled in the midst of transport. Medicine following and e-families mull over the area of phony things and keep the stock system free of fraudsters. The sharp names on the drugs can in like manner clearly advantage patients, for instance, by enabling securing of the group install, instructing buyers with respect to portions and end dates, and ensuring the validity of the remedy.

- *The business area of trade and supply chain management (SCM)*

  IoT can provide a couple of focal points in retail and SCM undertakings. For example, through RFID-arranged equipment and adroit control that monitor the present things ceaselessly, a vendor can improve various functions, for example, creating customized inspection of items receiving, steady seeing of storage items, following requirements, or the acknowledgment of robbery. Moreover, the IoT can facilitate making the figures from the trade store offered for propelling the collaborations of the entire store arrange. In case creators are acquainted with the storage products and arrangements figures from traders, they could make as well as send the correct measures of equipment, hence avoiding the condition of overage or lower production rate.

- *The industry for manufacture*

  By interfacing things in a sequence of advancement, either throughout implanted twisted gadgets or utilizing novel elements and data transporters that can speak among a carefully following framework structure and management systems, creation methods could be streamlined, and the entire period of things, throughout the exchange can be watched.
- *Industry of processing*

  In various factories of the lubricate as well as gas manufacturing units, flexible formation are sometimes used that believe potential results for fitting screw with innovation and procedures united by distinguishing/accelerating that is composed of the technology of IoT system as well as fuse remote seeing of oil staff in essential inland and offshore errands, compartment following, after of bore string parts of long channel, checking and supervising of permanent tools, etc.
- *Environment checking*

  The consumption of remote conspicuous apparatus and previous IoT propels in developed innovations as well as common management is a champion among the very talented business segment parts later on. There will be an extensible utilization of remote unmistakable devices in the earth all around arranged undertakings around the globe.
- *The industry of transport*

  IoT provides resolutions for section amassing and payment systems at check gates, vetting of explorers, and sacks involvement of commerce founders as well as the items relocated by the widespread burden structure which assist in the protection techniques of the lawmaking bodies and the shipping business, for fulfilling the extending requirement for safety in the sphere. Checking blocked streets throughout smartphones of the customers as well as game plan carried out by smart transportation frameworks will make the transportation of product and people progressively compelling.
- *Application in the area of farming*

  There are many rules formed for the tractability sake of the agrarian natural world and the improvements for them which need the utilization of advanced technologies such as IoT, creating promising the progressing area of creatures, for instance, in the midst of flare-ups of irresistible affliction. Further in many situations, the governments give sponsorship according to the number of animals in a gathering as well as their distinctive needs, to cultivation with cows and domestic animals such as sheep and goats. Because the confirmation of the number is not predictable, so there is a chance of some fake values. Extraordinary conspicuous confirmation methods can facilitate to limit such deception. Thus with the usage of ID technique, creature ailments can be checked, outlined, as well as turned away.

- *Commerce set up of media and amusement*

  The readiness of technology progressions in IoT will facilitate the cuff information collection reliant on territories of all the customers. Such information social event could happen by addressing technology, to observe which intelligent devices will be accessible in a particular territory, as well as approaching to those a (cash type) recommendation to assemble intelligent media movie about a particular event. Close sector communication marks may be joined to blurbs to give more idea by partner the peruser to a uniform resource identifier (URI) to concentrate on that contain better knowledge meaningful to the distribution.

- *Assurance industry*

  Habitually, the foreword of IoT development is viewed as a momentous assault on the security of persons. In any case, occasionally individuals are anxious to do business security for a predominant organization or a cash-related preferred standpoint. One model is vehicle security. If assurance customers will be glad to recognize mechanical recording devices in their smart vehicles, which can trace expanding velocity, speediness, and distinctive factors, and pass on this knowledge to their back up plan, they are expected to go to receive a more affordable price or payment system. For this circumstance the advancement generally helps in hindering considerable scale upkeep errands or considers significantly more affordable perceptive help before a scene occurs.

- *Recycling*

Remote advances in IoT can be effectively used to force the ability and satisfactoriness of various important town and state level geological surveys that also include the conduct of vehicle verification to measure undeviating air superiority, the collection of biodegradable products, the reuse of package of items and mechatronic parts, and the relocation of electronic waste (RFID is mechanized to discriminate mechatronic sub-subdivision of computing devices, smartphones, and other consumer hardware stuffs to assemble the reuse of these items and reduce e-waste).

9.6 Potential of Internet of Things in the pharmaceutical industry

The Web of Things is changing the drug store industry at a fast pace. IoT can possibly upgrade practically everyone in the procedures of the pharma business going from clinical preliminaries, medicate revelation, assembling, and store network to remote patient observing.

For IoT applications in the inventory network the board has turned out to be mainstream venture zones for some enterprises. Warehousing is a significant territory for the pharmacy business. IoT applications are put in the capacity zone which help stock things and transmit essential data (item area, stock subtleties) and report irregularities, for example, lost items legitimately to distribution centre chiefs' handheld gadgets/dashboards. Remedial estimates happen progressively—essentially improving the speed, precision, and productivity of the picking procedure.

Another territory where IoT innovations can include esteem is the capacity of temperature-touchy items. Dynamic and latent temperature lumberjacks connected to coolers in distribution centers and at different destinations consistently verify the warmth. Further, a different IoT arrangement may associate such gadgets, think about their estimations beside their security charts, and brief them to produce cautions in the event of warmth fluctuations.

Technology that uses the IoT can facilitate pharmaceutical medicine makers distantly monitoring conditions continuously by inserting digital hubs on the following hardware with self-start and closing components in distribution centers, motors, or consignments utilizing cell phones and miniature smart devices. Following the development of medication stock at each point can conceivably spare store network members billions of rupees. At the point when connected to bundling, the IoT bears a few preferences, as well as two side correspondence, following and updated status. It is particularly important in business sectors where fake medications routinely enter the esteem chain. 2D standardized tags, RFID labels, and brilliant bundling marks make it conceivable to follow every handshake in the store network, from assembling to apportioning. The outcome is a finished computerized impression. Advanced advances, for example, GPS area and condition checking manage the cost of ongoing permeability and security amid transport.

IoT innovations empower setting temperature-detecting labels with shipments. Labels ceaselessly record temperatures and ecological conditions. The information is transferred to the cloud. Accounts are in split-second access to the control room. For pharma organizations the IoT broadens deceivability in basically every region of the business—from improvement to assembling, transport, dispersion, administering, and utilization. Constant data, when combined with cutting-edge examination motors, can turn into the reason for making quicker, progressively precise choices; uplifting efficiencies, confirming item quality, and guaranteeing administrative consistency. The danger of doing nothing must be assessed against changing client and administrative desires and market elements. All is good and well for pharmacy organizations to quicken their usage and use of IoT stages and arrangements.

Stages in the pharmaceutical industry where IoT can be applied

- *Manufacturing:* Though bunch fabricating is the present practice in the pharma business, we can send computerization and control methods for controlling men and materials developments to keep up security and adequacy by improving permeability in different assembling exercises.

- *Warehousing:* It is an essential subject of any association to maintain a strategic distance from deficiencies, abundance inventories, and wastages at parent organization just as every one of the areas; it is available at different spots. Here, the innovation turns into a basic choice to have online deceivability through a network of different areas for better administration of stock accessibility, office advancement, preparatory activities, and so forth.
- *Transportation:* In a store network process, opportune accessibility of stocks at buyer point is a noteworthy testing task. Be that as it may, by receiving new age IoT methods, there is a plausibility to have a control on the equivalent by having sufficient energy-to-time observing of developments of stocks/vehicles through the GPS framework. This will empower us to screen the vehicle and the driver's execution and direct us toward streamlined choices as for rehashed development of stocks, decreased time to achieve advertise, and at last cost sparing.

9.7 Literature review of Internet of Things in pharmacy

Various IoT-related technologies such as RFID and WSN are the main technologies used to monitor real-time environmental conditions in the pharmaceutical domain. These technologies can be successfully adapted in the medical sector at an affordable cost. Angels in [1] discussed the significance of RFID technology in the SCM mainly in logistics management. It provided various data based on the introduction to RFID, various case samples, and other implementation guidelines based on published articles. Kelepouris et al. [2] analyzed the crucial traceability requirements and demonstrated the use of such technology in the pharmaceutical industry. It discussed the general information flow and architectural model of supply chain control system. A new RFID-based application architecture for the pharmacy industry was introduced by Yue et al. in [3]. It provided a guideline to various organizations that adopted RFID and enhance the speed at which RFID application can be used in the pharmaceutical supply chain. A software-defined architectural framework for item-level traceability and coordination of the entire pharmaceutical domain was proposed by Barchetti et al. [4]. Here, three distinct levels of RFID were described in detail. The benefit of electronic product code (EPC)-level technologies was discussed, and its significance in the pharmacy sector was highlighted. An RFID-based wireless technique has been developed by Moreno et al. [5] to trace pharmaceutical medicines. The model does container-level coordination and tracking of different routes to the repository to detect any anomalies in route optimization and create a response immediately if any issue occurs. RFID technology along with its correlation to IoT has been explained by Jia et al. [6]. It explains the significance of data processing tools such as RFID, GPRS, and WSN while analyzing the challenging issue of RFID technology. Chuan-Heng et al. [7] highlighted an anticounterfeit code for marine items categorization for monitoring and traceability in other countries. A temperature coordination model for chilled marine products has been developed by Xiao et al. [8], which is based on wireless constrained network combined with compressed transmission to enhance the effectiveness of the medium. In [9], the authors presented a WSN model to provide tracking of patients, localization and coordination services of patients, and medical staffs of various nearby clinical organizations. Passive IoT-integrated sensor technology has been developed in [10] to facilitate localization of equipments in medical centers. As RFID tags can work under the reader coverage sector, the application of RFID technology is restricted to patient and devices management and tracking in less geographic environments. In [11] a wireless localization model network tracks the geographic position of patients in various indoor surroundings and also to track their physical status is presented. Location-aware wireless constrained network to monitor patients using a ranging technique based on dynamic, and mobility adaptive filter is presented. A smart mobile communication technique using 6LoWPAN standard was introduced in [12] to coordinate and track the health status of patients and provide efficient medical service to them. A smart pillbox called MedTracker was developed in [13] to continuously monitor the medication of patients. It tracks various aspects of patients such as medical errors and anomalies and nonadherence issues. In [14] the authors presented a smart IT-based pillbox attached with a camera that is based on the medicine bag model. Medicine bags are attached with matrix barcode that is used to interact and manage the pillbox with patients that perform confirm as well as remind functionalities. An IPB (intelligent pill box) [15] is designed in correlation with a MBS (Medicine Bag System). The IPB is responsible to send the appropriate medicine bag out of the MBS in the required time. Suppose, the medicine bag is not collected by the patient then the IPB will send a notification to the caregivers through Skype.

9.8 Benefits of using Internet of Things in the pharmaceutical industry

IoT has influenced the logistics function too in a big way and has presented many opportunities to the pharma industry—the way it could improve its logistics operations.

An IoT-led ecosystem is bound to bring in more efficiency in the business operations especially in the supply chain. IoT is the panacea for the risks that are inherent in supply chain especially the ones that could result in costly disasters. Some of the benefits that can be realized by the implementation of IoT include in the following.

- *Improvement in the visibility of warehouse operations*

  The needs of the hour are real-time visibility into warehouse operations without which it would be difficult to track products besides unviable wastage of warehouse space and also the temperature variations that could be an issue while handling sensitive products. IoT-enabled smart warehouses having sensors and readers would increase the visibility of operations in the warehouse and presents the data seamlessly to the warehouse managers by reporting data about the products and their condition so that business can take a supported decision.

- *Overcome short supply of drugs*

  Implementation of digital technologies such as IoT ensures that there is a stable supply of drugs so that the needy patients could get their drugs on time. IoT helps the managing optimum inventories based on the business rules that the companies define so that planned decisions can be taken about manufacturing and also assist in sending out the relevant products to the market.

- *Enhance supply chain security*

  IoT helps to make the supply chain security more robust by facilitating bidirectional communication between the information seeker and the device, tracking of the inventory with respect to the location and the current status of the package in transit. Supply chain security can be ensured through tags in RFID format, bar codes in 2D, and smart labels for the purposes of packaging. The drug inventory movement at every check points can be easily tracked by the IoT. It can practically save supply chain stakeholders from losing millions of dollars while ensuring that genuine products are delivered to the customers while deterring drug counterfeiters.

  In order to reduce product recalls due to product degradation in transit, it is important to track the stored medicines' temperature during transportation to make sure that they stay stable by ensuring the temperature remains within suitable ranges till they reach their destination. IoT-enabled packaging continually tracks environmental circumstances in cold chains during transit—assuring efficacy and quality of the products with the usage of environmental sensors inserted into product packages/pallets in shipment containers. This would reduce wastage of drugs due to the increase of temperature in the containers, ensuring compliance with the regulatory requirements besides assuring the quality of the products.

- *Drug theft during transit*

The pharmacy industry has experienced losses due to the theft of drugs in transit which resulted in the losses to the companies and also deprived patients of the timely receipt of drugs. New age digital technologies such as condition monitoring and GPS location ensure security and visibility at real time during transport. IoT enables the logistics team to analyze and compare data on the progress of the cargo from anywhere. With the implementation of IoT solutions, it is possible to reduce the wastage of products due to damage by decreasing the time to market and by lowering transportation costs.

9.9 Patient-centric Internet of Things

IoT refers to the interconnection of the devices having the ability to send or receive data over networks. It has the ability to connect your waking up time to your coffee maker that in turn can begin your morning brew just before 5 minutes of your waking up time. It connects your GPS module and calendar to help you find the best route to a destination before you are even settled in the car. The development in the field of in-home sensors is considered to be one of the most widely known recent advances in the domain of patient-centric IoT. Users can freely live an independent life by taking the advantages of sharing insight into their day-to-day lives with their concerned health-care professionals and caregivers. Such kinds of intervention can be very helpful in the case of an emergency and strategies that govern the daily care of an individual. These sensors rely on interconnectivity between devices and individuals. Researches of Monash University have developed a system that can monitor health vitals of elderly patients in their home. An array of technologies such as light sensors, motion sensors, and vibration sensors are used in this system. There are also TV monitoring with Bluetooth activity, and sensors for detecting the movement of doors and windows, pressure of bed, and even embedded sensors in key-rings to detect the time of arrival and departure from the home.

These sensors are heavily customized for individual requirements and are not at all designed in a generalized purpose. Whenever the system encounters the elderly patient's break from the routine and abnormal behavior trends in the behavior, the system pushes an alert notification to the patient's caretaker, thus being ahead of all types of emergency conditions and availing highly personalized care to the users.

Like any other typical sector, the health-care sector also involves an ecosystem of the devices collecting organized information, transferring them to centralized analytics components, and thereby using the intelligence component to make efficient and profitable decisions. The second type of ecosystem has devices that are self-capable of autonomously sharing data with one another and thereby taking decisions. This focuses on the former.

Patient-centric health care deals with patients' health vitals such as weight and height, on-going activity archives such as exercises and food intakes, sophisticated medical indicators such as blood sugar levels and blood pressure.

The information is as follows:

- *Collected*—Done through the wearable devices and implanted devices on the patient's body.
- *Transmission to centralized analytics proponent*—Transferred from the individually managed and maintained a cloud-based server to highly specialized IoT-monitoring platform maintained and managed by hospital doctors.

- *Analyzed* by the analytics engine usually in correlation with the legacy data obtained from the same device or other devices.
- *Translated* into recommended actions—Anything from mentioning how many calories one needs to burn based off of his calorie consumption to inform a doctor about the tests that need to be done by the patient.

9.9.1 Patient-centric versus patient-centered information

The patient-centered information revolves around the patient's preferences and needs to meet his health-care concerns. It mostly depends on the communication between the patient and his caretaker with the health-care professional. Starfield mentions that patient-centered information depends on the core elements such as interaction or communication and obedience toward health-care recommendations.

The patient-centric information is produced by the patients themselves plus the data mined from the electronic health records, which mostly contains genomic information, which helps to figure out the predisposition of a disease.

The modern forms of wireless technologies are used to collect the vitals of patients and thereby filtering out a clinical document to fetch the patient-centric data. The usage of such patient-centric data including genomic data in emergency care is still not in order. Genomic information is still not included in the medical records, and the electronics health records are used by only a few health-care professionals.

9.10 Body area network overview

A BAN (body area network), also called WBAN (*wireless BAN*) or a *body sensor network* or an *medical BAN*, is a network of computation enabled devices that can be attached on the body, can be mounted on the body in a definite position, or can be attached with devices that people carry in various positions such as in bags or in cloth pockets. In general, networks consist of a combination of miniaturized body sensor units and a single body central unit though there is a trend of making miniaturized devices. Smart devices, such as Tab and iPad, play a significant role in being a data gateway, data hub and providing a graphical user interface (GUI) to manage and view BAN applications. WBAN development started in 1995 revolving around the concept of using WPAN (wireless PAN) technologies to device communications on, around, and near the human body. After 6 years BAN referred to systems where communication was done, on, within, and in the immediate human body proximity. A WBAN system uses WPAN to cover longer ranges. Wearable devices can be easily connected to the Internet through different kinds of gateway devices. This will enable the health-care professionals to receive the health vitals of a patient through Internet irrespective of his/her location.

The development in the field of low-power enabled integrated circuits, physiological sensors, and wireless communication has given birth to an entirely new era of WSN used for various purposes such as crops, monitoring of traffic, health, and infrastructure. BAN field is a form of multifaceted field that allows continuous and cheap monitoring of health-related parameters with real-time health-related updates from information sources through the Internet. A range of highly specific and artificially intelligent sensors be embedded with a wireless wearable BAN, which could be utilized for early detection of health conditions. It aims at implanting tiny biosensors in the human body that are extremely free to wear and don't hamper the normal activities of human. These sensors once implanted can start collecting the health information irrespective of the patient's location. The collected information is then transmitted to a centralized location. These sensors can transmit the real-time data of patient's health to doctors in different parts of the world. Whenever discrepancy occurs, the physicians would let the patient know about it through the system by conveying through alarms or messages. At present, the amount of data that can be provided over the net and the energy required to keeping the sensors running are extremely limited. This technology is in its initial stage. Continuous efforts are being made to improve the architecture and robustness of the system. Once the system is deployed, it will prove itself a revolutionary invention in the field of health care that can open new areas such as telemedicine and mHealth.

9.10.1 Challenges faced by body area network

Challenges associated with WBAN technology could include the following:

Data quality: Data generated and aggregated through BANs plays an important role in the patient monitoring process. Thus it is vital that this data is of a very good standard that ensures that the decisions made are based on the optimal information possible.

Data management: Since BAN generates a huge volume of data, there is an urgent need to track and manage these raw data effectively.

Sensor validation: There are instances where many pervasive-sensing nodes suffer from inherent connection and hardware anomalies such as unreliable network links and limited energy resources. This sometimes results in erroneous data getting transmitted back to the user. In the pharmaceutical domain, it is very important that these sensory nodes are very well validated in advance.

Data consistency: There are data that resides on several nodes and multiple mobile devices. This wireless patient notes required to be analyzed seamlessly. In a BAN, data related to patients may be segmented over a number of nodes and may be fragmented across several networks. The quality of health care of a patient starts degrading if the mobile device of a clinical expert does not have all the required data.

Security: Security is an important feature of a WBAN transmission in the medical domain. A secure BAN system must ensure that the patient data are not intermixed. Apart from this, generate data from WBAN should have restricted and secure access. Essential security requirements of WBAN include authentication, confidentiality, availability, integrity, and reliability.

Interoperability: WBAN models in the pharmacy sector should ensure reliable and seamless data transmission across various networks such as Bluetooth and ZigBee. Also, the systems must provide scalable and uninterrupted data transfer over time.

System devices: The sensors in WBAN should have the characteristics of low weight, small in size, energy efficient, and easy to reconfigurable. Also, the storage nodes should facilitate remote storage and viewing of data of patients.

Invasion of privacy: If the applications of WBAN move beyond secure health-care usage then a certain section of people might feel the WBAN technology as a threat to freedom. Social acceptance is very crucial in such a scenario.

Interference: The wireless transmission link used for body sensors must minimize the interference and maximize the coexistence of sensor devices available within the environment. It is very important for large-scale application of WBAN in the pharmaceutical domain.

Cost: In today's scenario, every consumer is expecting low-cost medical monitoring solutions with very high productivity. Implementation of WBAN needs to be cost optimized to offer appealing perspectives to consumers.

Constant monitoring: In a pharmacy model, users may require varying levels of monitoring and tracking. The monitoring management influences the quantity of energy needed and the life cycle of the BAN before the power source is reduced.

Constrained deployment: The chief attributes of WBAN should include lightweight, wearable, and not intrusive. The daily activities of the user should not be changed. In simple words the technology should be very transparent to the user.

Consistent performance: The performance of WBAN should be very consistent. Sensor measurements should be precise and calibrated even if the WBAN is deactivated. The wireless links should also be reliable and fault tolerant.

9.11 Internet of Health Things

Internet of Health Things (IoHT) is an interconnection of uniquely identifiable devices (particularly that are used in the medical field) over the networks. They prove to be very helpful to provide real-time information and localization about different assets concerning the medical environment. Through this system the user can easily monitor the resources remotely and through automated fashion. This not only saves a huge amount of time by ensuring way better quality of patient care but also ensures the safety of the patient. Now with the help of IoHT, the management of medical facilities has become more efficient due to the uninterrupted access to information and data related to patient's health. Personalization and flexibility are the most important features of an IoHT system. An IoHT framework is depicted in Fig. 9.3. The specialization and size of the facility is not a barrier for the system to be adapted in that environment. The connection, modification, and integration of the IoHT system with existing technologies are also feasible and easy.

9.11.1 Advantages of Internet of Health Things

Followings are the advantages of using IoHT architecture in different entities:

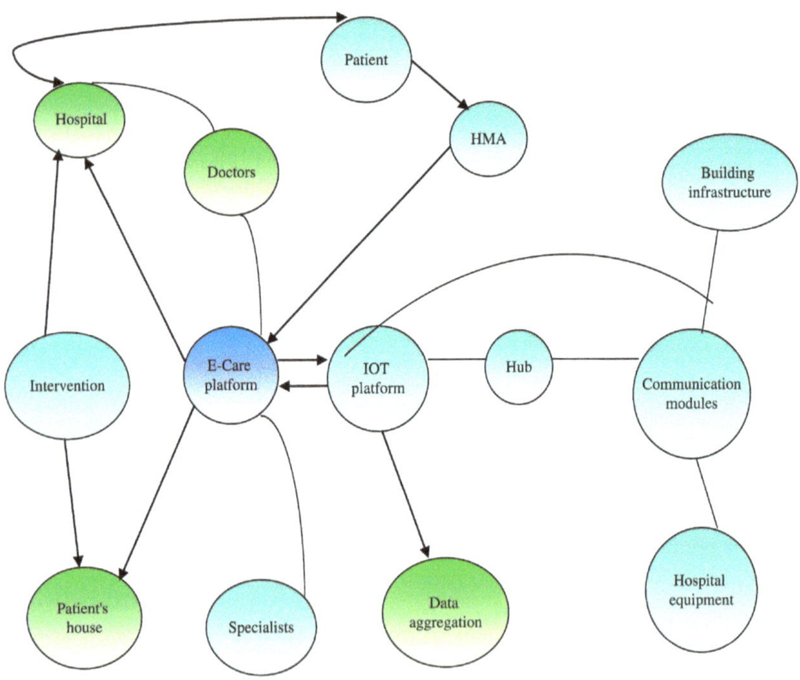

FIGURE 9.3 Operation scheme of IoHT platform. *IoHT*, Internet of Health Things.

For patients

- Easy and fast notification to the medical personnel about the patient through the device.
- Saving the health parameters of the patient through continuous monitoring.
- Detection of emergency situations and remote health monitoring provides increased safety.
- It enables faster diagnosis and treatment through effective and faster access to medical facilities.
- Provides comprehensive care to the patient when not inside the medical facilities.
- Automated transfer of patient's data followed by effective analysis.
- Distant medical consultations.
- Automated reminders.

For medical staff

- Uninterrupted access to patient's latest health data and previous medical history.
- Algorithms are so robust that they are capable of finding any kind of health discrepancy.
- Patients having orientation difficulties can be easily traced (e.g., dementia patients).
- Easy and quick tracing of the equipment and devices.
- Patients' health information can be easily accessed at any time through the mobile application.
- Easy access to the patient's database through a web browser.

For managers and IT staff

- It provides the feature of geo-fencing that provides notification whenever the device is beyond a defined area.
- Access to resources and data can be controlled.
- Automatic or remote control of lighting systems, which can save energy and regulate patients' circadian rhythm.
- Resources consumed over a definite period of time can be analyzed.
- Detection of patient's movements through various alerts generated due to sensor readings.
- Nature of maintenance is predictive/preventive.
- Enables quick transfer and access to the data in case of an emergency.
- Real-time error and fault detection.
- Automated maintenance system.

9.12 Analysis of medical nursing system using Internet of Things in the pharmaceutical domain

This work is an effect of integrating technologies such as NFC, WSN, and RFID to create an automated medical nursing system. In order to realize how medical nursing system several visits to Poh-Ai Hospital and a nursing home in Taiwan were done. A medical nursing system consists of five subsystems designed both for pharmacies and the nursing homes. This system not only helped in promoting the circumstances in a nursing home but also upgraded drug supply accuracy.

9.12.1 Discussed work

The overview of medical nursing system has been shown in Fig. 9.4. It is further divided into five subsystems, namely, ESS (environmental sensing system), IMS (identity management system), MS (medication system), BS (biomedical system), and POS (personal orientation system).

9.12.2 Identity management system

Every patient's respective family will be required to provide the basic health data of the patient such as gender, name, photo, and drug prescription. Then, the concerned staff will utilize the HF (high frequency) reader to generate two cards: one will be for MS and the other will be for the identity. IMS is a computer program coded in Microsoft Visual C# that has been illustrated in Fig. 9.5. It provides a GUI to enter the patient's data in the computer system and transmitting every data to the cloud server for future use.

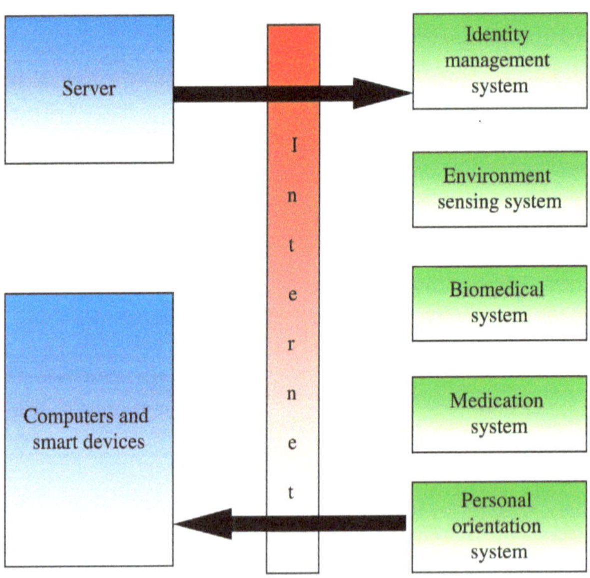

FIGURE 9.4 Framework of medical nursing system.

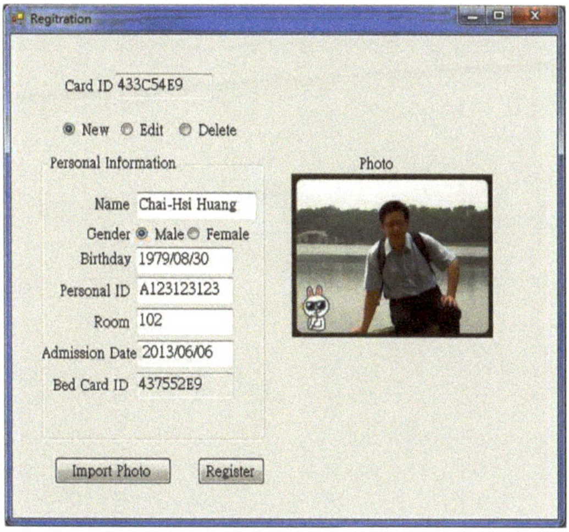

FIGURE 9.5 Identity management system.

9.12.3 Environmental-sensing system

Three sensors following ZigBee protocol, that is, humidity, temperature, and brightness sensors, will be installed in ESS. The utility of the ESS is to improvise the issues of the living environment of every resident residing in the nursing home. The programming of the system enables it to sense the sensor reading a particular number that is inputted by the user end as illustrated in Fig. 9.6. Following which, WSN transfers data to the centralized computer of the nursing home through ZigBee protocol. On receiving the data the system transfers them to the server deployed in the cloud which allows doing follow-up usage. Installation of a lot of ZigBee sensors due to their low cost and low rate in different areas of the nursing home can help one to achieve precise sensing of residents' living circumstances such as humidity, temperature, and brightness. The central computer follows a certain value for controlling it. Whether decided to switch cooling on, heating, dehumidifying machine or the increasing/decreasing the brightness in order to adjust the ease in the environment.

9.12.4 Biomedical system

In the third phase, that is, BS a Mobile Biomedical System (MBS) was designed by the amalgamation of IMS to make the job of the nursing home staffs much easier. The system is capable of sensing a range of values such as the pressure of blood, heartbeat and a saturation of oxygen in the blood (SpO_2). This data is sent through Bluetooth to transfer every measurement into smart mobile devices and thereby recording it. The nurse uses MBS to monitor the resident's physical health conditions. First, the nurse uses

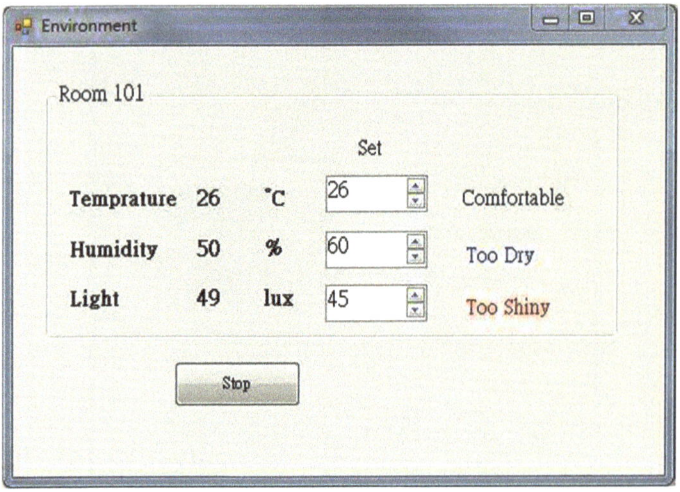

FIGURE 9.6 Monitoring condition of room.

NFC-mobile to detect the bed card of a resident which was built in the IMS phase. The system knows about which patient is going to be calculated by this action. Second, in the action of measuring, the nurse will just have to press the button present on the application activity written on NFC-mobile to activate MBS and in turn start measuring. The final calculated and measured data are transferred to mobile devices via Bluetooth and are displayed on the screen of smart devices as illustrated in Fig. 9.7. After the send button is clicked, the information will be transferred to the server host through a wireless network. The design of MBS helps to solve the issue of inconvenience faced by the nursing cart. As shown in Table 9.1, the MBS has advantages of size and operation way.

9.12.5 Medication system

The MS has been designed as an effort to improve the conditions of controlling drug processes and other advances in the nursing houses that were visited. The MS is divided further into two portions, that is, pharmacy and nursing home.

9.12.5.1 In pharmacy

Useful patient information such as side effects, indications, and pictures of drugs are inputted into the database by the pharmacy staff as illustrated in Fig. 9.8. It is done to make the follow-up operations easier and convenient. The pharmacy staff has to choose the data of the drug from the list entered into the computer in phase 1 for every patient separately and save them as

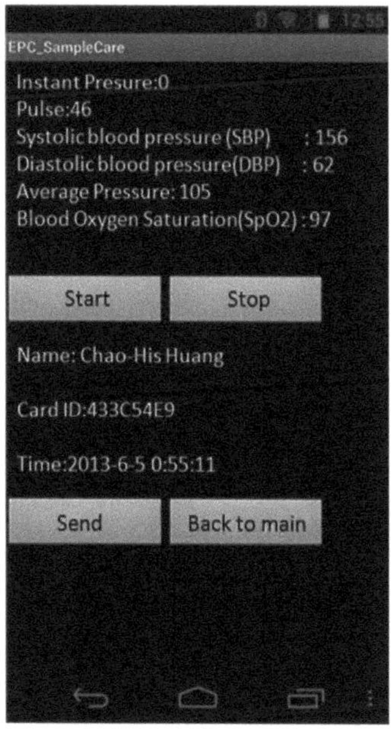

FIGURE 9.7 MBS measurement result.

TABLE 9.1 Comparing table of nursing cart and Mobile Biological System (MBS).

Nursing cart	Compare constant	MBS
Large	Size	Small
30 min	Battery	6 h
High	Cost	Low
Manual	Operating	Mobile device
By hand	Data record	Wireless protocol
Low	Interaction with patient	High

illustrated in Fig. 9.9. The main aim of this phase is to bind the resident and the drug prescription.

After the pharmacy staff finish packing the drug and inputting the drug detains in the information system, an HF Tag having all the data is applied. All these data are sent to the server. The nursing home staff then collects the drug bag having an HF Tag which consists of all the information about the drug.

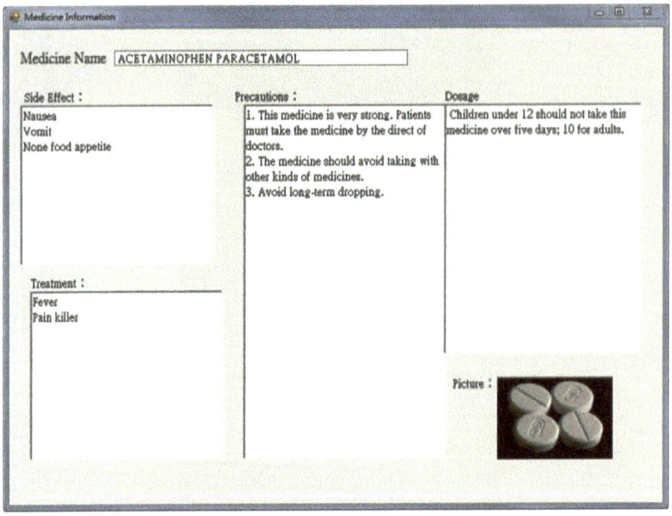

FIGURE 9.8 Inserting data into medication system.

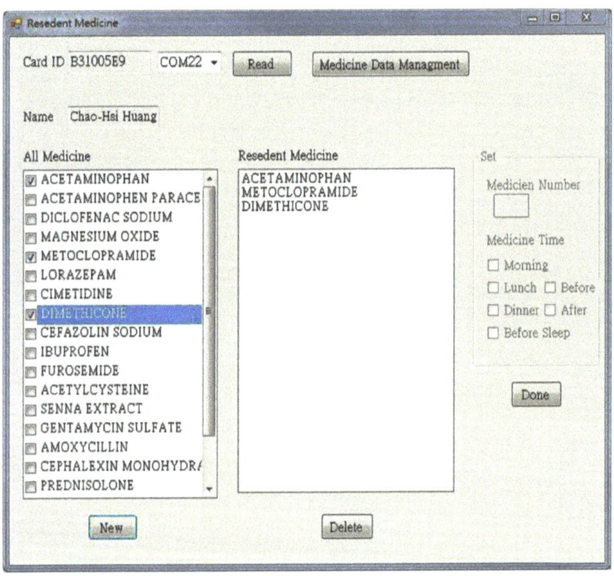

FIGURE 9.9 Sample showing drugs selection for patient.

9.12.5.2 *In nursing home*

- After the nursing home staff collects the drug bag, it is verified by them by scanning HF Tags.

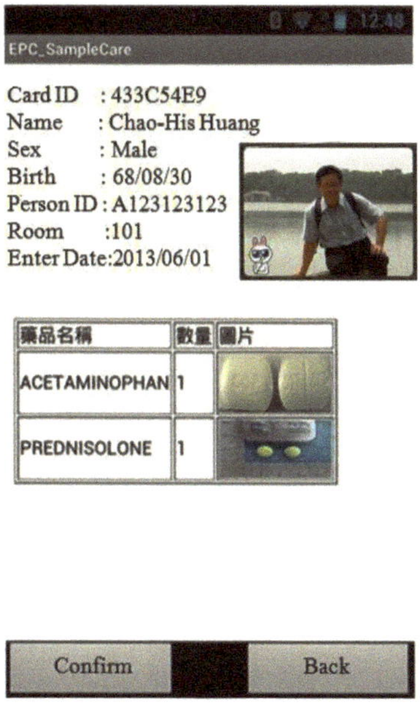

FIGURE 9.10 After sensing the bed card.

- During the time of dispensing of a drug the nurse would use the NFC-mobile to detect a patient's bed, HF Tag. The NFC-mobile screen then displays the patient's photo and name with the drug name, its information, and pictures demonstrating how the patient should consume the drugs as illustrated in Fig. 9.10. These data are used by the nurse to verify that the drug that they brought for the patient is correct.
- After finishing the drug dispensing the nurse clicks on the upload button present on the screen. The system then sends the completion time of medication to the cloud database. This information can be used for checking later on.

9.12.6 Personal orientation system

The patient's escape from the nursing home is a severe threat to the safety of the patient. This system has been designed to keep check of a patient's escape. Suppose, the staff is not attentive and the patient escapes from the room at that point of time then the system will detect this event and activate an alarm.

Every patient's room door is equipped with a ultra high frequency (UHF) reader. It is used to identify the opening of the door. The system prompts a message when a patient leaves his room. This is how the POS is capable of detecting the runaway possibilities as shown in Fig. 9.11.

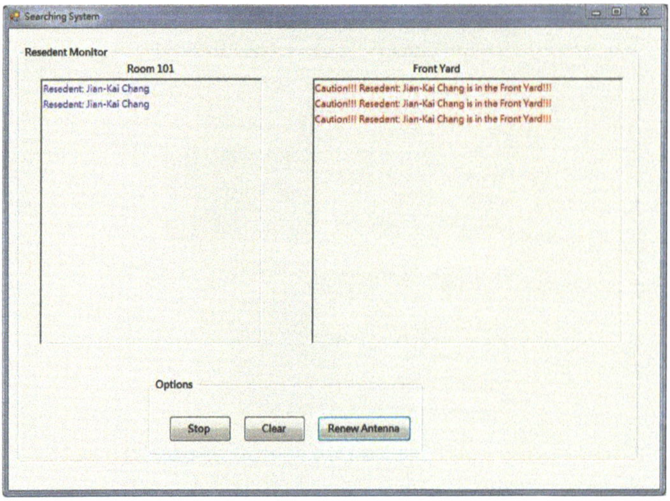

FIGURE 9.11 Personal orientation system.

9.13 Conclusion

IoT refers to the internetwork of several nodes integrated with sensors, actuators, electronics equipment and network connectivity to facilitate connectivity, and easy transfer of data in a real-time environment. Based on its application zones, IoT finds potential in the pharmaceutical sector. By the use of IoT in pharmacy domain, patients can be remotely monitored and automate drug discovery can be done. Similarly, various other tasks can be performed in this sector. The effective integration of pharmacy and IoT has opened many opportunities for massive social applications for mankind and industry. In this chapter, we have discussed various aspects of integration of IoT in the pharmaceutical sector. Several aspects related to the role of IoT in pharmacy have been addressed here. A sample case study related to IoT in pharmacy domain has been illustrated here. In this case study a smart system for medical nursing based on WSN, NFC, and RFID technology has been discussed. This system not only promotes nursing home conditions but also upgrades the drug supply accuracy.

References

[1] R. Angeles, RFID technologies: supply-chain applications and implementation issues, Inform. Syst. Manage. 2 (1) (2005) 51–65.

[2] T. Kelepouris, RFID-enabled traceability in the food supply chain, Ind. Manage. Data Syst. 107 (2) (2007) 183–200.

[3] D. Yue, X. Wu, M. Hao, A cost-benefit analysis for applying RFID to pharmaceutical supply chain, IEEE, 2011, doi:978-1-61284-311-7/11.

[4] U. Barchetti, A. Bucciero, M. De Blasi, L. Mainetti, L. Patrono, Traceability in the pharmaceutical supply chain, in: IEEE International Conference on RFID-Technology and Applications, 2010, pp. 17–19.

[5] A. Moreno, I. Angulo, H. Landaluce, A. Perallos, Easily deployable solution based on wireless technologies for traceability of pharmaceutical drugs, in: IEEE International Conference on RFID-Technologies and Application, 2011, pp. 252–258.

[6] X. Jia, Q. Feng, T. Fan, Q. Lei, RFID technology and its applications in Internet of Things (IoT), 2012, pp. 1282–1285, doi:978-1-4577-1415-3/12.

[7] S. Chuan-Heng, L. Wen-Yong, Z. Chao, L. Ming, J. Zeng-Tao, Y. Xin-Ting, Anticounterfeit code for aquatic product identification for traceability and supervision in China, Food Control 37 (2014) 126–134.

[8] X. Xiao, Q. He, Z. Fu, M. Xu, X. Zhang, Applying CS and WSN methods for improving efficiency of frozen and chilled aquatic products monitoring system in cold chain logistics, Food Control 60 (2016) 656–666.

[9] A. Redondi, M. Chirico, L. Borsani, M. Cesana, M. Tagliasacchi, An integrated system based on wireless sensor networks for patient monitoring, localization, and tracking, Ad Hoc Netw. 11 (2013) 39–53.

[10] A.A.N. Shirehjini, A. Yassine, S. Shirmohammadi, Equipment location in hospitals using RFID-based positioning system, IEEE Trans. Inf. Technol. Biomed. 16 (6) (2012) 1058–1069.

[11] D. De Donno, L. Catarinucci, L. Tarricone, RAMSES: RFID augmented module for smart environmental sensing, IEEE Trans. Instrum. Meas. 63 (7) (2014) 1701–1708.

[12] D. De Donno, L. Catarinucci, L. Tarricone, A battery-assisted sensor-enhanced RFID tag enabling heterogeneous wireless sensor networks, IEEE Sens. J. 14 (4) (2014) 1048–1055.

[13] T.L. Hayes, J.M. Hunt, An electronic pillbox for continuous monitoring of medication adherence, in: 28th IEEE EMBS Annual International Conference New York City, Aug 30–Sep 3, 2006.

[14] H.-K. Wu, C.-M. Wong, P.-H. Liu, S.-P. Peng, X.-C. Wang, C.-H. Lin, K.-H. Tu, A smart pill box with remind and consumption confirmation functions, in: 2015 IEEE 4th Global Conference on Consumer Electronics (GCCE), 2015.

[15] S.-C. Huang, H.-Y. Chang, Y.-C. Jhu, G.-Y. Chen, The intelligent pill box-design and implementation, in: Proceedings of the IEEE International Conference on Consumer Electronics, May 26–28, Taiwan, 2014.

Chapter 10

Smart pills: a complete revolutionary technology than endoscopy

Subhashree Sahoo[1], Amiya Bhusan Bagjadab[1] and Sushree Bibhuprada B. Priyadarshini[2]

[1]Sambalpur University Institute of Information Technology, Burla, India, [2]Institute of Technical Education and Research, Bhubaneswar, Siksha 'O' Anusandhan Deemed to be University, Bhubaneswar, India

10.1 Introduction

Rampant technological advancements in healthcare have protected countless lives and enhanced the standard of living for many years. Not solely has technology modified exposure for patients as well as the families; however, it's additionally had an enormous effect on healthcare processes and also the activities of healthcare professionals. In modern days, technology plays a very important role in every trade moreover as in our personal spheres. Among the manufacturing technologies, the current one exhibits an important role in healthcare, which is surely vital in all respects. Such a merger is accountable for the enhancement of as well as preserving innumerable lives throughout the planet.

Medical technology is a crucial area wherever invention has a supreme role, while considering health. Fields such as biotechnology, prescribed drugs, information technology, the event of medical appliances as well as instrumentation, and a lot of possess all created vital contributions to up the health of individuals throughout the world. Starting from "small" innovations such as adhesive bandages and ankle joint braces to broader, a lot of novel automations, such as Magnetic Resonance Imaging (MRI) appliance, artificial organs, and robotic prosthetic limbs, possess beyond any doubt framed out of this world impact on medication. Out of all the brilliant innovations and developments in medical technology, it becomes very easy for the healthcare practitioners to search tactics to enrich their practice, such as enriched surgical methods, better diagnosis, and enhanced patient care.

Emergence of Pharmaceutical Industry Growth with Industrial IoT Approach.
DOI: https://doi.org/10.1016/B978-0-12-819593-2.00010-8

Improving the standard of living is one in every of the most advantages of group action novel innovations for drugs. Healthcare technologies such as minimally invasive surgeries, higher tracking frameworks, and greater well-off instrumentality scanning are permitting patients to pay diminished time in recovery and longer experiencing a healthy life style.

10.2 Introduction to endoscopy

Diagnosing a patient with inflammatory intestines diseases such as Crohn's disease (CD) is a method that involves multiple steps. If the results of physical examination, blood tests, and stool tests counsel that a person's symptoms are being caused by inflammation within the digestive tract, then the physician might refer the patient to a gastroenterologist. Gastroenterologists are physicians specializing in the health of the digestive tract system; they're specially trained to perform a special form of the diagnostic technique referred to as endoscopy. Endoscopy procedures enable the gastroenterologist to check within a person's digestive tract employing a special instrument referred to as an "endoscope" associating with medical instrument that could be a long, thin tube with a small camera and light hooked up to the top of it, in which pictures of the digestive tract appear on a screen for analysis. Endoscopes will be inserted into the body by help of a natural gap, such as the mouth and down the throat, or through the bottom. An endoscope may be inserted by a tiny low cut (incision) created within the skin through which the keyhole surgery gets performed.

Fig. 10.1 depicts the overview of an endoscope within a human body. The endoscope is inserted into the human's body through his mouth. End of the endoscope contains a light and a tiny camera to project the picture of the intestine on the screen for further analysis.

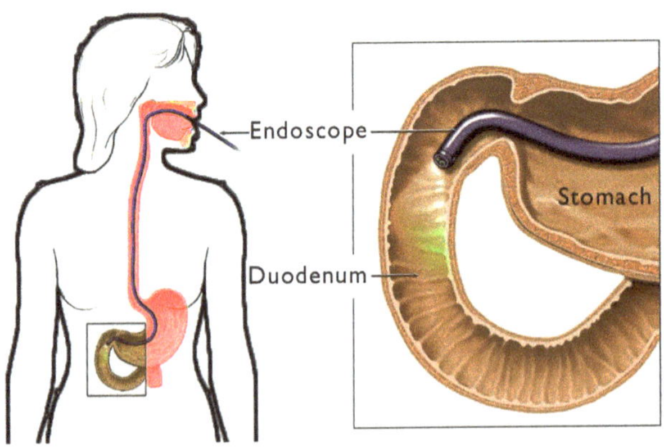

FIGURE 10.1 Overview of an endoscope.

10.3 Why endoscopy?

An upper endoscopy is employed to diagnose and, sometimes, treat conditions that have an effect on the upper part of our digestive framework, as well as the esophagus, abdomen, and beginning of the tiny intestine (duodenum). Doctors many times recommend endoscopy to assess the following:

- Pain in stomach
- Gastritis
- Ulcers
- Problem in swallowing
- Growth in colon
- Digestive tract bleeding
- Biopsy
- Changes in bowel habits
- Vomiting blood
- Unexplained weight loss
- Persistent abdominal pain
- Chest pain
- Presence of blood in stools

Sometimes doctors may also recommend endoscopy for the following reasons described in the below sections.

10.3.1 Investigating signs and symptoms

The endoscopy might facilitate the doctors who confirm what inflicts digestive symptoms, such as nausea, abdominal pain, vomiting, difficulty swallowing, and gastrointestinal hemorrhage.

10.3.2 Diagnosing

The concerned doctor might apply endoscopy to gather tissue samples (biopsy) to check for diseases and situations, such as anemia, bleeding, diarrhea, inflammation, or the cancers of the entire digestive system.

10.3.3 Treating

The doctor passes special tools through the endoscope for the purpose of treatment of issues in concerned digestive system, such as stretching a narrow esophagus, clipping off a polyp, or eliminating a remote entity. The endoscopy gets usually conglomerated with alternative approaches, such as an ultrasound. The ultrasound investigation is also connected to the endoscope to form specialized pictures pertaining to the wall of the esophagus or abdomen. Endoscopic ultrasound (EUS) may facilitate the doctor to produce pictures of hard-to-reach organs, such as the pancreas. Novel endoscopes employ high-definition videos to afford clarity in pictures.

10.4 Types of endoscopy

There are several types of endoscopy available:

- Upper gastrointestinal endoscopy
- Upper gastrointestinal endoscopy and dilation
- Colonoscopy
- Endoscopic retrograde cholangiopancreatography (ERCP)
- Bronchoscopy
- Percutaneous endoscopic gastrostomy (PEG)
- Flexible sigmoidoscopy
- Transbronchial biopsy
- Cystoscopy
- Hysteroscopy
- EUS

10.4.1 Upper Gastro-Intestinal (GI) endoscopy

Upper gastrointestinal endoscopy is often called as upper endoscopy and esophagogastroduodenoscopy. Physicians employ upper GI endoscopy for treating symptoms that have an effect on the esophagus, stomach, and upper gut or small intestine [1].

Upper GI endoscopy will facilitate realize the reason behind unexplained symptoms, namely,

- continuous heartburn,
- bleeding,
- nausea,
- vomiting,
- pain,
- problems swallowing, and
- unexplained weight loss.

Upper GI endoscopy helps for distinguishing various kinds of diseases, such as

- gastroesophageal reflux disease,
- ulcers,
- cancer,
- inflammation or swelling,
- precancerous abnormalities, namely, Barrett's esophagus,
- celiac disease,
- strictures or narrowing of the esophagus,

- blockages,
- tumors,
- infections of the upper GI tract, and
- CD in the upper GI tact.

During the process of upper GI endoscopy, the doctor might

- draw tiny samples of either tissue, cells, or fluid in the upper gastrointestinal tract for testing;
- finish any hemorrhage; and
- carry out different methods, such as opening up strictures.

10.4.1.1 Risks of upper GI endoscopy

The risks associated with an upper GI endoscopy are very low, however, might incorporate

- bleeding from the location wherever the physicians takes the tissue samples or discard a polyp;
- hole within the lining of concerned upper gastrointestinal tract; and
- abnormal reactivity concerned to the sedative, as well as respiration or heart issues.

10.4.1.2 Medications

Endoscopy is frequently done as an out-persistent method. Patients are prompted not to drive to his arrangement as the narcotics can take as long as 24 hours to wear off. Explicit directions will be given by the staff at the clinic where the method will be performed. For 8 hours preceding the methodology, patients are not allowed to certainly eat or drink anything aside from possibly little measures of water until one-and-a-half hours before the system. This limits the danger of yearning (sucking or motivation) of gastrointestinal substance into the aviation routes and lungs. It likewise guarantees that the upper gastrointestinal tract is unfilled to increase ideal perspectives on the walls and mucosa.

10.4.2 Colonoscopy

Colonoscopy represents an examination employed to observe alterations or abnormalities within the colon as well as rectum. Throughout the colonoscopy, a long, versatile tube (colonoscope) gets inserted toward the rectum. A small video camera at the tip of the concerned tube permits the diagnoser to look at the inside of whole colon. Fig. 10.2 illustrates a scenario of colonoscopy.

If desired, polyps or different forms of tissues having abnormality may be discarded by the scope throughout a colonoscopy. Before colonoscopy, intravenous (IV) liquids are administered, and the patient is put on a screen for consistent observing of heart beat, blood pressure, and oxygen in the

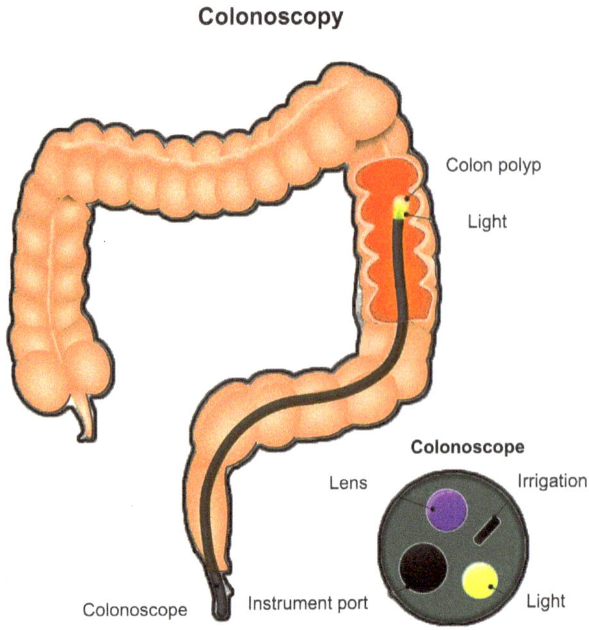

FIGURE 10.2 Demonstration of colonoscopy.

blood. Medications (narcotics) generally are given through an IV line so the patient ends up tired and loose, and to experience less pain. If necessary, the patient may get extra dosages of prescription during the technique. Colonoscopy regularly creates a sentiment of pressure, cramping, and swelling in the stomach area; in any case, with the guide of prescriptions, it is commonly all around endured and inconsistently causes serious pain. Patients will lie on their left side or back as the colonoscope is gradually best in class. When the tip of the colon (cecum) or the last segment of the small digestive tract (terminal ileum) is achieved, the colonoscope is gradually pulled back, and the covering of the colon is cautiously analyzed. Colonoscopy more often than not takes 15 minutes to 1 hour. In the event of the whole colon, the doctor may choose to attempt colonoscopy again sometime in near future with or without an alternate bowel planning or may choose to arrange an X-beam or CT of the colon [1].

The physician often recommends colonoscopy to detect any changes related to the following reasons.

10.4.2.1 Examining intestinal signs and symptoms

The colonoscopy will facilitate the physician in exploring possible reasons of abdominal pain, rectal harm, chronic constipation, diarrhea, and different intestinal issues.

10.4.2.2 Test for more polyps

If a person's had a polyps prior, physicians may recommend a follow-up colonoscopy to test for more polyps, if any, which reduces the risk associated with colon cancer.

10.4.2.3 Risks of colonoscopy

The risks associated with colonoscopy are given below:

Unpropitious reaction to the sedative employed throughout the examination.
Bleeding from the location wherever a tissue sample (biopsy) was chosen or a polyp or different abnormal tissues were removed.
A tear within the colon or rectum wall (fissure).

10.4.3 Endoscopic retrograde cholangiopancreatography

Endoscopic retrograde cholangiopancreatography is often called ERCP that is a medical procedure which combines both the procedures of upper GI endoscopy and X-rays to treat the problems associated with bile and pancreatic ducts [1,2]. Fig. 10.3 portrays an illustration of ERCP.

Physicians perform ERCP during the time a person's pancreatic ducts get either narrowed or blocked due to the following problems:

- Gallstones

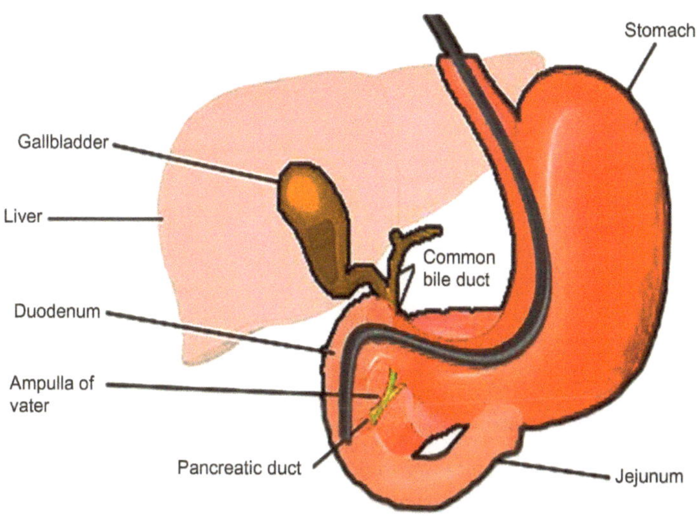

FIGURE 10.3 ERCP illustration. *ERCP*, Endoscopic retrograde cholangiopancreatography.

- Infection
- Intense pancreatitis
- Acute pancreatitis
- Trauma
- Surgical problems in the bile or pancreatic ducts
- Pancreatic pseudocysts
- Cancers in the bile ducts
- Cancers in the pancreas

There are two common treatments for ERCP that are listed in the following sections.

10.4.3.1 Sphincterotomy

In this technique, small cut is made in papilla of Vater to enlarge the size of opening of pancreatic duct. This method is performed to improve the drainage of the pancreatic duct and removed the stones present there.

10.4.3.2 Stenting

A stent is a little plastic tube that is set and left in a blocked or narrowed duct to remove blockage. The narrowing should be extended before the stent is put. A few stents are intended to go out into the digestive system following half a month when they have done their work. Different stents must be evacuated or changed following 3–4 months.

10.4.3.3 Risks of endoscopic retrograde cholangiopancreatography

There are several risks associated with ERCP procedures some of them are listed below:

- Pancreatitis
- Infection of either the bile ducts or gallbladder
- Much hemorrhage
- An abnormal reaction to the sedative, and metastasis or cardiac issues
- Hole within the bile or duct gland ducts, or within the small intestine close to the opening where the bile and duct gland ducts empty into it
- Tissue injury from X-ray contact
- Death, though such complication is very rare

10.4.4 Bronchoscopy

Bronchoscopy is a procedure through which the physician performs issues related with the respiratory system. Doctors use an instrument known as a bronchoscope through a patient's nose or mouth and down the throat to reach the lungs. The bronchoscope is made of versatile fiber-optic materials and

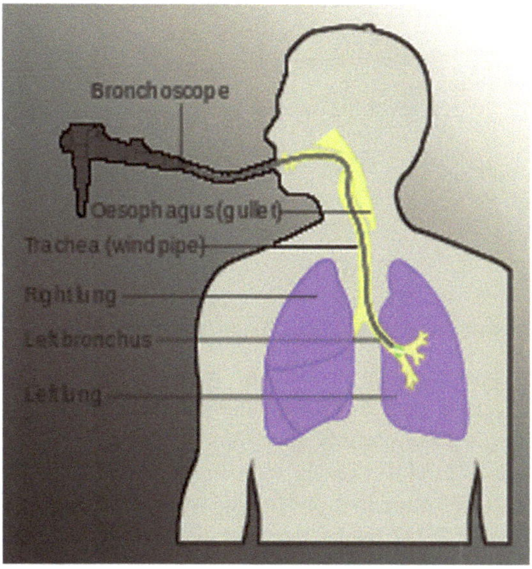

FIGURE 10.4 Bronchoscopy illustrations.

includes a light source and a camera on the top. Fig. 10.4 show an illustration of bronchoscopy.

Before the procedure, patients are advised to stop taking aspirin such as blood thinning medications. Bronchoscopy is typically done in a hospital room or in an emergency clinic working room. The whole method, including procedure and recuperation time, ordinarily takes around 4 hours. Bronchoscopy itself more often than not keeps going around 30 minutes to 1 hour. A patient will be approached to sit or lie back on a table or a bed. The patient gets associated with a screen so the physicians' team group can follow patient's pulse, blood pressure, and oxygen level during the procedure. The patient is given medication through a vein to enable to unwind. He may feel tired, yet regardless he will be wakeful. A desensitizing medicine will be splashed in his throat and conceivably through the nose. This prescription, called an analgesic, numbs the area. It decreases choking and coughing as the bronchoscope is set into the throat [1,2].

Bronchoscopy used to diagnose the following diseases:

- Lung disease
- Tumor
- Chronic cough
- Infection

10.4.4.1 Risks of bronchoscopy

There are some medical complications associated with bronchoscopy:

- Bleeding
- Infection
- Trouble breathing
- A low blood oxygen level during test
- Fever
- Coughing up blood
- Heart attack
- Lung collapse

10.4.5 Percutaneous Endoscopic Gastrostomy (PEG)

PEG represents percutaneous endoscopic gastrostomy, a system where an adaptable feeding tube gets set through the patient's abdominal wall as well as into the stomach. PEG permits nourishment, liquids, or potentially medicines to be placed straightforwardly into the stomach, bypassing the mouth and throat. It provides a method of feeding when oral intake is not suitable. Typically mild sedation is used by the physicians while performing PEG procedure. It doesn't require general anesthetic [2]. Fig. 10.5 demonstrates a scenario of PEG tube insertion in stomach.

10.4.5.1 Techniques of PEG

There are two techniques available for PEG techniques. In the first technique, the stomach wall gets distinguished and procedures get utilized to guarantee that there exists no such organ between the anterior stomach walls and the skin. In this connection, digital pressure gets exerted to the stomach wall that gets visible indenting the front gastric wall by the physicians.

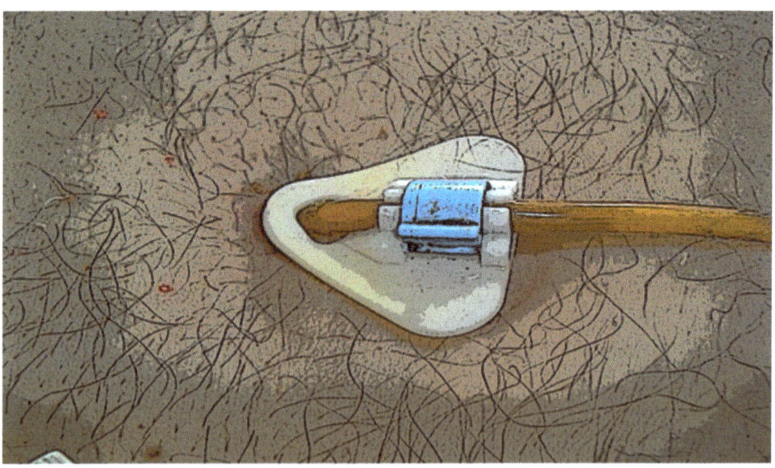

FIGURE 10.5 PEG tube inserted in stomach.

In transillumination method the light transmitted from the endoscope inside the stomach can be seen through the stomach wall. A little needle gets passed into the stomach prior to the bigger cannula gets passed inside the patient's stomach.

In other techniques an angiocath is utilized to cut the stomach divider through a little entry point. A delicate guidewire is embedded through this and hauled out of the mouth. The feeding tube gets joined to the guidewire and pulled along the mouth, throat, stomach, conjointly out of the cut [2].

10.4.5.2 Contraindications of PEG

Following are the reasons for which PEG method to be used:

- Failure to carry out the esophagogastroduodenoscopy
- Uncorrected coagulopathy
- Peritonitis
- Untreatable huge ascites
- Bowel obstruction
- Gastric wall neoplasm
- Abdominal wall infection
- Gastric mucosal abnormalities

10.4.5.3 Complications of PEG

- Cellulites around the gastrostomy site
- Hemorrhage
- Gastric ulcer
- Perforation of bowel
- Puncture of the left lobe of the liver resulting in liver capsule pain
- Gastrocolic fistula
- Gastric separation
- Buried bumper syndrome
- Leakage of stomach content around the PEG tube site
- Pain at the PEG site

PEG tubes can keep going for a considerable length of time or years. Maybe nonetheless, on the grounds that they can separate or end up stopped up over broadened timeframes, they ought to be supplanted. Physicians can lucidly discard or replace a tube without tranquilizers or anesthesia, in spite of the fact that physicians may pick to utilize sedation and endoscopy now and again. PEG sites close rapidly when the tube gets evacuated, hence, coincidental dislodgment needs.

10.4.6 Flexible sigmoidoscopy

Flexible sigmoidoscopy is a procedure that enables the doctor to examine the body part and also the lower (sigmoid) colon. The sigmoidoscope is a

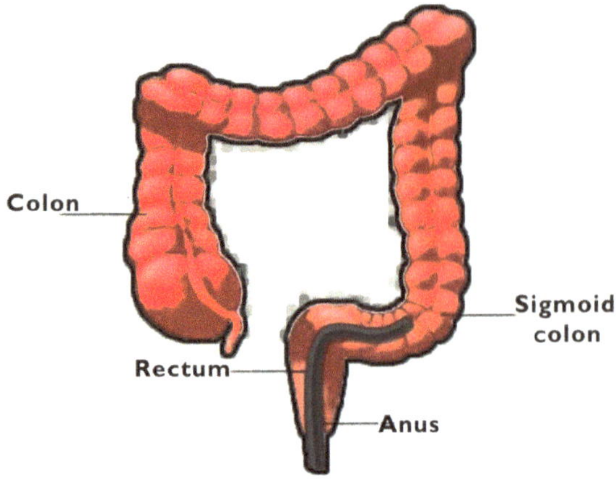

FIGURE 10.6 Demonstration of sigmoidoscopy.

versatile tube 60-cm long and regarding the thickness of a bit finger. It's inserted gently into the anus and is advanced slowly into the body part as well as the lower colon. It's a correct and easy methodology of investigation for body part hemorrhage, modification in intestine habit, and body part symptoms such as pain and diarrhea. Endoscopy is also an element of colon screening and surveillance for carcinoma [1].

10.4.6.1 Contraindications of flexible sigmoidoscopy

Following are the reasons for which flexible sigmoidoscopy method to be used. Fig. 10.6 demonstrates a scenario of sigmoidoscopy.

10.4.6.1.1 Investigate intestinal signs and symptoms

The flexible endoscopy test will facilitate the doctor to explore possible causes of abdominal pain, body part hurt, changes in bowel habits, chronic symptom, and different intestinal issues.

10.4.6.1.2 Screen for colon cancer

Sigmoidoscopy represents the one possibility for colon cancer screening; however, there are alternative choices that permit visualization of the entire colon. Sigmoidoscopy could sometimes be most popular over an endoscopy as a result of the preparation for flexible sigmoidoscopy and also the test itself may take less time. In addition, anesthetic is usually not needed. There is less risk of direct damage such as perforation with flexible sigmoidoscopy compared with colonoscopy.

- Unexplained weight loss
- Pain in abdomen

10.4.6.2 Preparation of flexible sigmoidoscopy

A trained medical professional performs a versatile flexible sigmoidoscopy throughout a workplace visitor at a hospital or a patient center. One usually doesn't want sedatives or anesthesia, and also the procedure takes concerning 20 minutes. For the work, patients will be interrogated to reside on a table whereas the doctor inserts an endoscope into his anus and slowly supervises it through his rectum and into his colon. The scope pumps air into his large intestine to provide the physician a far better read. The camera transfers a video image of his intestinal lining to a track, permitting the doctor to investigate the tissues lining in his colon and rectum. The doctor might raise him to maneuver many times on the table to regulate the scope for higher vision. Once the scope reaches his colon, the doctor withdraws it and investigates the liner of his colon once more. Throughout the procedure, the doctor might take away polyps and transfer them to a research laboratory for testing. Colon polyps are general in adults and are harmless in many situations. On the contrary, most colon cancer begins as a polyp; thus, removing polyps earlier is an efficient way to stop cancer. If the doctor finds abnormal tissues, he or she might carry out a biopsy. If the physician found polyps or different abnormal tissues throughout a versatile flexible sigmoidoscopy, the concerned doctor might recommend the patient to come for an endoscopy.

10.4.6.3 Complications of flexible sigmoidoscopy

Complications associated with flexible sigmoidoscopy are very rare but followings are some of the complications associated with it.

- Severe pain
- Fever
- Rectal bleeding

10.4.7 Cystoscopy

Cystoscopy could be a process that enables the physician to investigate the liner of the bladder; therefore, the tube that holds excreta out of body (urethra). A hollow tube (cystoscope) equipped with a lens is inserted into the urethra and is slowly advanced into the bladder. Cystoscopy is also carried out in a test room, employing a topical anesthetic jelly to numb to urethra. Or it is going to be accomplished as an outpatient strategy, with sedation. An alternative choice is to own a cystoscopy within the hospital throughout anesthesia. There are two types of cystoscopy available [3]:

1. *Flexible cystoscopy*
2. *Rigid cystoscopy*

Below are the listed reasons for which cystoscopy method to be used.

10.4.7.1 Examining causes of signs and symptoms

The signs along symptoms will embrace blood within the excreta, incontinence, active bladder, and painful excreting. Cystoscopy is capable of even facilitating to confirm the reason behind frequent urinary tract infections. On the contrary, cystoscopy usually isn't accomplished whereas you have got a lively urinary tract infection.

10.4.7.2 Diagnosing bladder diseases and conditions

It helps one to detect bladder cancer, bladder stones, and bladder inflammations.

10.4.7.3 Treat bladder diseases and conditions

Those kinds of small bladder tumors can be easily removed through cystoscopy by passing special tools through the cystoscope to treat certain conditions.

10.4.7.4 Diagnose an enlarged prostate

By the help of cystoscopy, doctors can detect the narrowing of the urethra at the place where it gets passed through the prostate gland. This helps one to detect an enlarged prostate gland.

10.4.7.5 Preparation of cystoscopy

A simple outpatient cystoscopy will paus fore 15 minutes once wiped out of a hospital having sedation or anesthesia. Cystoscopy involves 15 minutes to the half-hour. Patients are interrogated to empty the bladder. Afterwards, patients lie on a table on his back. Patients probably are positioned together with their feet in stirrups posture conjointly having knees bent. Patients may or might not want a sedative or anesthetic. If they get a sedative, they will feel asleep and relaxed throughout the cystoscopy; however, they still have some hangover. If they receive an anesthetic agent, they don't remain anxious throughout the procedure. Each kind of medication could also be given through a vein in their arm. The doctor can insert the cystoscope and for that a desensitizing jelly is employed to the urethra to help stop pain once the cystoscope is inserted. When waiting some minutes for the desentization, the concerned physician can fastidiously push the cystoscope into the epithelial duct, thus exploiting the tiniest scope attainable. Larger scopes are required for tissue samples or passing surgical equipment into the bladder.

The physician can investigate urethra and bladder. The cystoscope contains a lens on the tip that works sort of a telescope to amplify the inner surfaces of the urethra along with the bladder. A doctor would possibly place a special video camera over the lens to project the pictures onto a video screen. The bladder is stuffed with a sterile solution. The solution inflates the bladder and permits a doctor to induce a far better look within. Since the bladder fills, the patient will feel the requirement to urinate. He will be inspired to do this once the procedure is over.

Tissue samples can be taken. A physician would possibly take tissue samples for research laboratory testing or carry out varied alternative methodologies throughout the cystoscopy [1].

10.4.7.6 Complications of cystoscopy

Complications of cystoscopy can include

- an inability to urinate after cystoscopy,
- heavy blood clots present in urine,
- abdominal pain and nausea,
- chills,
- fever,
- pain or burning during urination,
- infections,
- pain,
- bladder perforation,
- narrowing of the urethra, and
- urinary retention.

Fig. 10.7 illustrates a scenario of cryptoscopy in the case of both males and females.

10.4.8 Transbronchial endoscopy

Transbronchial biopsy is performed by pulmonologists to analyze central and diffuse lung illnesses. Contrasted and open lung biopsy, transbronchial biopsy has lower bleakness and mortality. Biopsy of the lung used to be performed by methods for open careful strategies until 1963, when Anderson performed bronchoscopic lung biopsy with an inflexible bronchoscope. This methodology has two fundamental uncommon entanglements: pneumothorax and serious pneumonic dying.

10.4.8.1 Preparation of transbronchial bronchoscopy

A lung authority (pulmonologist) prepared to play out a bronchoscopy that showers a topical or neighborhood sedative in mouth as well as throat. This will initiate hacking at first, which will stop as the sedative works. At the point during the time the territory feels "thick," this is adequately numb. Patients might be afforded an IV narcotic to enable to unwind. Such medicine may make him drowsy and ought to lessen any tension they may have about the strategy. The methodology can likewise here and there be accomplished through utilizing general anesthesia, amid which are oblivious and

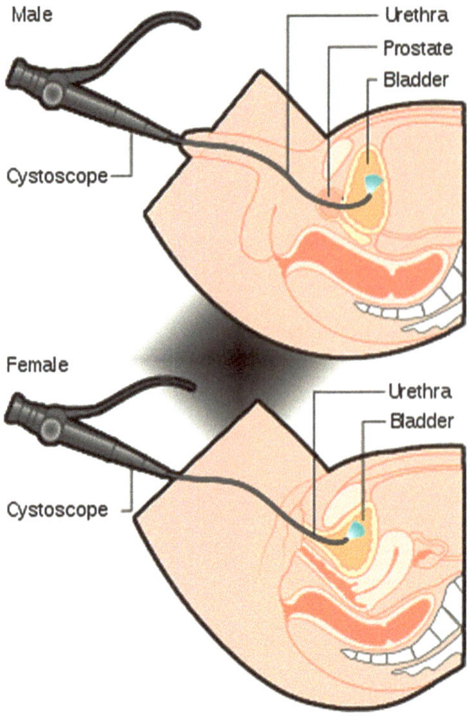

FIGURE 10.7 Visualization of cystoscopy in male and female.

torment free. On the off chance that the bronchoscopy is performed by means of the nose, a sedative jam gets embedded into one nostril. At the point when the nostril is numb, the extension gets embedded by the nostril until this goes through the throat into the trachea along with bronchi. As a rule, an adaptable bronchoscope gets utilized. Such an instrument is a cylinder that is under 1/2 in. wide and around 2 ft long. Since the bronchoscope gets utilized to look at the aviation routes of the lungs, tests of the lung discharges might be acquired to transfer for research facility examination. Saline liquid might be utilized to flush the zone and to gather cells may be that ought to be dissected through a physician. The transbronchial biopsy technique is performed utilizing a small forceps that goes through a channel of the bronchoscope into the lungs. Patients will be prescribed to inhale out gradually as the pulmonologist gets a little example of lung tissue. Such progression is generally rehashed until a few examples of tissues have been obtained for investigation. Sporadically ongoing chest X-beams (fluoroscopy) are utilized amid the bronchoscopy to assist direct the forceps to the ideal territory of lung.

Local anesthesia is utilized to loosen up the throat muscles. One may experience liquid running down the back of the throat and feel to cough or muffle until the sedative produces results. Notwithstanding the anesthesia, one may have vibes of weight or mellow pulling as the tubes travel through the trachea. Numerous patients experience a sentiment of suffocation when the tubes are in the throat, yet there is no danger of suffocation. Attempt to try to avoid panicking. On the off chance that patients cough during the test, increasingly soporific will be included. An X-beam is frequently taken after the bronchoscope is evacuated. At the point when the soporific wears off, the throat might be scratchy for a few days. Thereafter, in the test, the cough reflex will return in 1–2 hours, at that point ordinary eating and drinking are permitted [3].

Followings are the reasons for which transbronchial bronchoscopy is to be used:

- Bronchial abnormalities
- Tumors
- Endobronchial mass
- Adenoma (tumor)
- Infection
- Inflammation of the lungs pertaining to allergy-type reactions
- Rheumatoid lung disease
- Vasculitis
- Alveolar abnormalities like alveolar proteinosis
- Granulomas

10.4.8.2 Complications of transbronchial bronchoscopy

There are several types of complications associated with it.

- Disordered heart rhythm (i.e., arrhythmias)
- Heart attack
- Diminished blood oxygen (i.e., hypoxemia)
- Nausea and vomiting
- Sore throat
- Muscle pain
- Breathing problems
- Depressed heart rate
- Change in blood pressure
- Kidney damage

Figs. 10.4 and 10.8 illustrate the process of transbronchoscopy and transbronchoscopy needle aspirations respectively.

10.4.9 Hysteroscopy

Hysteroscopy could be a process that makes the doctor able to look within the uterus for diagnosing and treating the reasons of abnormal hemorrhage.

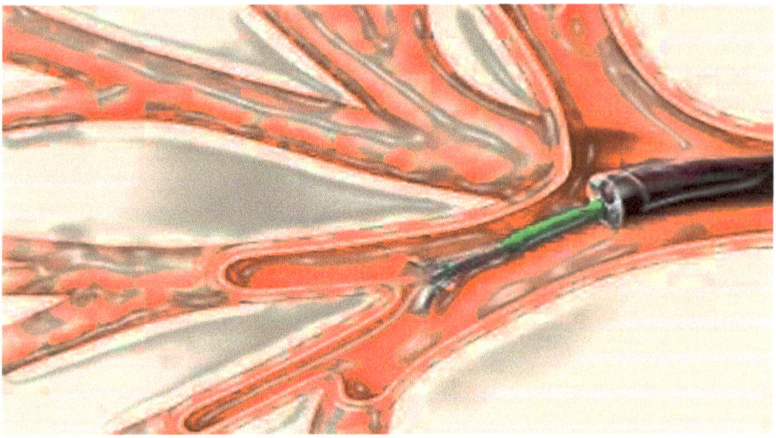

FIGURE 10.8 Transbronchoscopy needle aspirations.

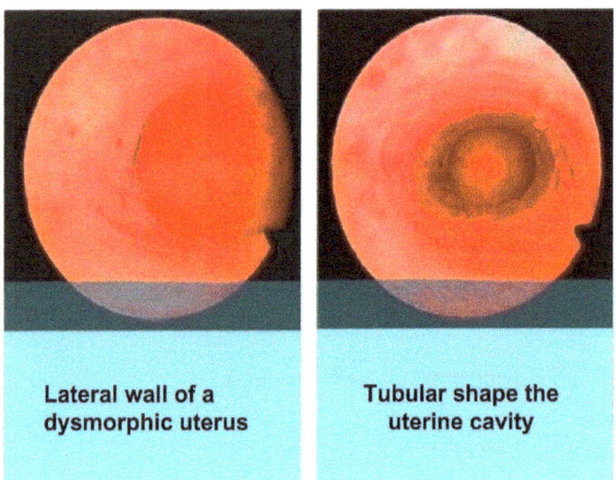

FIGURE 10.9 Hysteroscopy visualization.

Endoscopy gets completed employing a hysteroscope, a narrow, lighted tube that gets injected into the vagina to look at the cervix as well as within the uterus. In this connection, endoscopy will be either diagnostic or operative. Diagnostic endoscopy is employed to check issues of the uterus. Diagnostic endoscopy gets further accustomed to ensuring the outcomes of various tests, such as hysterosalpingography (HSG). HSG is X-ray dye checks, which is accustomed to check the uterus and fallopian tubes. Moreover, diagnostic endoscopy will typically be exhausted in an office setting [3]. Fig. 10.9 shows a scenario of hysteroscopy and Fig. 10.10 portrays the images of hysteroscopy.

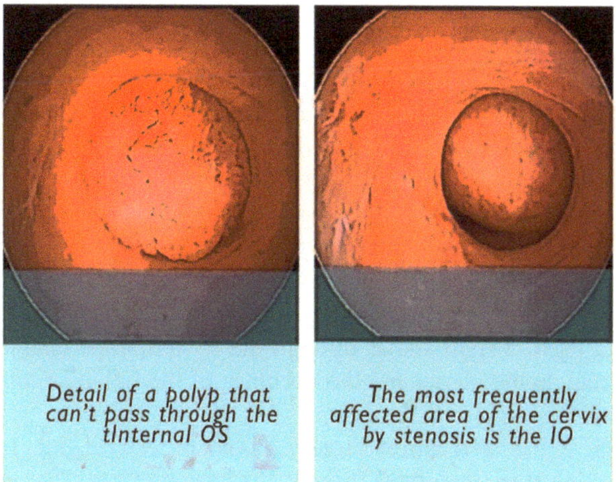

FIGURE 10.10 Hysteroscopy images.

Further, endoscopy will be employed with different methodologies, such as laparoscopy, or prior procedures such as dilation and curettage. In the case of laparoscopy, your physician can insert an endoscope (a slender tube fitted with a fiber-optic camera) into your abdomen for speculating the skin of your uterus, ovaries, and fallopian tubes. The endoscope gets inserted by incision created through or below your navel. Operative endoscopy is employed to correct an abnormalcy that has been tracked throughout a diagnostic hysteroscopy. If an abnormalcy gets tracked throughout the diagnostic endoscopy, an operative endoscopy will be carried out at the same time, averting the requirement for a second surgery. Throughout operative endoscopy, little appliances accustomed right the state is inserted by the hysteroscope.

Hysteroscopy is also used in the following situations:

- Discard adhesions that will occur due to infection or from past surgery.
- Examining the explanation for continual miscarriage once a lady has quite two miscarriages in an exceeding row.
- Locating an intrauterine device.
- Perform sterilization, within which the hysteroscope gets employed to put little implants into a woman's fallopian tubes as a permanent sort of contraception.

10.4.9.1 Preparation of hysteroscopy

Hysteroscopy can help one to recognize the reason for a heavy or long menstrual stream, just as seeping between periods or subsequent to the menopause. Endometrial removal represents one strategy where the hysteroscope, alongside different instruments, is utilized to obliterate the uterine coating so as to treat a few reasons for heavy bleeding. Otherwise called Asherman's syndrome, uterine adhesions are groups of scar tissues that can shape in the uterus and might prompt alterations in menstrual flow just as infertility. Hysteroscopy can enable your specialist to find and expel the adhesions. The system itself happens in the accompanying request:

The specialist can extend your cervix to permit the hysteroscope to get attached. The hysteroscope is attached through the vagina and cervix into the uterus. Carbon dioxide gas or a fluid arrangement is afterwards inserted into uterus, by the hysteroscope, to make it grow and to gather up any blood or bodily fluid. Then, a light shone through the hysteroscope enables the specialist to speculate the uterus and the openings of the fallopian tubes into the uterine cavity.

At length, if the medication surgery has to be carried out, little instruments are inserted into the uterus through the hysteroscope. The time it needs to carry out hysteroscopy can go from under 5 minutes to over 60 minutes. The duration of the methodology depends upon whether it is diagnostic or operative and whether an additional strategy, for example, laparoscopy, is accomplished in the meantime. When all is said in done, in any case, diagnostic hysteroscopy takes reduced time than operative [3].

10.4.9.2 Advantages of hysteroscopy

- Lesser hospital stay
- Lesser recovery interval
- Reduced pain medication needed during surgery
- Averting hysterectomy
- Rejection of "open" abdominal surgery

10.4.9.3 Complications of hysteroscopy

There are several complications associated with hysteroscopy. Some of them are listed below. Hysteroscopy may be a comparatively safe methodology. On the contrary, like any variety of surgery, problems are attainable.

- Risks related to anesthesia
- Infection
- Heavy hemorrhage
- Injury to the cervix, uterus, bowel, or bladder
- Intrauterine scarring
- Reaction to the substance accustomed enlargement of uterus
- Accidental damage to the womb
- Accidental damage to the surgery
- Excessive bleeding during or after surgery
- Infection of the womb
- Feeling faint

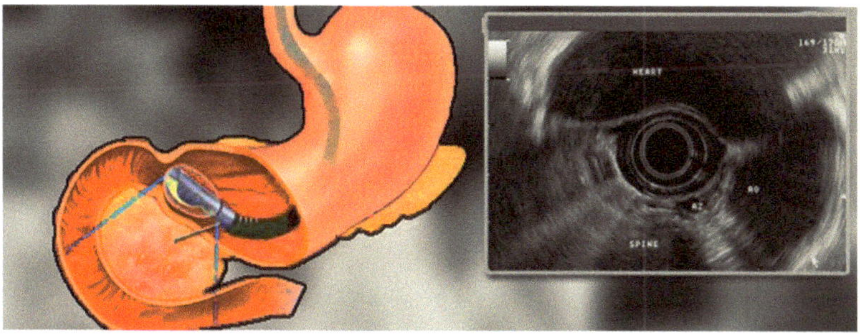

FIGURE 10.11 Endoscopic ultrasound visualization.

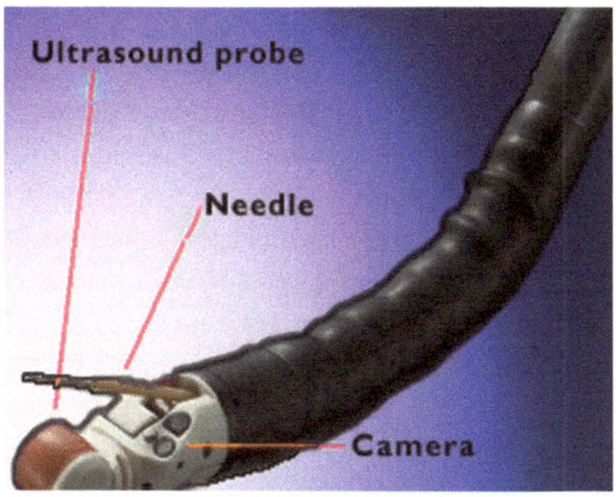

FIGURE 10.12 Visualization of an endoscope.

10.4.10 Endoscopic ultrasound

EUS is a technique that enables a specialist to acquire pictures and data about the digestive tract and the encompassing tissue and organs, including the lungs. Ultrasound testing utilizes sound waves to make an image of inside organs. Amid the method, a little ultrasound gadget is introduced on the tip of an endoscope. An endoscope is a little, flexible tube with a camera appended. By embeddings the endoscope and camera into the upper or the lower stomach–related tract, the specialist can acquire high-quality ultrasound pictures of organs. Since the EUS can draw near to the organ(s) being inspected, the pictures got with EUS are regularly more precise and detailed than the pictures given by customary ultrasound that must go from the outside of the body. Fig. 10.11 illustrates a scenario of EUS visualization, and Fig. 10.12 illustrates a visualization of endoscope.

For EUS pertaining to the higher duct, a probe gets injected into the esophagus, stomach, and small intestine throughout a methodology known as esophagogastroduodenoscopy. Among alternative usages, it permits for televising for cancer in pancreas, esophagus, or gastric cancer, similarly as benignant tumors of the upper epithelial duct. This conjointly permits for a diagnostic assay of any focal lesions detected within the upper epithelial duct, like passage tuberculosis. Such thing is often accomplished by inserting a needle along the abdomen lining into the desired target. Ordinarily this procedure is carried out to spot various malformations and much within the digestive fluid ducts and exocrine gland ducts.

EUS gets carried out with sufferers who are insensible. The endoscope has undergone the mouth as well as toward the suspicious space by the esophagus. From numerous locations between the esophagus and small intestine, organs among and outside the epithelial duct will be viewed to determine if they're abnormal and that they get biopsied through a method known as "fine needle aspiration." Various organs such as the liver, pancreas, and adrenal glands get simply biopsied. In addition, the gastrointestinal wall will be imaged to determine whether it's very thick, thus indicating inflammation or malignancy. Such strategy is very susceptible to the detection of pancreatic cancer significantly in sufferers who are thought to possess a mass or presence with jaundice. Such activity in staging sufferers with pancreatic cancer gets restricted to native metastases; but, together along the Computerized Tomography (CT) scan that offers data on zonal metastases, this affords a superb imaging modality for diagnosing and staging of exocrine gland cancer. EUS may also be employed coupled with ERCP. Further, the ultrasound investigation gets employed to find gall stones that can have migrated into the common duct. Such incidence could cause an obstruction of the drain used by the liver and exocrine gland, which can cause lower back pain, jaundice, along with redness.

EUS is performed to detect the following conditions:

- Evaluate how profoundly a tumor enters your abdominal wall in esophageal, gastric, rectal, pancreatic, and lung cancer.
- Decide the degree (organize) of cancer, if present.
- Decide whether cancer has spread (metastasized) to your lymph hubs or different organs.
- Give exact data about nonlittle cell lung cancer growth cells, to manage treatment.
- Assess abnormal discoveries from imaging tests, for example, growths of the pancreas.
- Guide emptying of pseudocysts and other anomalous accumulations of liquid in the stomach area.
- Grant exact focusing for conveying drug legitimately on the pancreas, liver, and different organs.

10.4.10.1 Complications of endoscopy ultrasound

The risks associated with EUS are listed below:

- Bleeding
- Infection
- Tearing in the intestinal wall or throat
- Pancreatitis
- Fever
- Chest affliction
- curtailment of breath
- Black or too dark-color stool
- Acute or persistent abdominal pain

10.5 Smart pills

The expression "smart pills" refers to smaller than normal electronic gadgets that are formed and planned in the shape of pharmaceutical containers yet perform exceedingly advanced functions, for example, detecting, imaging, and medication conveyance. They may incorporate biosensors or picture, pH or compound sensors. When they are gulped, they travel along the gastrointestinal tract to catch data that is generally hard to get and they are effectively dispensed with from the system. Their arrangement as ingestible sensors makes them particular from implantable or wearable sensors. Many endoscopy methodologies and a large number of colonoscopy methods are likewise performed to analyze or screen for cancer. Traditional, rigid endoscope utilized for these strategies are awkward for patients and may cause inward wounding or lead to disease as a result of reuse on various patients. Smart pills wipe out the requirement for obtrusive systems: wireless communication permits the transmission of continuous data; advanced batteries and locally available memory make them valuable for long term sensing from inside the body [4,5].

Smart pills have altered the finding of the gastrointestinal issue and could replace ordinary symptomatic methods, for example, endoscopy. Usually, an endoscope is embedded into a patient's throat, and in this manner the upper and lower gastrointestinal tract, for demonstrative purposes. There is a danger of aperture or tearing of the esophageal coating, and the patient faces inconvenience during and after the system. A smart pill, is that as it may, can without much of a stretch be gulped and moved to catch pictures, and requires negligible patient readiness, for example, sedation. The implicit sensors permit the estimation of body fluid and gases in the gut, giving the doctor a multidimensional image of the human body.

Core body temperature, nearby pH, and internal pressure are essential markers of patient prosperity. Although a thermometer can give an exact perusing during normal checkups, the observation of experts in high-intensity circumstances requires an increasingly precise internal body temperature sensor. An ingestible chemical sensor can record sharpness and pH levels along the gastrointestinal tract to screen for ulcers or tumors. Sensors additionally can be incorporated with prescriptions to follow compliance [6].

Medicine can be ineffective if the medication payload isn't conveyed at its proposed spot and time. Since an oral medication goes through a broad pH range, the pill encapsulation could break down at the wrong time. Notwithstanding, a smart pill with environmental sensors, a feedback algorithm, and a drug release procedure can offer ascent to smart pill conveyance systems. This can guarantee ideal medication conveyance and anticipate inadvertent overdose. The scaling down of electronic parts has been vital to the smart pill design. As distributed computing and wireless communication stages are coordinated into the social healthcare, the utilization of smart pills for observing imperative signs and prescription consistency is probably going to increment. In the long haul, smart pills are relied upon to be an essential segment of remote patient monitoring and telemedicine [2].

10.6 Purpose of WCE

The first wireless capsule endoscopy was performed in 1999, and the US Food and Drug Administration affirmed its utilization in the United States in 2001. The M2A container (mouth to rear-end) was the most readily accessible pill camera and was in the long run renamed as PillCam SB. The capsule is ingested and transmits pictures at two to six images for every second through the span of 8–12 hours until the battery terminates. It creates 512 by 512-pixel, high-resolution pictures that permit itemized assessment of the gastrointestinal mucosa. A prepared gastroenterologist at that point surveys the pictures. Battery life can be a restricting variable during wireless capsule endoscopy, and 16.5% of studies are incomplete because of battery lapse. Wireless capsule endoscopy is an analytic system. It can just limit sores in the throat, stomach, little gut, and colon, however, can't be utilized for biopsy or treatment. It is utilized frequently for repetitive and cloud gastrointestinal seeping after traditional endoscopic methods have neglected to distinguish a draining source. It very well may be a valuable report for confining a sore before angiography, medical procedure, or further endoscopic methods. Wireless capsule endoscopy can be utilized to evaluate the throat, stomach, small digestive system, and colon. It is ingested simply like some other medicine and goes through the throat into the stomach. It at that point goes through the pyloric sphincter into the duodenum, jejunum, and ileum. The capsule continues through the ileocecal valve into the cecum. It at that point progresses through the colon and is discharged during defecation. By and large, the patient will observe the entry of the capsule; however, plain stomach films can be utilized to evaluate total section of the video case.

The most widely recognized sign is for cloud gastrointestinal drain thought to be situated in the little gut after upper and lower endoscopic

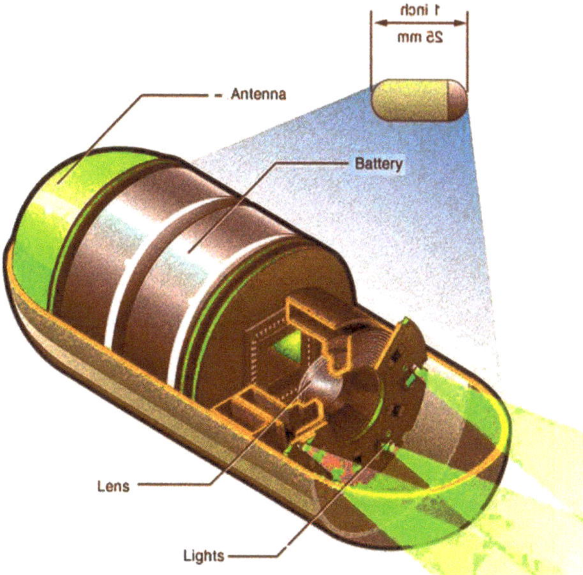

FIGURE 10.13 Interior design of smart pills.

methods neglected to discover a draining source. A video capsule has a 35%–77% discovery rate of cloud gastrointestinal drains. Different signs for small bowel capsule endoscopy are as per the following: conclusion of Crohn's malady and assessment of Crohn's illness action, finding of celiac sickness, and assessment of hard-headed celiac ailment, polyposis disorder reconnaissance, small-digestive tract tumors, for example, neuroendocrine tumors, or carcinoid tumors. Wireless capsule endoscopy is demonstrated to assess the throat for esophageal varices screening, Barrett's throat screening, and esophagitis recognizable proof. Colon case endoscopy is demonstrated for colon malignancy screening in patients with a past deficient colonoscopy, patients that have real dangers for a colonoscopy itself, and patients that can't endure sedation. Fig. 10.13 shows the interior design of smart pills.

Below given some of the reasons for which WCE is to be used to detect [7]

- iron deficiency;
- unexplained weight loss;
- unexplained tract bleeding;
- detecting tumors, polyps, and cancer;
- detecting CD;
- detecting celiac disease;
- screening for ulcers.

10.6.1 Accuracy of WCE

The accuracy of wireless capsule endoscopy can differ by the point of the examination and the gadget utilized. According to a recent report from University Hospital Ghent in Belgium, wireless capsule endoscopy can effectively analyze dynamic seeping in the small digestive system in around 58%–93% of cases. At the point when Crohn's sickness is used to analyze, wireless capsule endoscopic is viewed as predominant at identifying early provocative injuries contrasted with every single other methodology. It is 26% more exact than an X-beam, 16% more precise than a barium think about, 25% more precise than colonoscopy, and 21% more precise than a figured tomography (CT) filter.

Likewise, wireless capsule endoscopy is between 83% and 89% exact in accurately distinguishing celiac infection. Wireless capsule endoscopy is a vital method for assessing the gastrointestinal tract when conventional endoscopic procedures have fizzled. It is protected, effortless, has no hazard for contamination, and does not require sedation. It has a higher indicative yield than numerous different modalities for assessing the intestinal lumen and can confine sores. The deficiencies of wireless capsule endoscopy are its absence of therapeutic capabilities. Its high diagnostic yield and low complication rate settle on it an engaging decision for evaluation of the intestinal lumen when shown regardless of the failure to acquire tissue tests and give treatment.

10.6.2 Technology of WCE

Wireless capsule endoscopy or smart pills consist of three components:

- Capsule
- Data recorder belt/Smart wearable
- Workstation

10.6.2.1 Capsule

The wireless capsule is a dispensable device, estimating 11 × 26 mm (somewhat bigger than a substantial nutrient container) and weighting 3.7 g. The 2 vault, chamber molded case is made of a biocompatible plastic with a smooth surface that permits the peristalsis of the intestinal tract to propel the container through the lumen. The wireless capsule contains a complementary metal oxide silicon chip camera, a lens, an illuminating light-discharging diode, and a vitality source and a radiotelemetry transmitter. The case battery life is around 8 hours, which is adequate for imaging the small digestive tract. At the point when the battery control is exhausted, the transmitter changes the case to the guide mode. In this mode, it transfers the data to the recorder about the area of the case enabling workstation or system to follow the capsule in its pathway for 10 extra hours. The expendable capsule is built of extraordinarily fixed biocompatible material impervious to stomach-related liquids. The capsule transmits video signs and information progressively. The capsule is ingested by the patient and gives video pictures of the GI mucosa amid its travel all through the GI tract at a rate of two pictures for every second. The system can gain around 50,000 pictures. The capsule is normally discharged after roughly 8–72 hours [8,9].

10.6.2.2 Data recorder belt/Smart wearable

The wearable worn by the patient around the abdomen gets the signals transmitted by the container through a variety of sensors put on the patient's body. The sensor exhibit [antenna] is involved eight indistinguishable, 4 cm measurement sensors appended to the skin by dispensable adhesive pads. It gets the pictures from the capsule and sends signals to an information recorder. The sensors are associated with the recorder by an adaptable coaxial link.

The information recorder is a walkman-size battery-worked unit that gets the information transmitted by the capsule. It involves a receiver, processor modules, and a hard disk drive to store the information. Eight nickel-metal 6-V battery-powered batteries are utilized for the task. The information recorder is prepared for the task once the sensor exhibit and batteries and information recorder are altogether associated. A blue flickering light demonstrates that the recorder is capturing the information. The information recorder can download around 50,000 pictures to the workstation.

10.6.2.3 Workstation

System processes the information downloaded from the information recorder. The workstation is an altered standard PC intended for handling of the information into a video motion picture and introduction. The output enables doctors to pursue the way gone by the capsule, see the sores, and spare critical pictures and short video cuts. The video motion picture is included edges and can show from 1 to 50 images for every second. It has stopping and switching abilities. More often than not, the motion picture is viewed as 5–10 images for every second except higher or lower velocities can be chosen to see the pictures. The doctor endoscopy survey time is around 1.5 hours. Proceeded with advancement and improvement of WCE has brought about the improvement of a blood-detecting calculation that utilizes bleeding design, enabling the doctor to concentrate more on bleeding areas [10].

10.6.3 Preparation of WCE

10.6.3.1 Day of before WCE

After lunch on the day preceding the capsule endoscopy examination, patients are advised to begin the fluid eating routine as recommended by the doctor. From midnight the night prior to the capsule endoscopy, it is advised to eat or drink aside from fundamental drug with a taste of water. It is advisable not to take any medicine starting 2 hours before experiencing the capsule endoscopy. Patients should keep away from smoking 24 hours before the capsule endoscopy. Male patients with a great deal of chest/stomach hair should shave their mid-region 6 in. (15 cm) above and beneath the navel upon the arrival of the examination.

10.6.4 Working of WCE

Smart pills works as follows: when a person ingests or takes smart pills, it goes through the throat into stomach. The pills contain a sensor-enabled chip, a small camera, light, and a battery. The pills contain a 1 mm^2 sensor that is coated in two digestive metals: copper and magnesium.

Ingesting copper and magnesium is not harmful because these two ingredients are already present in our regular multi-vitamin. When the sensor-enabled pills are reached into the stomach, the stomach acid or electrolytes activate the chip. The chip then sends the signals to patients' wearable patch. The patch records all the information captured by the pills as the anxiety levels, stress levels, blood pressure of the patients and stores the data. After the patch receives the data from the device, it encrypts the data and transmits it to the app on a smartphone. Then the data is encrypted once more by the app and transmits to the doctors mobile.

As the pills contain a camera, when the pills travel through the esophagus, stomach, or inside the human body, it takes photographs very rapidly almost two to– six images per second with a high resolution of pictures 512 by 512 pixel. The devices take images continuously until the battery life expires, which is almost for 8 hours. Then the capsule (smart pills) is passed through the natural bowel movement and flushed away. The pills then send these photographs to the patient's wearable patch where the images are stored and then are reviewed by the physician later or transmit the images to the computer system where the images get monitored by the physician for diagnosing diseases.

10.7 Conclusions

The current chapter details the different types of endoscopy and the risks and complications associated with each type. Then we discussed the wireless capsule endoscopy (WCE) or smart pills, how this technology works and how it is beneficiary for patients rather than the traditional endoscopic method. As discussed above, all the complications and problems associated with the traditional endoscopic method will be overcome by wireless capsule endoscopy method or smart pills. Moreover, smart pills will revolutionize the healthcare industry than traditional endoscopy method. Thereafter, the concept of smart pills and its contribution in the healthcare industry were briefly discussed.

References

[1] <https://www.medicinenet.com/capsule_endoscopy/article.htm#introduction>.

[2] K.D. Robertson, R. Singh, Capsule endoscopy, in: StatPearls [Internet], StatPearls Publishing, Treasure Island, FL, 2019. Available from: <https://www.ncbi.nlm.nih.gov/books/NBK482306/#>. [Updated 2019 Feb 28].

[3] <https://www.nhs.uk/conditions/endoscopy/>.

[4] S.V. Zanjal, G.R. Talmale, Medicine reminder and monitoring system for secure health using IOT, Procedia Comput. Sci., 78, 2016, pp. 471–476.

[5] <https://www.linkedin.com/pulse/digital-pills-iot-healthcare-smart-medicine-future-poonam-as>.

[6] C. McCaffrey, O. Chevalerias, C. O'Mathuna, K. Twomey, Swallowable-capsule technology, IEEE Pervasive Comput. 7 (1) (2008) 23–29. Available from: https://doi.org/10.1109/mprv.2008.17.

[7] S.B. Baker, W. Xiang, I. Atkinson, Internet of things for smart healthcare: technologies, challenges, and opportunities, IEEE Access, 2017, Digital Object Identifier 10.1109/ACCESS.2017.2775180,

[8] R. Goffredo, D. Accoto, M. Santonico, G. Pennazza, E. Guglielmelli, A smart pill for drug delivery with sensing capabilities. 2015, 978-1-4244-9270-1/15/$31.00 ©2015 IEEE.

[9] N. Fawaz, D. Jansen, Enhanced telemetry system using CP-QPSK band-pass modulation technique suitable for smart pill medical application ,2008, 978-1-4244-2829-8/08 ©2008 IEEE.

[10] R. Goffredo, A. Pecora, L. Maiolo, A. Ferrone, E. Guglielmelli, and D. Accoto, A swallowable smart pill for local drug delivery, 2016, 1057-7157 © 2016 IEEE.

Further reading

D. DeMeo, M. Morena, Medication adherence using a smart pill bottle, 2014, 978-1-4799-6740-7/14 ©2014 IEEE.

Chapter 11

BioSenHealth 2.0—a low-cost, energy-efficient Internet of Things–based blood glucose monitoring system

Vikram Puri[1], Raghvendra Kumar[2], Dac Nhuong Le[3], Sandeep Singh Jagdev[4] and Nidhi Sachdeva[5]
[1]Duy Tan University, Da Nang, Vietnam, [2]CSE Department, LNCT Group of College, Jabalpur, India, [3]Faculty of Information Technology, Haiphong University, Haiphong, Vietnam, [4]Ellen Technology (Pvt). LTD, Punjab, India, [5]Fairleigh Dickinson University, Vancouver, BC, Canada

11.1 Introduction

Since the last few years, the health-care industry has seen a major issue due to an increase in the number of patients with long-term illness and their long-term diagnosis. Many systems have been proposed to reduce the number of patients in hospitals due to chronic diseases [1–3]. Millions of people face the irregular heartbeat called *arrhythmia*, according to the report of the center of disease control [4]. Usually, people aged 65 or more have high chances of arrhythmia. Many autonomous health systems have been proposed to monitor major health problems including diabetes. According to the World Health Organization (WHO), the number of diabetic patients shots to more than 422 million and 1.5 million people died due to diabetes in 2014 [5]. WHO considered diabetes under the category of top 10 causes of death, and this number has been rapidly increasing during the period from 1990 to 2014 and will be two to three times more by 2030 [5]. A person of any age and gender could face diabetic disorder. Impact of diabetes is not only limited to the developed countries but can also be seen in the developing countries. In the United States, 200,000 people [6] under the age of 20 suffer from diabetes. Moreover, diabetes is directly related to other major health problems including heart attack, liver or kidney failure, and some other mortal diseases [7].

Emergence of Pharmaceutical Industry Growth with Industrial IoT Approach.
DOI: https://doi.org/10.1016/B978-0-12-819593-2.00011-X

Fall and its effects cannot be forgotten because it causes serious health injuries, bone fractures, and neck and serious head injuries [8]. These injuries take a long time for healing, which may cause high cost and reduce the quality of life [9]. Diabetes, atherosclerosis, heart attack, and elder people are correlated, and diabetes is recognized as a peril for fall [10]. Unfortunately, no permanent cure is present for diabetes, and some available solutions are not appropriate [11]. To overcome this issue, the UK Prospective Diabetes Group [12] has proposed a solution called continuous glucose monitoring (CGM) that helps to minimize the long-term effects such as macrovascular and microvascular complications.

Internet of Things (IoT) is a crisscross of physical devices' network combined with sensors, cloud services, and network, which enables data sharing and storage [13]. IoT has transformed the health-care industry into a new futuristic level. About 40% of IoT technology is dependent on the health-care industry and developing $117 IoT-connected health-care market [14]. The synergy of the medicine and IoT, such as the Internet of Medical Things (IoMT) [15], will move toward better health-care services including reduced uncertainties, improved patient experience as well as management of remote medical care, and drugs management. Fig. 11.1 represents the general architecture of IoMT.

Wearable health-monitoring device connected through the Internet or IoT is a more reliable and effective solution for long-term illness and for the health-care centers' access to extract the diagnostic and treatment [16] information of the patient. Although the existing medical treatment devices for the healthcare monitoring are insufficient to fulfill the basic need of treatment and inefficient to fetch the accurate values through the sensors which cannot analyze accurate membership values [17]. Existing health-care system is focused on diet control and recommendation of drugs, which is another problem in the IoT-enabled health-monitoring architecture [18]. Currently, people are using existing proposed systems for the food nutritional consultation, which are unable to provide accurate information to the patients on the basis of real-time physiological information and sometimes make the patient's situation worse. Diabetic patients take dietary advice from the expert when their blood glucose level is increased or decreased. Moreover, it is a tough situation for the health-care experts to extract the patient's risk factor information and compile a meaningful perception for the medication, diet, or dietary food. Usually, compiled information is very unpredictable due to the current condition, drugs, or food because it varies from patient to patient. Perception and treatment both have complexity and unpredictability.

IoT enabled many futuristic technologies to include deep sensing, cloud computing, a static or mobile sensor network that assists to maintain the quality of life. Sensed data is stored at cloud server that helps to process the real-time actions or feedback at anytime, any location. For instance the

FIGURE 11.1 IoMT architecture. *IoMT*, Internet of Medical Things.

IoT-based glucose system works autonomously, and it injects insulin inside the body when blood sugar is detected high.

Although diabetes IoT monitoring system has some pros, namely, real-time monitoring at remote location and data storage for the action or response, but it also has cons especially security. Data transferred inside the network is not secured because it is not properly encrypted and processed by unauthorized third parties [19]. In addition, many proposed IoT health-care systems are not able to handle distributed storage, data processing, and data analytics.

A proper approach to overcome this challenge is to apply a fog layer with a gateway to manage energy dissipation at sensor nodes [20]. Fog layer works on the upper end of the smart gateway to maintain the quality of service (QoS). In this chapter a wearable sensor is used to fetch the blood glucose data and send this processed data to the smart gateway through the Wi-Fi module. The main processing (i.e., blood sugar algorithm) is done at the smart gateway to secure data through the encryption technique. Moreover, fog helps to manage the overload on the cloud server via preprocessing of the data and also assists other complex services including failed detection, synchronization at database, and mobility [21]. For example, if there is any interruption between the smart gateway and cloud server, the real-time processing always maintains with them. With the addition of the advanced services in fog, it will not only overcome many uncertainties but also intensify QoS.

In this chapter a smart IoT-based diabetes monitoring system is proposed. To improve the accuracy in blood glucose detection and diagnosis, the proposed system is not only capable to detect the blood sugar but also able to process the body temperature and nearby environmental parameters including the temperature, humidity as well as barometric pressure. The proposed system data is encrypted with the use of fog layer at the smart gateway. The encryption and decryption of data are used at the time of sending and receiving, respectively. The proposed sensor node is flexible, energy efficient to monitor the vital signal and offers advanced services such as cloud storage and fault detection.

The remainder of this chapter is organized as follows: Section 11.2 includes related studies. Section 11.3 provides methodology including the architecture of proposed work and the circuit diagram. Section 11.4 discusses and analyzes the outcomes. Section 11.5 concludes the work.

11.2 Related studies

Many studies and efforts have been proposed for real-time health-care monitoring systems through the use of IoT. Verma [22] proposed a remote health-care monitoring system using the abstraction of fog computing and applied in smart homes. For the real-time processing of patient data, data transmission based on the event triggering is used, and temporal mining concept is deployed to analyze health index of a valetudinarian. Data of 67 patients are collected through a monitoring system in 30 days to check the validation of the proposed system. Rathore [23] proposed a real-time emergency health-care system to overcome uncertainties such as processing of big data and real-time processing. In addition, the proposed work consists of a smart building (data analysis building) illustrated by deployed model and architecture and is used for decision-making. Numerous health-care datasets and real-time data are considered to check the efficiency of the proposed model. Yang [24] developed intelligent health-care platform called "iHome" that improved the user experience and efficiency regarding the service. iHome is based on the open-source intelligent pharmaceutical packing box with extended data connectivity to make a reliable connection between the sensor and related services. Proposed platform integrates IoT devices within health-care services to check the feasibility on trial basis. Jara [25] presented an IoT-based health-care solution that is categorized into two ways: (1) manage patients' profile through radio frequency identification (RFID) and (2) make internal connectivity between the patient's sensor node, nurses/caretaker application to manage the health card, patient's web portal. This solution is based on personal devices for the patients to calculate the insulin dosage. In [26] the author discussed benefits of m-health things (m-IoT) architecture to manage the glucose level. The architecture utilizes laptop as a smart gateway to make a connection between the sensor node and the cloud server. Dubey [27] presented an architecture based on data mining and data analytic techniques on raw data that is collected from the sensors. It also uses it for the validation and evaluation of fog data. In [28] the author illustrated the role of IoT in the health-care industry and technological feasibility regarding their implementation and futuristic approach. A framework has been proposed to check the validity of implementing IoT in the health-care industry. Gope [29] discussed and proposed a highly secured Internet-enabled health-care system called BSN-Care (body sensor network-care) to maintain the privacy of patients' data and also highlighted the major security loopholes of BSN-based health-care platforms. In Ref. [30] an IoT-based health-care approach is proposed for the rural and poor people. Health parameters included in this system are blood pressure, blood sugar (diabetes), and uncertain growth in the body parts. Farahani [31] illustrated the holistic IoT-based health-care system architecture to deal with chronic diseases.

Alfian [32] proposed a self-managed health-care solution for diabetic patients to deal with their chronic condition. The proposed solution has fused two different platforms such as Apache Kafka and MongoDB for the data streaming and storing patient data from deployed sensors, respectively. Bluetooth low energy is worked as a bridge to fulfill the gap between sensor and server. Diabetes dataset is tested through the machine learning classification technique.

11.3 Methodology

11.3.1 Architecture

An overview of IoT [33] based blood glucose system is illustrated with the aim of showing main part of sensor strip in the system. Fig. 11.2 represents the architecture of proposed system. The proposed work is categorized into three different parts: (1) sensor strip circuit, (2) smart gateway, and (3) cloud processing. Different parts are discussed in the following subsections.

11.3.1.1 Sensor strip nodes

Sensor strip circuit node consists of glucose strip, transconductance amplifier, and microcontroller (ATmega328), and Wi-Fi module.

- *Glucose strip*: Glucose strips are the disposable plastic strips, which play a vital role to monitor and control blood glucose level. These strips are

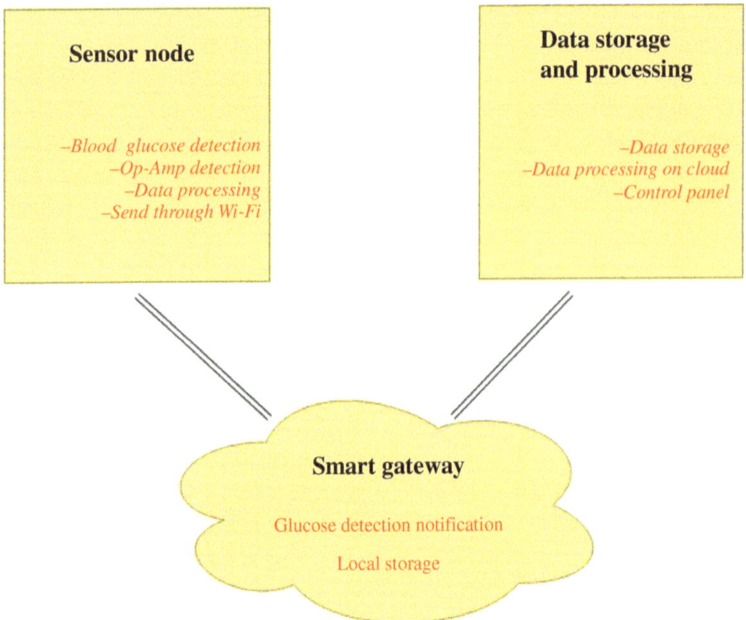

FIGURE 11.2 Architecture of IoMT. *IoMT*, Internet of Medical Things.

the primary component to check the blood glucose level. Strips will be on work mode when blood sample is placed on it. Enzyme is present on the strip called glucose oxidase that is react with blood sample to form gluconic acid in the presence of glucose in the blood. In the next step, strip is inserted inside the glucose meter. It provides current to the strip through the terminals and able to compute the current between the terminals of the strip. Electric current between the terminals is directly depended on the concentration of glucose present in the blood. Three electrodes (see Fig. 11.3) present in the strip are as follows:

- *Working electrode*: Electrons are generated during the chemical reaction between the strip enzyme and blood glucose.
- *Reference electrode*: Hold the constant voltage with respect to working electrode to make chemical reaction feasible.
- *Counter electrode*: It helps to supply enough current to the working electrodes.

- *Transconductance amplifier*: Transconductance amplifier (LM358 [34]) is a high gain, frequency compensated operational amplifier (Op-Amp) designed for extended range of voltage over single-input power supply. LM358 consists two independent Op-Amp in dual-inline package as well as surface mount device (SMD) chip size package, and current drain is

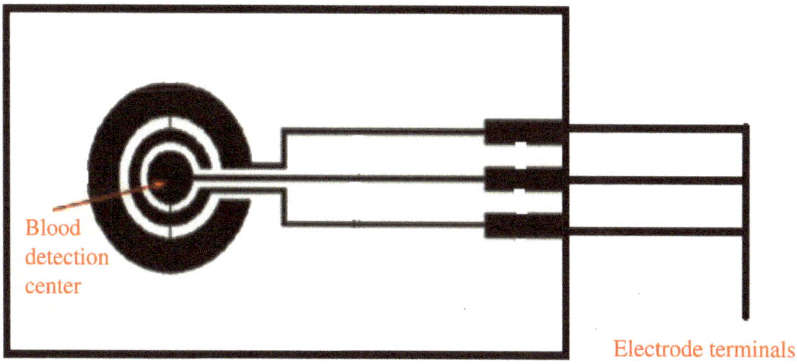

FIGURE 11.3 Internal circuit of glucose strip.

separated from the power supply voltage magnitude. Table 11.1 represents the technical specifications of LM358.

Output from test strip is in form of current; for processing of these values, we need to convert it into voltage. Here, transconductance makes a bridge between the sensor and microcontroller. The main purpose behind the use of Op-Amp is to convert current into voltage. Output of amplifier is fed back to input so current is converted into voltage. Gain required at output can be adjusted by value of resistors used.

- *Microcontroller (ATmega328)*: ATmega328 [35] is an 8-bit reduced instruction set computer (RISC)-based microcontroller. It consists of 14 input or output (I/O) pins, 6 pulse-width modulation, and 6 analog input pins and also supports serial programming through the in-serial programming (ISP). In addition, three timer or counter with compare mode interrupts universal synchronous asynchronous receiver transmitter (USART), watchdog timer, serial peripheral interface, inter-integrated circuit are integrated inside the microcontroller. Table 11.2 represents technical specification of ATmega328.

There are several of reasons to choose ATmega as compare to other microcontrollers:

- Power-saving mode.
- 32 kB memory extended up to many applications.
- For autonomous resetting, watchdog timer can be used.
- Controller executes instruction rapidly because of RISC architecture.
- Operating at high temperature because of embedded temperature sensor.

The main purpose behind the Arduino Nano boards (ATmega328) is to fetch data from the deployed sensor, encrypting and sending this processed data to the ESP8266(Wi-Fi module). It also controlled the input devices,

TABLE 11.1 LM358 technical specification.

Serial no.	Specification	Values
1.	Single supply	3–32 V
2.	Dual supply	1.5–16 V
3.	Current drain	500 μA
4.	Offset voltage	2 mV
5.	Unity gain	1 MHz
6.	DC voltage gain	100 db

TABLE 11.2 ATmega specification.

Serial no.	Specification	Values
1.	Bit processor	8 bit
2.	Instruction	20 million IPS at 20 MHz
3.	Memory	32 kB
4.	RAM	2 kB SRAM
5.	Programming memory	1 kB EEPROM
6.	Operating frequency	200 MHz (max)
7.	Write/erase cycle	10,000 flash and 100,000 EEPROM
8.	Data retention	20 years/100 years at 85c/25c
9.	Operating voltage	1.8–5.5 V
10.	Multiplier	On-chip two cycles

ESP8266 module via universal asynchronous receiver transmitter (UART) through the serial communication pins. Microcontroller is responsible for the major power consumption of sensor node or device. In addition, microcontroller chosen is very important for deploying any sensor node in terms of energy efficiency. In our proposed glucose monitoring system, 16- and 32-bit microcontrollers are not suitable as compare to 8-bit microcontroller. According to the experimentation analysis [36], 8-bit microcontrollers are more reliable and efficient as compare to 32-bit microcontroller in regards to processing and data access time. For instance, 32-bit microcontroller needs 33 cycles for fetching 1 B of data although 8-bit microcontroller only need 12 cycles for same data. During algorithm performance, 8-bit microcontroller consumes 70 B of stack, while 32-bit microcontroller requires 192 stacks. In terms of current consuming, 8-bit microcontroller only consumes 36.1 μA, and 32 bit microcontroller needs 48.1 μA.

In our proposed work, we have clearly mentioned that 8-bit microcontroller has a number of advantages in respect to data processing cycle, consume memory, and perform many tasks without any lag in the real-time monitoring. The main feature in the ATmega328 is sleep mode to conserve battery power and support nearby communications. Therefore it perfectly fits our proposed work.

11.3.1.2 Wi-Fi (Module)

ESP8266 (see Fig. 11.4) is a serial communication–based low-cost Wi-Fi module that equipped with full TCP/IP stack and designed for the low-power consuming devices. It has on-board chip for the data processing and antenna that move toward into a new low-cost connectivity market and helps to enable the integration with different sensors, microcontrollers through inbuilt general-purpose-input-and-output pins. Due to on-built chip and antenna, it consumes less power consumption as compare to the other wireless connectivity modules such as Bluetooth, xbee module. With the use of software support, ESP8266 enhances its features. UART bus consumes less power as compare to the I2C or parallel communication.

ESP8266 consists of AT Command firmware that is enabled to access the input/output access. AT commands are already packed with the software development kit version 1.1.1 of ESP Firmware, and Table 11.3 represents some foremost commands to access the ESP8266 module.

11.3.1.3 Smart gateway

A smart IoT gateway is a solution-oriented technique based on software to overcome the uncertainties between sensor data and information uses by the high-level application [37].

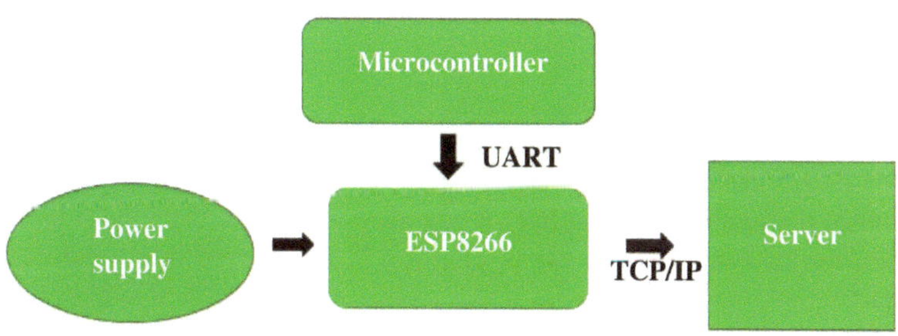

FIGURE 11.4 ESP8266 architecture.

TABLE 11.3 Commands for ESP8266 module.

Serial no.	Commands	Descriptions
1.	AT + RST	For resetting ESP8266 command
2.	AT + CWMODE	Change the Wi-Fi mode either STA mode or API mode
3.	AT + CWJAP	Connect the Wi-Fi module with SSID and Password
4.	AT + CIPSTATUS	Checking for the status of connection
5.	CIFSR	Fetch IP address

API, Application Programming Interface; *SSID*, service set identifier

With the magnification in the communication technologies such as wired and wireless, device gets smarter and can be controlled from anywhere such as homes, industrial units, hospitals, and working areas. Smart gateway is the major component for the blood glucose detection. It acts as bridge to fulfill the gap between the node and back end and is also able to enable the connectivity between them, protocol conversion that includes IPV4, IPV6 mechanisms. Data transmission and reception plays an important role in the smart gateway and works on the two things: (1) it stored the blood glucose values and execute the first-order IIR filtering on it, (2) it sorted and mapped encrypted data stored on the CSV format. Fig. 11.5 represents the step-by-step working of smart gateway and cloud server.

For analyzing and processing data from blood glucose detection system, big data analyzer is used. This analysis is based on the decision-making models, and also these steps are executed in the cloud. At cloud, three steps are as follows: (1) create model, (2) train and test model, and (3) blood glucose detection.

11.3.1.4 Cloud server

For the implementation of IoT, several things are required in the software and hardware synchronization. Blynk is an iOS- and android-based platform to control the applications via IoT. It can also use for the data and presented in the form of visualization. Blynk platform is categorized into three different parts:

- *Blynk app*: It provides the visualize interface for your device.

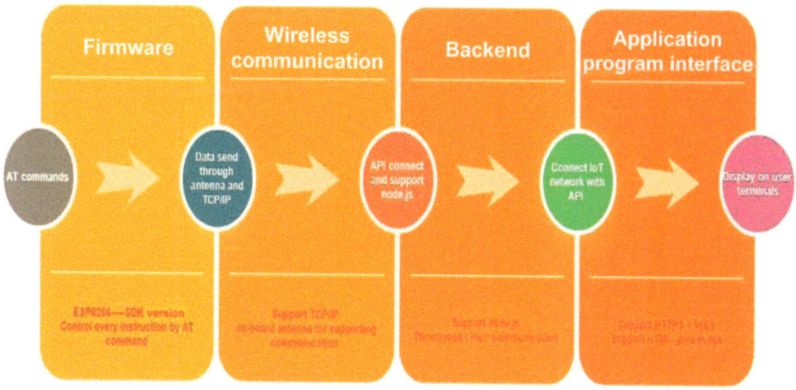

FIGURE 11.5 Smart gateway architecture.

- *Blynk server*: It makes a connection between the sensors and user terminal device.
- *Blynk libraries*: It helps to make a connection between popular hardware platforms and server for processing all input and output commands.

Application programming interface (API) is an interface, which provides the communication to the sensors within IoT environment. In addition, Blynk allows you to build an application around the data collection from the sensors. In addition, it also provides real-time data fetching, storage, and processing and sends visualize data on user terminal. Blynk helps to stamp the incoming data with the data and time and also have sequential ID.

11.3.2 Circuit diagram

In the proposed system (see Fig. 11.6), glucose strip is used to measure the blood glucose. The main principle behind the glucose strip is chemical reaction between the enzyme present on the strip and blood sample, and resultant outcome is gluconic acid. Op-Amp plays a dual role at our proposed system. First, it helps to convert the current into voltage, and second, it provides the amplification to the conversion data for easy processing. Op-Amp is directly connected to the glucose strip for providing the sufficient voltage to the strip terminals and fetch current data from it. The gain of the output value can be varied according to the requirement via variable resister present at the OP-Amp. Output pin of LM358 is connected to the analog pin of Arduino Nano. Reasons behind to choose Arduino Nano are as follows:

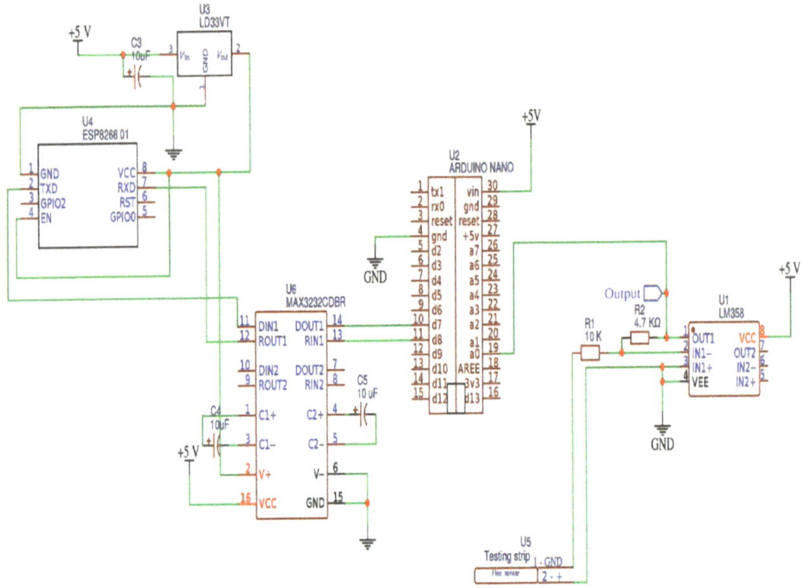

FIGURE 11.6 Circuit diagram of proposed glucose meter.

- ATmega328 is worked as main board and its 8-bit microcontroller consumes less current as compare to the 32-bit microcontroller.
- Arduino Nano boards have in-build USB that is can be used for the debugging purpose.
- Due to the ATmega328, it supports multiple frequencies such as 8, 16, and 20 MHz.

Arduino Nano fetched the analog output from the LM358 and processed the data according to the threshold value. Arduino Nano is further connected to ESP8266 through serial communication. ESP8266 supports AT commands for resetting, starting connection, and sending data. ESP8266 works in two modes: (1) awake mode and (2) sleeping mode. In sleeping mode, ESP8266 consumes very less power as compare to the awake mode. Smart gateway is implemented with the aid of Raspberry PI. It makes a gateway more secure and reliable. At the last step, Blynk IoT server is used to store and visualize data fetched through the smart gateway from ESP8266 (see Figs. 11.7 and 11.8) module. The main focus of this proposed system is the reliable, secure, and energy efficiency of the sensor node.

Algorithm 1 represents the workflow of our proposed work. In this system, Arduino Nano, LM358 operates at the 5 V and ESP8266 at 3.3 V. Voltage send from the LM358 to Arduino Nano varies between 0 and 5 V, and it is directly depended on the concentration of blood glucose. More the

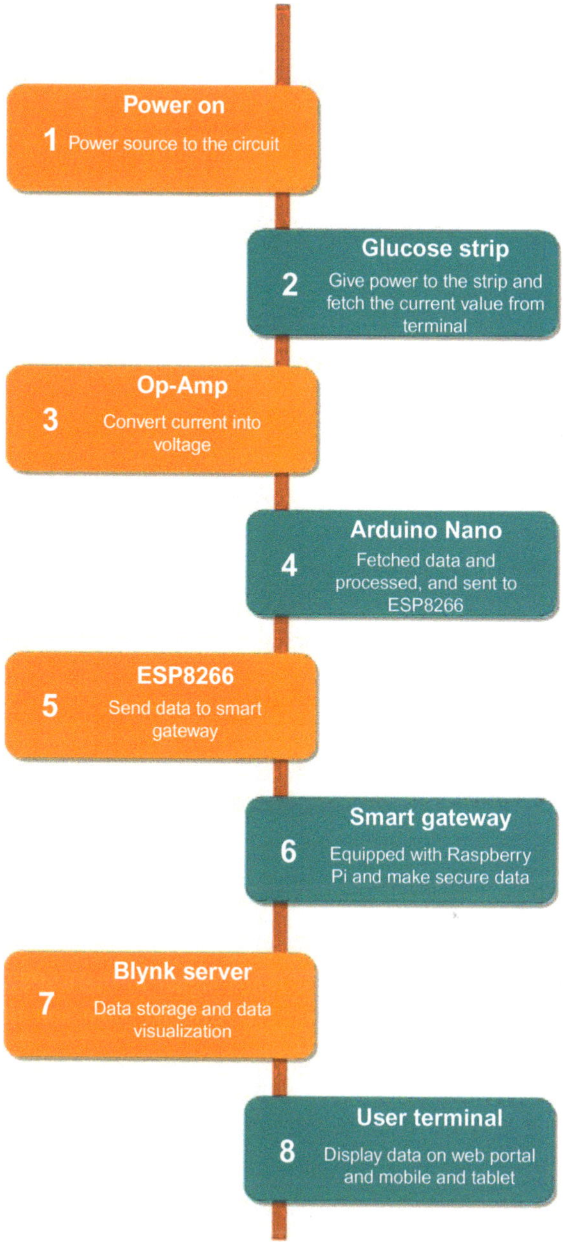

FIGURE 11.7 Flow diagram of the proposed work.

concentration of blood glucose is equal to more current generated. More current generated means conversion of voltage is also high. Arduino Nano processes LM358, and results lie between 0 and 1023 V according to the input

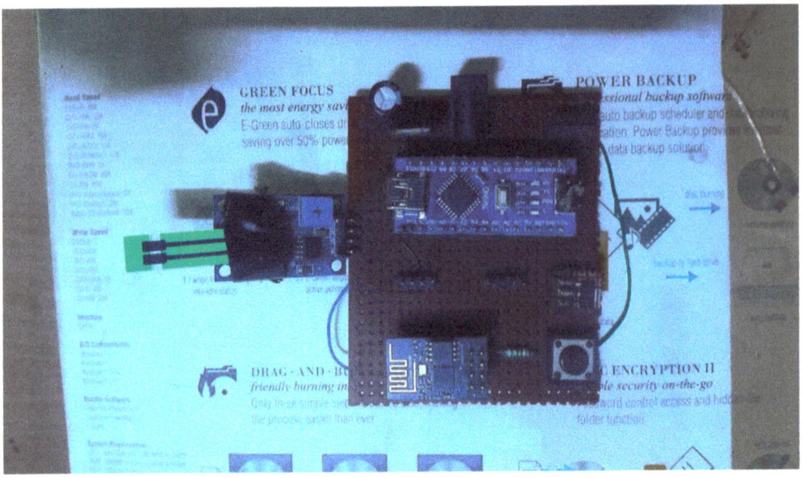

FIGURE 11.8 Proposed blood glucose monitoring system.

TABLE 11.4 Blood sugar level decision.

Checking conditions	Decision
Process value < threshold value	Blood sugar is low
Process value > threshold value	Blood sugar is high
Process value = or nearly lies threshold value	Blood sugar is normal

voltage signal. Threshold value is prefixed to check the blood sugar level is high, normal, or low. Table 11.4 represents the decision of blood sugar detection.

After the decision, Arduino Nano sends signal to the ESP8266 and processes to make a connection between the Arduino Nano and ESP8266, the steps are described as follows:

Step 1: Send AT command to the ESP8266.
Step 2: Checking the OK message is received or not if OK is received means handshaking between both devices are done and if not, means retry again.
Step 3: Send AT + CWJAP = "Wi-Fi name" and "Password."
Step 4: If received OK means you got the IP address for further process and ready to send data. If not means, retry again.

Next step send data to Blynk server through smart gateway.

ALGORITHM 1 Proposed blood glucose monitoring

```
thre← Threshold value
sen← Sensor value
op-amp← Operational Amplifier
micro← microcontroller processing value
micror← microcontroller receiving value
micros← microcontroller sending value
esp← esp8266 processing and sending values
server← server value
loop
  Blood sample on the enzyme
  processing current value on op-amp
  if sen > thre then
    send data for conversion
    change current values to voltage
    for i = 1 to n do
      collection of voltage values and micro
      if micros > micror then
        match the API with server
        send esp
        blynk process data and visualize data
      else
        nosend esp
      end if
    end for
  end if
end loop
```

11.4 Result and discussion

For the analysis and discussion of IoT-based blood glucose monitoring, we have used two different tools: (1) MATLAB 2018a and (2) Blynk server. Before implementation on the Blynk server, we examine the data on MATLAB 2018a. I5 processor with 8 GB RAM and 2 GB graphic card is used for processing of data. Results are categorized into following two cases:

- monitoring blood sugar level before meal and
- monitoring blood sugar level at time of sleeping.

Our proposed system tested into two different platforms. First, we tested at MATLAB 2018a to predict the values. After testing, we send data to Blynk server through smart gateway.

1. *Implementation of data on MATLAB*

For data analyze in MATLAB, connected our blood glucose monitoring to PC with the use of USB cable, which is used as bridge to fill the gap between sensor node and computer. For the glucose monitoring system, glucose monitoring considered for two situations: (1) before meal and (2) at sleeping time.

Figs. 11.9 and 11.10 represent blood glucose before meal and at time of sleeping respectively. In the case of before meal, the blood glucose is lies between 14 and 36, which means that our proposed system maintains reliability and accuracy. At the time of sleeping, blood sugar level is almost constant and lies between 17.7 and 17.9. Results that come from MATLAB are before and after meal and also on the time of sleeping.

2. *Implementation of data on Blynk server*

In our implementation, we are using Blynk server. With the use of Blynk server, we can store and fetch data from blood glucose strip using hypertext protocol over the Internet. Figs. 11.11 and 11.12 represent the blood glucose monitoring before meal and at sleeping time, respectively.

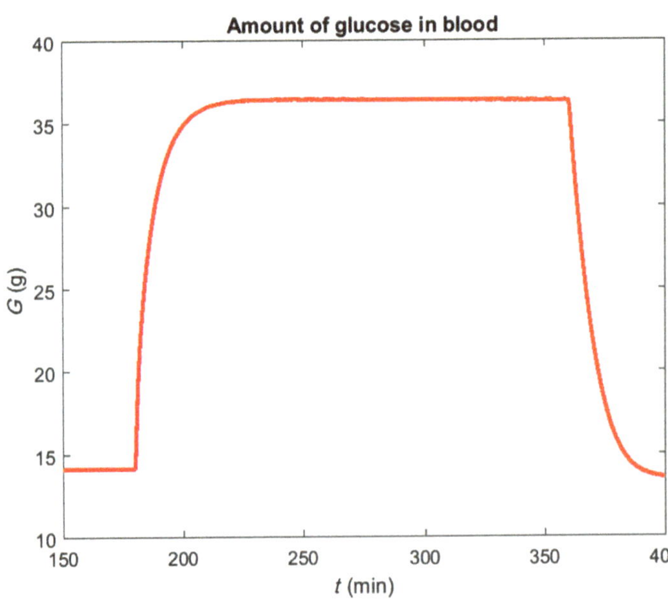

FIGURE 11.9 Blood glucose monitoring before meal.

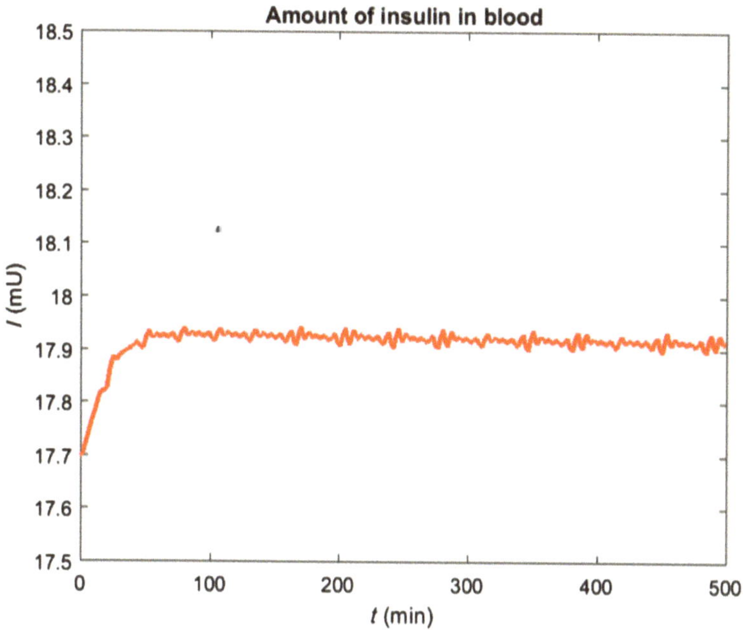

FIGURE 11.10 Blood glucose monitoring at the time of sleeping.

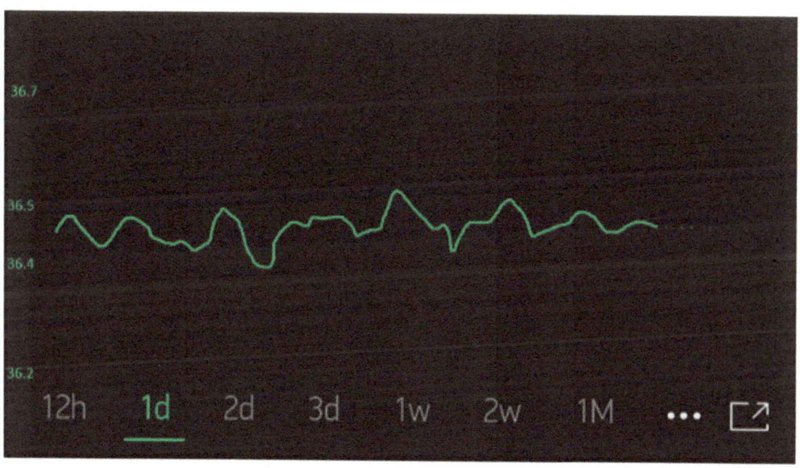

FIGURE 11.11 Blood glucose monitoring before meal at Blynk server.

11.5 Conclusion

In this chapter, we presented an IoT-based real-time CGM system. The deployed IoT-based architecture is a fulfilled system from design to the implementation such as implemented from sensor node to the back end

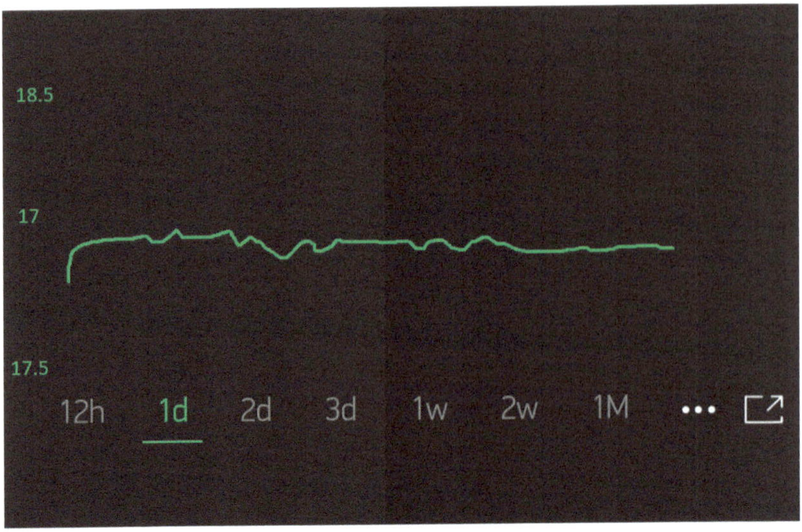

FIGURE 11.12 Blood glucose monitoring at the time of sleeping (Blynk server).

server. In the implemented IoT system, anytime, anywhere monitoring can be done through the smartphone or tablet via patient caregiver or doctor. Sensor nodes are able to fetch the blood glucose data, processing and send wirelessly to the cloud server through the smart gateway efficiently as regards to energy dissipation. Moreover, choosing microcontroller is more beneficial to reduce overall power consumption of the sensor node and make it energy-efficient node. With the assistance of ESP8266, patient smart devices, such as smartphone and tablet, visualize the data fetched from the sensor node through gateway and cloud server and also provide notification service to alert in the critical condition. The results show that our proposed system is reliable, accurate, and energy efficient to monitor blood glucose level.

References

[1] C.T. Bau, R.C. Chen, C.Y. Huang, Construction of a clinical decision support system for undergoing surgery based on domain ontology and rules reasoning, Telemed. e-Health 20 (5) (2014) 460–472.

[2] A. Lanata, G. Valenza, M. Nardelli, C. Gentili, E.P. Scilingo, Complexity index from a personalized wearable monitoring system for assessing remission in mental health, IEEE J. Biomed. Health Inf. 19 (1) (2015) 132–139.

[3] T. Shemeikka, P. Bastholm-Rahmner, C.G. Elinder, A. Vég, E. Törnqvist, B. Cornelius, et al., A health record integrated clinical decision support system to support prescriptions of pharmaceutical drugs in patients with reduced renal function: design, development and proof of concept, Int. J. Med. Inf. 84 (6) (2015) 387–395.

[4] CDC report of Heart Disease and Stroke prevention. <https://www.cdc.gov/dhdsp> (accessed 10.05.19).

[5] World Health Organization report for diabetes. <https://www.who.int/en/news-room/fact-sheets/detail/diabetes> (accessed 10.05.19).

[6] National Institute of Health Report for type 1 and type 2 diabetes among the teens. <https://www.nih.gov/news-events/news-release/rates-new-diagnosed-cases-type-1-type-2-diabetes-rise-among-children-teens> (accessed 10.05.19).

[7] C.D. Mathers, D. Loncar, Projections of global mortality and burden of disease from 2002 to 2030, PLoS Med. 3 (11) (2006) e442.

[8] D.A. Sterling, J.A. O'Connor, J. Bonadies, Geriatric falls: injury severity is high and disproportionate to mechanism, J. Trauma Acute Care Surg. 50 (1) (2001) 116–119.

[9] CDC report about the adult facing falls. <https://www.cdc.gov/homeandrecreationalsafety/falls/adultfalls.html> (accessed 11.05.19).

[10] S. Volpato, S.G. Leveille, C. Blaum, L.P. Fried, J.M. Guralnik, Risk factors for falls in older disabled women with diabetes: the women's health and aging study, J. Gerontol. A: Biol. Sci. Med. Sci. 60 (12) (2005) 1539–1545.

[11] Solution for the diabetes cures. <http://apps.who.int/iris/bitstream/10665/204871/1/9789241565257eng.pdf> (accessed 11.05.19).

[12] P. King, I. Peacock, R. Donnelly, The UK prospective diabetes study (UKPDS): clinical and therapeutic implications for type 2 diabetes, Br. J. Clin. Pharmacol. 48 (5) (1999) 643–648.

[13] A. Zanella, N. Bui, A. Castellani, L. Vangelista, M. Zorzi, Internet of Things for smart cities, IEEE Internet Things J. 1 (1) (2014) 22–32.

[14] Internet of Things connected medical market. <http://www.mckinsey.com/industries/high-tech/our-insights/the-internet-of-things-sizing-up-the-opportunity> (accessed 11.05.19).

[15] V. Puri, J.G. Tromp, N.C. Leroy, C. Le Van, N.G. Nguyen, Analysis of telemedicine technologies, Emerging Technologies for Health and Medicine: Virtual Reality, Augmented Reality, Artificial Intelligence, Internet of Things, Robotics, Industry 4.0 Wiley Publication, 2018, pp. 153–161.

[16] S.R. Islam, D. Kwak, M.H. Kabir, M. Hossain, K.S. Kwak, The Internet of Things for health care: a comprehensive survey, IEEE Access 3 (2015) 678–708.

[17] T.C. Chiang, W.H. Liang, A context-aware interactive health care system based on ontology and fuzzy inference, J. Med. Syst. 39 (9) (2015) 105.

[18] I. Azimi, A.M. Rahmani, P. Liljeberg, H. Tenhunen, Internet of Things for remote elderly monitoring: a study from user-centered perspective, J. Ambient Intell. Humanized Comput. 8 (2) (2017) 273–289.

[19] T. Xu, J.B. Wendt, M. Potkonjak, Security of IoT systems: design challenges and opportunities, Proceedings of the 2014 IEEE/ACM International Conference on Computer-Aided Design, IEEE Press, 2014, pp. 417–423.

[20] F. Bonomi, R. Milito, J. Zhu, S. Addepalli, Fog computing and its role in the Internet of Things, Proceedings of the First Edition of the MCC Workshop on Mobile Cloud Computing, ACM, 2012, pp. 13–16.

[21] L.M. Vaquero, L. Rodero-Merino, Finding your way in the fog: towards a comprehensive definition of fog computing, ACM SIGCOMM Comput. Commun. Rev. 44 (5) (2014) 27–32.

[22] P. Verma, S.K. Sood, Fog assisted-IoT enabled patient health monitoring in smart homes, IEEE Internet Things J. 5 (3) (2018) 1789–1796.

[23] M.M. Rathore, A. Ahmad, A. Paul, J. Wan, D. Zhang, Real-time medical emergency response system: exploiting IoT and big data for public health, J. Med. Syst. 40 (12) (2016) 283.

[24] G. Yang, L. Xie, M. Mäntysalo, X. Zhou, Z. Pang, L. Da Xu, et al., A health-IoT platform based on the integration of intelligent packaging, unobtrusive bio-sensor, and intelligent medicine box, IEEE Trans. Ind. Inf. 10 (4) (2014) 2180–2191.

[25] A.J. Jara, M.A. Zamora, A.F. Skarmeta, An Internet of Things–based personal device for diabetes therapy management in ambient assisted living (AAL), Pers. Ubiquitous Comput. 15 (4) (2011) 431–440.

[26] R.S. Istepanian, S. Hu, N.Y. Philip, A. Sungoor, The potential of Internet of m-health Things "m-IoT" for non-invasive glucose level sensing, 2011 Annual International Conference of the IEEE Engineering in Medicine and Biology Society, IEEE, 2011, pp. 5264–5266.

[27] H. Dubey, J. Yang, N. Constant, A.M. Amiri, Q. Yang, K. Makodiya, Fog data: enhancing telehealth big data through fog computing, Proceedings of the ASE Bigdata & Socialinformatics 2015, ACM, 2015, p. 14.

[28] S. Tyagi, A. Agarwal, P. Maheshwari, A conceptual framework for IoT-based healthcare system using cloud computing, 2016 Sixth International Conference-Cloud System and Big Data Engineering (Confluence), IEEE, 2016, pp. 503–507.

[29] P. Gope, T. Hwang, BSN-Care: a secure IoT-based modern healthcare system using body sensor network, IEEE Sens. J. 16 (5) (2015) 1368–1376.

[30] V.M. Rohokale, N.R. Prasad, R. Prasad, A cooperative Internet of Things (IoT) for rural healthcare monitoring and control, 2011 Second International Conference on Wireless Communication, Vehicular Technology, Information Theory and Aerospace & Electronic Systems Technology (Wireless VITAE), IEEE, 2011, pp. 1–6.

[31] B. Farahani, F. Firouzi, V. Chang, M. Badaroglu, N. Constant, K. Mankodiya, Towards fog-driven IoT eHealth: promises and challenges of IoT in medicine and healthcare, Future Gener. Comput. Syst. 78 (2018) 659–676.

[32] G. Alfian, M. Syafrudin, M. Ijaz, M. Syaekhoni, N. Fitriyani, J. Rhee, A personalized healthcare monitoring system for diabetic patients by utilizing BLE-based sensors and real-time data processing, Sensors 18 (7) (2018) 2183.

[33] M.A. Al-Taee, W. Al-Nuaimy, A. Al-Ataby, Z.J. Muhsin, S.N. Abood, Mobile health platform for diabetes management based on the Internet-of-Things, 2015 IEEE Jordan Conference on Applied Electrical Engineering and Computing Technologies (AEECT), IEEE, 2015, pp. 1–5.

[34] LM358 technical details. <http://www.ti.com/lit/ds/symlink/lm358-n.pdf> (accessed 12.05.19).

[35] ATmega328 8 bit micro controller suitability. <https://www.mouser.com/pdfdocs/Gravitech_ATMEGA328_datasheet.pdf> (accessed 12.05.19).

[36] 8 bit micro controller or 32 bit micro controller. <http://www.atmel.com/images/45107a-choosing-a-mcu-fredriksen_article_103114.pdf> (accessed 12.05.19).

[37] Y.H. Lee, S. Nair, A smart gateway framework for IoT services, 2016 IEEE International Conference on Internet of Things (iThings) and IEEE Green Computing and Communications (GreenCom) and IEEE Cyber, Physical and Social Computing (CPSCom) and IEEE Smart Data (SmartData), IEEE, 2016, pp. 107–114.

Index

Note: Page numbers followed by "*f*" and "*t*" refer to figures and tables, respectively.

A

Adaptive boosting, 33
AllJoyn, 99
Alpha wave, 25–26, 25*f*
Ambient Assistive Living (AAL)
  AAL frameworks, 217–219, 227–228, 231
  AAL homes, 234–235
Anticounterfeit technology in pharmaceutical industry, 101
Apache Drill, 115
Apache Hadoop, 113–114
Apache Mahout, 114
Apache Spark, 113–114
Application programming interface (API), 315
Arduino Nano, 315–316
ARIMA (autoregressive integrated moving average), 156
Asperger syndrome (AS), 58–59
  difference between autism spectrum disorder and, 59–60
Audio analytics, 117
Augmented Dickey–Fuller test, 156
Autism
  cause of ASDs, 58
  food impacts on, 56*t*
    artificial ingredients, 55
    corn, 55
    diary, 55
    gluten, 55
    sugar, 55
  foods associated therapies for, 54–55
  pervasive developmental disorders (PDDs), 57
  pharmacological involvements, 51–55, 54*t*
    in autism spectrum disorder (ASD), 52*f*
    for chronic neuroinflammation, 52–53
    to control and enhance behaviors, 51–52
    for hormonal changes and imbalance, 52–53
    for immune system, 52–53
    for mitochondrial dysfunctions, 52
    for stabilizing free radicals, 52–53
  pollen allergy in autistic kids, 56–57
  prevalence in Pakistan, 56–57
  repetitive behavior, 58
  symptoms, 57–58
  types of autism spectrum disorder, 57*f*
Autism schools, in Pakistan, 75–76, 77*t*
Autism spectrum disorder
  diagnosis and treatment in, 60
  difference between Asperger syndrome (AS) and, 59–60
  recommendations
    for autistic child discipline, 76–79
    diet plan, 78
    for nutritional interventions, 78–79
  types of, 57*f*
Autism spectrum disorder, case study
  at age of 5.8 years, 62–66
    ability to follow multistep instructions, 65–66
    attention control, 63
    behavior, 63
    expressive language, 64–66
    language skills, 64
    medical and developmental history, 63
    play skills, 63
    receptive language, 64
    recommendations, 65–66
    social interaction, 63
  applied behavior analysis therapy assessment report, 68–75
  diagnosis, 61
  general behavior up to 2.5 years, 61
  herbal treatment, 61
  parental family medical history, 60
  patient history, 60–61
  therapies, pharma, and Internet of Things, 61–62
  at 8 years of age, 66

Autism spectrum disorder, case study (*Continued*)
changes in behavior, 66–67
use of Internet of Things, 66–67, 67*f*
at 10.5 years of age, 68
Autistic disorder, 58
Autoregressive (AR) method, 29–30
Azathioprine, 52–53
Azure IOT, 148–150, 148*f*

B

Band-pass filter, 27
Barcode, 138
Beta wave, 25–26, 25*f*
Big data
advantages of, 97, 97*f*
handling of pharma data, 113–117
tools for analytics, 114–117
linking with Internet of Things (IOT) technology, 94–96, 95*f*
in pharmaceutical industry, 92–94
platforms, 95–96
Biomedical system, 267–268
Bipolar montage, 27
Blind source separation (BSS), 26–27
Bluetooth low energy (BLE), 137–138
Blynk platform, 314–315
Body area network (BAN), 261–263
Brain-computer interface (BCI)-based applications, 26–27
Brain injury detection, 37
Brain signals
kinds of waves, 25–26
representation of, 25*f*
types of, 25–26
event-related potential (ERP), 25–26
neural oscillation, 25
somatosensory evoked potential (SSEP), 25–26
Bronchoscopy, 282–284, 283*f*
risks associated with, 283–284
BSNCare (body sensor network-care), 308–309
Burg's method, 29–30

C

Capacitance, 169–172
Capacitive sensor, 169–172
CareSmart Seniors Consulting Inc., 233–234
Childhood disintegrative disorder. *See* Heller's syndrome
Cloud-based software architecture, 98
Cloud computing, 6, 88, 246
Cloudlet, 136
CoAP interface, 98
Cold chain, definition, 133
Cold chain pharmaceutical market, 133
Cold chain pharmaceuticals
common sensor and devices for, 138
conditional monitoring and predictive maintenance of containers, 158
deployment considerations, 158–159
Internet of Things (IOT)-based cold chain solution, 133, 136, 137*f*
approach for building, 146–150
benefits of processing IOT data at edge, 140–142
conceptual framework, 138–146, 139*f*
cyber attacks, 141
demerits of edge server, 141
deployment model, 140–141, 140*f*
device management, 140
distributor, 139
edge server, 140
experiments and results, 148–150
implementation, 146
IOT Edge, 134
local edge with cloud platform connectivity, 141–142, 142*f*
manufacturer, 138–139
monitoring, 140
network latency issues, 141
pharmacist, 140
prediction of orders, 140–141
warehouse worker, 139
issues in, 159
logistics process, 133–135, 135*f*
problems, 134
product demand forecasting, 154–157
role of edge computing and analytics, 135, 137–138
benefits, 136
edge failure, issue of, 159
edge layer and its components, 147*f*
experiments and results, 148–150
important trends, 136
role of containers, 150–153, 150*f*, 151*f*, 152*f*, 153*f*
status with updated modules, 148*f*
study
literature review, 135–138
objective, 134–135
track and trace technology, 157–158

Colonoscopy, 279–281
risks associated with, 281
signs and symptoms, 280
test for polyps, 281
Connected homes, 224–225
Containers in Internet of Things edge, 150–153
architecture, 151*f*
orchestration, 150, 152*f*
prime function of, 151
tools, 151
popular, 149*f*, 150, 150*f*, 152*f*, 153*f*
ContQuest, 99
Counterfeiting, 101
CuBLAS library, 98–99
Cumulocity, 99
Cystoscopy, 287–289
bladder diseases and conditions, 288
complications of, 289
diagnosing bladder diseases and conditions, 288
enlarged prostate, diagnosis of, 288
preparation of, 288–289
signs and symptoms, 288

D

Data analytics, 7
Decision tree, 32–33, 33*f*
Deep brain recording, 23
Delta wave, 25–26, 25*f*
Diabetes IoT monitoring system, 307–308
analysis and discussion of, 319–320
data analyze in MATLAB, 320
glucose monitoring at time of sleep, 321*f*, 322*f*
glucose monitoring before meal, 321*f*
sending data to Blynk server, 320
architecture, 309–315
circuit diagram of proposed glucose meter, 315–319, 316*f*
cloud server, 314–315
glucose strips, 309–310, 311*f*
microcontroller (ATmega328), 311, 312*t*
sending data to Blynk server, 317*f*, 318–320
sensor strip circuit node, 309–313
smart IoT gateway, 313–314, 315*f*
transconductance amplifier (LM358), 310–311, 312*t*, 316
Wi-Fi module (ESP8266), 313, 313*f*, 314*t*, 316
blood sugar level decision, 318*t*
Digital health, 2*f*, 7
Discrete cosine transform (DCT), 29
Discrete Fourier transform (DFT), 28–29
Discrete wavelet transform (DWT), 26–27
Drug-delivery system, challenges in, 184–185
Dryad, 114–115

E

Economic load dispatch problem, 1–2
Elderly population, 221*t*
ambient assistive living systems, 226–227
distribution, 219–220, 223*f*
health conditions of, 217–219
IoT assistance, 225–226
AAL homes, 234–235
activity recognition framework, 230–231
AlarmNet, 235
CAALYX venture, 240–241
CareWatch framework, 236–237
CodeBlue remote sensors, 237–238
GERHOME, 237–238
I-LivingTM framework, 238
MITHouse, 239–240
ORCATECH, 239–240
ready-to-use products, 233–234
SOPRANO, 239–240
TAFETA venture, 240
TeleCARE, 240–241
wearable systems, 231–233
ZigBee information correspondence remote strategy, 237–238
requirements of activity recognition, 227–229
societal adaptions, 220–224
Electrochemical sensors, 169–173, 174*f*
Electroencephalogram (EEG), 21–22
acquisition techniques, 22–23
electrodes type used, 23
invasive, 23
noninvasive, 23
placements of electrodes, 23*f*
working process of, 22*f*
biomedical engineering applications, 22
channel selection techniques, 24–25
embedded technique, 24
filtering, 24
human-based techniques, 25
hybrid technique, 25
wrapper method, 24
classification of EEG signals based on features, 31–34

Electroencephalogram (EEG) (*Continued*)
adaptive boosting, 33
decision tree, 32–33, 33*f*
k-Nearest neighbor (*k*-NN), 31–32, 31*f*
linear discriminant analysis (LDA), 32, 32*f*
multilayer perceptron (MLP), 33–34
Naive Bayes classifiers, 34
evolution of channel selection algorithms, 24
feature extraction and classification from signal, 27–31
chaotic motions, 30
discrete cosine transform (DCT), 29
discrete Fourier transform (DFT), 28–29
feature extraction transformation model, 28, 28*f*
genetic algorithm (GA), 28, 30–31
Hurst exponent, 28, 30
random search method, 30–31
frequency of signal and power from artifact, 29–30
autoregressive (AR) method, 29–30
Burg's method, 29–30
Music method, 29–30
Pisarenko technique, 29–30
sample entropy, 30
time-frequency distribution analysis, 29
Yule–Walker method, 29–30
Internet-of-Thing application of, 35–40
brain injury detection, 37
hand movement trajectory reconstruction approach, 38–39
mental state recognition, 38
neuro-marketing, 39–40
object controlling, 37
product-based application, 39–40
seizure detection, 36
preprocessing of, 26–27
artifact removal, 26–27
blind source separation (BSS), 26–27
brain-computer interface (BCI)-based applications, 26–27
independent component analysis (ICA) technique, 26–27
use of basic filtering, 27
using finite impulse response filter, 26–27
wavelet transform (WT), 26–27
Electronic Product Code (EPC), 136–137
Endoscopic retrograde cholangiopancreatography, 281–282, 281*f*
risks associated with, 282
sphincterotomy, 282
stenting, 282
Endoscopic ultrasound (EUS), 295–297, 295*f*
complications of, 296–297
Endoscopy, 276, 276*f*
diagnosis for, 277
investigation of signs and symptoms, 277
purpose of treatment, 277–278
rationale for, 277–278
types of, 278
bronchoscopy, 282–284
colonoscopy, 279–281, 280*f*
cystoscopy, 287–289
endoscopic retrograde cholangiopancreatography, 281–282
endoscopic ultrasound (EUS), 295–297
flexible sigmoidoscopy, 285–287
hysteroscopy, 291–294
PEG, 284–285
transbronchial biopsy, 289–291
upper gastrointestinal endoscopy, 278–279
Environmental-sensing system, 267
e-Pharma, 213–214
Event-related potential (ERP), 25–26
Extranet, definition, 50–51

F

Fit Bit Pure Pulse, 179–180
Flexible sigmoidoscopy, 285–287, 286*f*
complications with, 287
contraindications of, 286
colon cancer screening, 286
signs and symptoms, 286
preparation for, 287
Fog computing, 89–90

G

Global Trade Item Number (GTIN), 101
GPGPU device, 98–99

H

Hadoop system, 98–99
Hand movement trajectory reconstruction approach, 38–39
H7, 179–180
Heller's syndrome, 59
symptoms after regression, 59
Holograms, 101

HRM-Tri, 179–180
Human brain, 21–22
lobes of, 21*f*
Hurst exponent, 28, 30
Hysteroscopy, 291–294, 292*f*, 293*f*
advantages of, 294
application of, 293
complications of, 294
preparation of, 293–294

I

Identity management system, 266
iHome, 308–309
iMedBox, 107
iMedPack, 107
Independent component analysis (ICA) technique, 26–27
Industry 4.0, 92, 102, 108–109
"Intelligent water drop" (IWD) algorithm, 1–2
categorization of, 1–2
main factors/variables in, 1–2
movement of water drops, 2
gravitational pull, effect of, 2–3
high-speed water drop, 3
velocity of water drop, 3
optimization methodologies, 3–4
for QOS performances, 3–4
related work, 4–7
simulation results, 7–15, 8*t*, 9*t*
end-to-end delay (E2D), 7–9, 10*t*, 11*f*
normalized routing load (NRL), 7–8, 12, 13*t*, 14*f*
packet loss ratio (PRL), 12, 15*f*, 15*t*, 16*f*
PDR results, 9*f*, 9*t*
simulation time, 7–8
throughput, 7–9, 12*f*, 12*t*, 13*f*
simulation setup, 8*t*
for ZRP optimization, 3–4, 7
Internet, definition, 50–51
Internet of Health Things (IoHT), 264–265
advantages of, 264–265
Internet of Medical Things (IoMT), 217–219, 306, 310*f*
Internet of Things (IOT) technology, 1, 195–198, 245–246, 306
advantages, 248–249
applications of, 6, 86, 196–198, 198*f*, 252–255
assurance industry, 255
automotive industry, 252
environment checking, 254
farming industry, 254–255
independent living, 253
industry of telecommunications, 252–253
manufacturing industry, 254
media and amusement, 255
medical and health-care industry, 253
pharmaceutical industry, 253
processing industry, 254
recycling, 255
solution in various fields, 164*f*
supply chain management (SCM), 253–254
transport industry, 254
architectural framework of, 250*f*, 307*f*
application layer, 252
gateways and networks layer, 251
management service layer, 251–252
smart device/sensor layer, 249–250
attribute selection, 126*t*
characteristics features of, 247–248
components
availability of big data, 246
cloud computing, 246
low power-based embedded systems, 246
networking connection, 246
connectivity technology, 88, 248
data handling, 248
devices, 49–51, 50*f*, 105–107, 247
changes, 248
flowchart for detection and classification of body movements, 125*f*
flowchart for treatment of biological waste, 109*f*
heterogeneous, 248
interoperability of, 97–100
patient's health-care monitoring, 99
enablers, 86
era of, 85–87
form of engagement, 247
infrastructure, 246*f*
interconnectivity, 247
intermittent connectivity, 88
journey of, 196
linking with big data, 94–96, 95*f*
with MANET, 7, 16–17
network requirements, 51
in pharma industry, 90–91, 200–202, 202*f*, 205*t*, 255–257
access to real-time visibility of warehouse, 103

Internet of Things (IOT) technology (*Continued*)
anticounterfeit technology, 101
automation of tasks in, 93
benefits, 258–259
blood pressure monitoring, 105
body temperature monitoring, 105
challenges, 103, 198, 209–213
classification, 122–124
cloud-based services, 90–91
controlling of environmental factors in drugs manufacturing, 204
data acquisition, 120
designing of digital drug administration interface, 100–101
electrocardiogram monitoring, 105
enhancement of processes, 90
facility management, 206
feature extraction, 120–122
framework for digitalization, 91*f*
glucose level monitoring, 105
health-care monitoring and emergency situation alerts, 104
heart sensors, 104–105
inventory management, 209
IoT_BM-PDA algorithm, 123*b*
literature review, 257–258
management with sensors, 210*f*
manufacturing of pharmaceuticals, 102, 202–203, 256–257
monitoring of production flow, 203–204
network topology, 199–200
objective connectivity, 202
to overcome short supply of drugs, 212
packaging optimization, 205
pharmaceutical formulations and drug delivery, 93–94
pharma logistics, 100–111
plant safety and security, 211–212
production process, 201
quality control of product, 204
road map, 198–200, 200*f*
security for supply chain, 212
supply chain, 206–208, 207*f*, 208*t*
theft of drugs during transportation, 213
tracking shipment of product, 201–202
transportation, 257
visual analytics, 118–119
warehouse operations, 206, 207*t*, 257
wheelchair management, 105
quirky machine-to-machine communication, 89–90
safety, 248
TCP/IP model, 201, 201*f*
thing-related services, 247–248
utility of, 87*f*
Internet protocol version 6 (IPV6), 86
IPV6 endpoint (virtual device), 98
IoT-enabled wearable health care, 163–166
antennas, 173–174, 173*f*
architecture of, 164–166, 165*f*
application layer, 166
communication layer, 165
data processing layer, 165–166
sensing layer, 164
challenges and future directions, 190–191
challenges in fabrication of, 172–173
design and development of, 169
in drug dispensing, 184–187
challenges, 185
data-enabled drug-administrative wearable devices, 186
forms and functions, 185–186
trend, 186–187
wearable drug-administrative device, 185
functions of wearable sensors, 175–178
for physiological parameters measurement, 166–168, 168*f*
biochemical parameters, 168
interfacing unit, 167
physical parameters, 167–168
revenue, 174*f*
safety and security issues, 187–188
taxonomy of, 166, 167*f*
types of wearable sensors, 169
blood glucose monitoring, 176*f*
blood pressure monitoring sensor, 175*f*, 182
body temperature sensor, 174*f*, 182
electrocardiogram sensor, 176*f*
glucose monitoring, 187
invasive sensors, 169, 170*t*
noninvasive sensors, 169, 170*t*, 174*f*, 175*f*
photoplethysmographic (PPG) sensors, 179–180
piezoresistive pressure sensor, 175–178
pulse oximetry sensors, 175*f*, 183–184
pulse sensor, 179–180
respiratory rate sensor, 174*f*, 180–181
wearable devices in pharmaceutical applications, 171*t*, 178–184
wireless communication protocols, 177*f*
wireless sensor network, architecture, 177*f*

for women safety, 188–190
working principles of wearable sensors, 169–172
IPB (intelligent pill box), 257–258

J

Jaspersoft, 115
Joint approximate diagonalization of Eigen-matrices method, 26–27

K

K-Nearest neighbor (*k*-NN), 31–32, 31*f*, 122–123
Kubernetes, 151–152

L

Laplacian montage, 27
Life expectancy, 217–219
Linear discriminant analysis (LDA), 32, 32*f*
Local edge with cloud platform connectivity, 141–142, 142*f*
accelerometer, 142
aggregated data, 144
cloud platform, 141–142
curated data, 144
cyber attacks, 141
data for, 143
deployment model, 141
device management, 140, 144
monitoring, 140
network latency issues, 141
prediction of orders, 140–141
sensor data from edge, 142
storage layer, 143
temperature and humidity sensor, 142
Location awareness, 136
Lyapunov exponent (LLE), 28, 30

M

MapReduce, 98–99, 113–114
Mass serialization, 101
MBS (Medicine Bag System), 257–258
Medication system, 268–271
MedTracker, 257–258
6-Mercaptopurine, 52–53
Metadata, 93
Microsoft HoloLens, 199
Mobile ad-hoc network (MANET) platform, 2*f*, 7
Mobile Biomedical System (MBS), 267–268, 269*f*, 269*t*, 270*f*
MODALITi (FraMewOrk for EmbeDded and CollAborative data anaLysIs with HeTerogeneous DevIces), 100
Multidimensional knapsack problem (MKP), 1–2
Multilayer perceptron (MLP), 33–34
Music method, 29–30

N

Naive Bayes classifiers, 34
Networking of Internet-linked smart devices. *See* Internet of Things (IOT) technology
Neuro-marketing, 39–40
Notch filter, 27
N-queen puzzle, 1–2
Nursing system, Internet of Things and, 265–271

O

Object controlling, 37
OMA-LwM2M, 100

P

Packet delivery ratio (PDR), 3–4
Patient-centric Internet of Things, 260–261
vs patient-centered information, 261
Pedigree, 101
PEG (percutaneous endoscopic gastrostomy), 284–285, 284*f*
complications of, 285
contraindications of, 285
techniques available for, 284–285
Personal orientation system, 271, 272*f*
Pharmaceutical industry, big data and, 85–97
Pharmaceutical manufacturers, 1
Pharma data, 96–97
handling of, 96–97, 113–117
for optimizing processes, 111–113
tools for analytics, 114–117
Pharma intelligence, 96–97
Pharma logistics, 100–111
Pisarenko technique, 29–30
Predictive analytics, 117
PUC-Rio Dataset, 124

Q

QR code, 138
Quality of service (QOS), 1, 307–308

R

Radio frequency identifications (RFID) tags, 86, 88–89, 100–101, 103–104, 107–110, 135–138, 142, 200–201, 257–258
Random search method, 30–31
Rett syndrome, 59
  symptoms, 59
Risperidone, 52–53
Robot path planning, 1–2

S

Seizure detection, 36
Smart band in healthcare, 170*t*
Smart city, 197–198
SmartDose wearable devices, 185
Smart pills, 297–298, 299*f*
Smooth trajectory planning, 1–2
Somatosensory evoked potential (SSEP), 25–26
Splunk, 115
Storm, 113–116
Structured data, 96
Swarm-inspired algorithms, 1–2

T

Text analytics, 116–117
Theta wave, 25–26, 25*f*
Track and trace technology, 101, 157–158
Transbronchial biopsy, 289–291, 292*f*
  complications of, 291
  preparation of, 289–291
Traveling salesman problem (TSP), 1–2

U

Universal Product Code scanning technique, 107–108
Unstructured data, 96, 111–112, 116–117
Upper gastrointestinal endoscopy, 278–279
  medications, 279
  process of, 279
  risks associated with, 279

V

Vaccination drugs, 133, 135
Vehicle routing problem, 1–2
Video analytics, 117
Visual analytics in pharma industry, 118–119
  gene expression, 118, 118*f*
  target discovery, 119, 119*f*

W

Wavelet transform (WT), 26–27
Wearable IoT (WIoT) devices, 104, 110, 306
  in healthcare. *See* IoT-enabled wearable health care
  related studies, 308–309
Wiener filter, 26–27
Wireless body area network (WBAN), 179–180, 261–263
  challenges associated with, 262–263
Wireless capsule endoscopy (WCE), 298–302
  accuracy of, 300
  preparation of, 302
  technology of, 300–301
    capsule, 300–301
    data recorder belt/Smart wearable, 301
    workstation, 301
  working of, 302

X

X-graph simulator, 7–8
Xively, 99

Y

Yule–Walker method, 29–30

Z

Zigbee, 107, 137–138, 267
Zone Routing Protocol (ZRP), 3–4
Z-Wave technologies, 137–138

FSC
www.fsc.org
MIX
Papier | Fördert
gute Waldnutzung
FSC® C083411

Kolibri 360 GmbH
Ferdinand-Jühlke-Straße 7
99095 Erfurt, Deutschland
produktsicherheit@kolibri360.de